T0227430

Radiation Protection in Medical Imaging and Radiation Oncology

Series in Medical Physics and Biomedical Engineering

Series Editors: John G Webster, E Russell Ritenour, Slavik Tabakov, and Kwan-Hoong Ng

Series in Medical Physics and Biomedical Engineering

Radiation Protection in Medical Imaging and Radiation Oncology

Edited by

Richard J. Vetter
Professor Emeritus, Radiation Safety Officer Emeritus, Mayo Clinic

Magdalena S. Stoeva
Chair, IOMP *Medical Physics World* Board

CRC Press
Taylor & Francis Group
Boca Raton London New York

CRC Press is an imprint of the
Taylor & Francis Group, an **informa** business

CRC Press
Taylor & Francis Group
6000 Broken Sound Parkway NW, Suite 300
Boca Raton, FL 33487-2742

International Standard Book Number-13: 978-1-4822-4537-0 (Hardback)

Visit the Taylor & Francis Web site at
http://www.taylorandfrancis.com

and the CRC Press Web site at
http://www.crcpress.com

Brief Contents

Contents

SECTION IV   **Radiation Protection in Diagnostic Radiology**   **69**

CHAPTER   5 ▪ Radiation Protection in Diagnostic Radiology   71

ELIZABETH BENSON and CORNELIUS LEWIS

CECILIA HINDORF and LENA JONSSON

IBRAHIM DUHAINI, HUDA AL-NAEMI, SHADA WADI-RAMAHI, NABAA NAJI, HANAN AWADA AL-DOUSARI, RIAD SHWEIKANI , and ALI AL-REMEITHI

CHAPTER 13 ■ Regulatory Structures and Issues in Latin America   275

SIMONE KODLULOVICH RENHA, LIDIA VASCONCELLOS DE SA , and ILEANA FLEITAS ESTEVEZ

KIN YIN CHEUNG

KEVIN NELSON and KELLY CLASSIC

List of Figures

List of Tables

About the Series

The Series in Medical Physics and Biomedical Engineering describes the applications of physical sciences, engineering, and mathematics in medicine and clinical research.

The series seeks (but is not restricted to) publications in the following topics:

- Artificial organs
- Assistive technology
- Bioinformatics
- Bioinstrumentation
- Biomaterials
- Biomechanics
- Biomedical engineering
- Clinical engineering
- Imaging
- Implants
- Medical computing and mathematics
- Medical/surgical devices

- Patient monitoring
- Physiological measurement
- Prosthetics
- Radiation protection, health physics, and dosimetry
- Regulatory issues
- Rehabilitation engineering
- Sports medicine
- Systems physiology
- Telemedicine
- Tissue engineering
- Treatment

The *Series in Medical Physics and Biomedical Engineering* is an international series that meets the need for up-to-date texts in this rapidly developing field. Books in the series range in level from introductory graduate textbooks and practical handbooks to more advanced expositions of current research.

The *Series in Medical Physics and Biomedical Engineering* is the official book series of the International Organization for Medical Physics.

THE INTERNATIONAL ORGANIZATION FOR MEDICAL PHYSICS

The International Organization for Medical Physics (IOMP) represents over 18,000 medical physicists worldwide and has a membership of 80 national and 6 regional organizations, together with a number of corporate members. Individual medical physicists of all national member organisations are also automatically members.

The mission of IOMP is to advance medical physics practice worldwide by disseminating scientific and technical information, fostering the educational and professional development of medical physics and promoting the highest quality medical physics services for patients.

A World Congress on Medical Physics and Biomedical Engineering is held every three years in cooperation with International Federation for Medical and Biological Engineering (IFMBE) and International Union for Physics and Engineering Sciences in Medicine (IUPESM). A regionally based international conference, the International Congress of Medical Physics (ICMP) is held between world congresses. IOMP also sponsors international conferences, workshops and courses.

The IOMP has several programmes to assist medical physicists in developing countries. The joint IOMP Library Programme supports 75 active libraries in 43 developing countries, and the Used Equipment Programme coordinates equipment donations. The Travel Assistance Programme provides a limited number of grants to enable physicists to attend the world congresses.

IOMP co-sponsors the *Journal of Applied Clinical Medical Physics*. The IOMP publishes, twice a year, an electronic bulletin, *Medical Physics World*. IOMP also publishes e-Zine, an electronic news letter about six times a year. IOMP has an agreement with Taylor & Francis for the publication of the *Medical Physics and Biomedical Engineering* series of textbooks. IOMP members receive a discount.

IOMP collaborates with international organizations, such as the World Health Organisations (WHO), the International Atomic Energy Agency (IAEA) and other international professional bodies such as the International Radiation Protection Association (IRPA) and the International Commission on Radiological Protection (ICRP), to promote the development of medical physics and the safe use of radiation and medical devices.

Guidance on education, training and professional development of medical physicists is issued by IOMP, which is collaborating with other professional organizations in development of a professional certification system for medical physicists that can be implemented on a global basis.

The IOMP website (www.iomp.org) contains information on all the activities of the IOMP, policy statements 1 and 2 and the 'IOMP: Review and Way Forward' which outlines all the activities of IOMP and plans for the future.

Preface

Radiation protection is a leading and rapidly developing area of medical physics and society as evidenced by the emphasis hospitals and medical organizations are placing on radiation protection culture. The interdisciplinary nature of radiation protection in medicine makes it a key discipline in ensuring the safety of the public.

Safety and quality assurance in the use of radiation in medicine aims to reduce unnecessary radiation risks while maximizing the benefits. Improvements in quality and safety in radiation medicine require a strong radiation safety culture. To better achieve the goals of strengthening radiation safety in healthcare and better protection of the patients from excessive or unnecessary radiation exposure, a concerted effort by all role players including the radiologists, referring practitioners, technologists, professional organizations, international bodies, and regulators is essential.

The International Organization for Medical Physics (IOMP) and the International Radiation Protection Association (IRPA) have worked together to produce this book on radiation protection in medical imaging and radiation oncology, intended for use both in countries that have well-developed medical and health physics disciplines and in countries that have fewer resources.

The IOMP represents more than 18,000 medical physicists worldwide and 81 adhering national member organizations. The mission of IOMP is to advance medical physics practice worldwide by disseminating scientific and technical information, fostering the educational and professional development of medical physicists, and promoting the highest quality medical services for patients.

The IRPA represents some 18,000 members from 50 associate societies representing 63 countries. IRPA's vision is to be recognized by its members, stakeholders, and the public as the international voice of the radiation protection profession in the enhancement of radiation protection culture and practice worldwide Awareness of the need for emphasis on radiation protection contributes significantly to the safety of healthcare providers, patients, and the public. Contributions are most evident in facility design, in monitoring of personnel and the patient care environment, and in development of procedures and practices for proper handling and limitation of radiation exposure. Medical health physicists are often challenged to maximize protection of personnel while minimizing the cost of resources necessary to keep radiation doses as low as reasonably achievable (ALARA). Advances in medical health physics will continue to be based on evidence gathered through basic and applied research. Periodic review of the evidence will help medical health physicists focus on the issues and advance the science.

Radiation Protection in Medical Imaging and Radiation Oncology focuses on the professional, operational, and regulatory aspects of radiation protection covering virtually all regions of the world. The theoretical background is based on current recommendations of the International Commission on Radiological Protection (ICRP) and is complemented by detailed practical sections and professional discussions by the world's leading medical and health physics professionals. Information is well organized into discreet chapters from basic protection to advanced imaging and treatment modalities. Chapters stand on their own with minimal cross referencing so that readers may focus on the chapters of greatest interest to them. This book is a valuable source of information for the medical physicist and related specialties targeting a reading level of Master of Science and above.

Richard J. Vetter
Magdalena Stoeva

Acknowledgments

We are grateful to the International Organization of Medical Physics (IOMP) for suggesting this book project and to the International Radiation Protection Association (IRPA) for their support. We are deeply indebted to the authors who contributed chapters to this book. All authors are connected to one or both of these organizations by virtue of their membership in the associate societies of these organizations. The authors are all active professionals who contributed time from their busy lives to write the material found in this book. We appreciate each one of them for sharing their expertise, for their adherence to the publisher's strict timelines, and for their response to our editorial suggestions. We also deeply appreciate the contributions by authors who represented international organizations in radiation protection including the International Atomic Energy Agency, International Commission on Radiological Protection, World Health Organization, and International Union for Physical and Engineering Sciences in Medicine.

We thank Taylor & Francis for the commitment and professional support of the management and production teams who devoted many hours to this project. We especially appreciate the expert guidance of our production leader, Francesca McGowan, Physics Editor at Taylor & Francis, who diligently arranged and led many teleconferences and provided us with guidance to keep the project moving. She worked hard behind the scenes and made many suggestions along the way to keep everyone on task. This project would not have been possible without her expert leadership and guidance.

The outcome of this project is the result of contributions from many, but any errors of omission or commission belong solely to the editors.

Richard J. Vetter
Magdalena Stoeva

Editors

Richard J. Vetter
Professor Emeritus,
Radiation Safety Officer Emeritus
Mayo Clinic
rvetter@mayo.edu

Richard J. Vetter, Ph.D., is Professor Emeritus and former Radiation Safety Officer for the Mayo Clinic in Rochester, Minnesota, USA. He received his B.S. and M.S. degrees in Biology from South Dakota State University and his Doctorate at Purdue University in 1970, and is board certified by the American Board of Health Physics and the American Board of Medical Physics. From 1970 to 1980, Dr. Vetter served as Professor of Health Physics at Purdue University where he taught health physics, conducted research, and served as Assistant Radiological Control Officer. He served as Radiation Safety Officer and Professor of Biophysics at the Mayo Clinic from 1981 to 2010. From 2010 to 2014 he served as the Government Liaison for the Health Physics Society.

Dr. Vetter is past Editor-in-Chief of the Health Physics Journal, past President of the Health Physics Society, past President of the American Academy of Health Physics, and author or coauthor of more than 230 publications, books, book chapters, and articles. He has served as a reviewer for 20 journals and granting agencies. He is a member of several professional organizations that reflect his interest in radiation safety and the use of radiation in medicine and research.

Dr. Vetter is a past member of the National Academy of Sciences Nuclear and Radiation Studies Board and a Distinguished Emeritus Member of the National Council on Radiation Protection and Measurements. He served as Vice Chair of the Nuclear Regulatory Commission Advisory Committee for Medical Uses of Isotopes and was a member of the Radiation Advisory Committee of the Environmental Protection Agency Science Advisory Board. He has served on the academic advisory boards for the medical and health physics training programs at three universities.

Dr. Vetter is a Fellow Member of the Health Physics Society and the American Association of Physicists in Medicine, and a recipient of the Health Physics Society's Founders Award. He is currently a member of the Executive Council of the International Radiation Protection Association.

Magdalena S. Stoeva
Chair IOMP Medical Physics World
Board
Associate Professor,
Medical Imaging Department,
Medical University-Plovdiv, Bulgaria
ms_stoeva@yahoo.com

Dr. Magdalena Stoeva holds Master's degrees in Medical Physics and Computer Science and a Ph.D. in Diagnostic Radiology/Medical Physics. She has 15 years teaching experience at the Technical University, Medical University, and International Medical Physics Center in Plovdiv, Bulgaria, and ICTP, Trieste, Italy.

Dr. Stoeva serves several active roles in the International Organization of Medical Physics: Chair of the Medical Physics World Board, Editor of IOMP Medical Physics World; Technical Editor of IOMP Journal Medical Physics International; Member of IOMP Publication Committee, Web Committee, and Women's Subcommittee; responsible for IOMP and IOMP-W websites; and a member of the IUPESM Administrative Council. Dr. Stoeva is a member Bulgarian Society of Biomedical Physics and Engineering, the European Society of Radiology, and EuroSafe.

Dr. Stoeva currently serves as Associate Professor in the Medical Imaging Department and the Translational Neurosciences Center, Medical University-Plovdiv, Bulgaria, and is responsible for teaching over 300 students in medical physics, medicine, dentistry, and post-doctoral in various topics related to Medical Physics and Equipment, Physics of Diagnostic Imaging, Radiation Protection, and Quality Control.

Dr. Stoeva actively participates in six international projects under TEMPUS and Leonardo da Vinci European Union programs. She is a co-author and leading IT developer for the projects European Medical Imaging Technology Training (EMIT) and European Medical Imaging Technology e-Encyclopaedia for Lifelong Learning (EMITEL) and is the leading developer of the website for the e-Encyclopedia of Medical Physics and related Multilingual Dictionary.

In recognition of Dr. Stoeva's work, she has received several awards, including: the 2004 Lenardo da Vinci Award to acknowledge achievements in education and training, the 2012 IOMP Award for Website Development, the 2012 International Union of Pure and Applied Physics (IUPAP) Young Scientist Award, and the 2013 Mayor of Plovdiv award for contributions in Education and Science.

Contributors

Author bios are provided in the order of the chapters within the book.

Kin Yin Cheung
President, International Organization for Medical Physics (IOMP), 2012–2015
Senior Medical Physicist, Hong Kong Sanatorium and Hospital
kycheung@hksh.com

Dr. Kin Yin Cheung is a Senior Medical Physicist and Radiation Protection Advisor at the Hong Kong Sanatorium and Hospital. He is also an Adjunct Professor at Tung Wah College, Hong Kong, and Adjunct Associate Professor at the Chinese University of Hong Kong. He is a Fellow of IOMP, Fellow of the Institute of Physics and Engineering in Medicine (IPEM), Fellow of the Hong Kong Institute of Education (HKIE), Honorary Member of Hong Kong College of Radiologists, and a Certified Medical Physicist. He has published more than 110 journal papers, abstracts, and book chapters in medical physics.

Renate Czarwinski
President,
International Radiation Protection Association (IRPA)
czarwinski@me.com

Ms. Czarwinski studied Physics and finalized a postgraduate study in Nuclear Safety and Radiation Protection. In 1996 she was appointed as Head of Radiation Protection at Workplaces Section in the German Federal Office for Radiation Protection. From 2007 to 2012, Ms. Czarwinski acted as Head of Radiation Safety and Monitoring Section of the International Atomic Energy Agency (IAEA) in Vienna. In 2012 Ms. Czarwinski received the HPS Landauer Lecturer Award, which is to honor prominent individuals who have made significant contributions to the field of radiation research and protection. Since 2004 she has been a Member of the Executive Council of the International Radiation Protection Association (IRPA) and was elected as President of IRPA in 2012.

Slavik Tabakov
President, International Organization
for Medical Physics (IOMP), 2015–
2018
Director of MSc Programmes,
MSc Medical Engineering and Physics,
MSc Clinical Sciences,
King's College, London
slavik.tabakov@emerald2.co.uk

Dr. Tabakov is IOMP President (2015–2018) and co-Editor of the IOMP Journal of Medical Physics International. Leading Education and Training Committees of IOMP, International Federation for Medical and Biological Engineering (IFMBE), and International Union for Physical and Engineering Sciences in Medicine (IUPESM) (2000–2012). Coordinator of seven international projects, including EMERALD and EMIT, awarded the EU Leonardo da Vinci Award. Leading developer of the concept of e-Learning in Medical Physics, the first educational Image Databases, e-books, e-Encyclopaedia, and Multilingual Dictionary of Terms in 29 languages. Awarded with the IOMP Harold Johns Medal.

Christopher H. Clement
Scientific Secretary
International Commission on
Radiological Protection (ICRP)
sci.sec@icrp.org

Mr. Clement worked for the Canadian Low-Level Radioactive Waste Management Office for more than a decade, then for nearly another decade for the Canadian Nuclear Safety Commission in several positions culminating in Director of Radiation Protection. Since 2008 he has been Scientific Secretary of the ICRP. ICRP is an independent registered charity established to advance radiological protection for the public benefit. He is responsible for day-to-day management of ICRP, and is the Editor-in-Chief of the Annals of the ICRP. In 2012, Mr. Clement was elected to the Executive Council of the International Radiation Protection Association.

Elizabeth Benson
Clinical Scientist and Laser Protection
Adviser
King's College Hospital, London, UK
Elizabeth.chaloner@nhs.net

Elizabeth Benson is a Clinical Scientist and Laser Protection Adviser at King's College Hospital NHS Foundation Trust. Her interests are ionizing and non-ionizing radiation protection. She is extensively involved in special interest group work in these areas for both the British Institute of Radiology (BIR) and the Institute of Physics and Engineering in Medicine (IPEM).

Cornelius Lewis
Director of Medical Engineering and
Physics
King's College Hospital, London, UK
cornelius.lewis@kcl.ac.uk

Cornelius Lewis graduated from Aston University in 1974. He undertook postgraduate research in particle radiotherapy and was awarded a doctorate from the University of Leeds in 1979. After a brief period of postdoctoral research at the Universities of Surrey and British Columbia, he joined the UK National Health Service as a physicist in diagnostic radiology and radiation protection at Mount Vernon Hospital, Middlesex in 1981. He subsequently worked at University College Hospital, St Peter's Hospitals, and the Institute of Urology before moving to King's College Hospital in 1991, where he is currently Director of Medical Engineering and Physics.

Cecilia Hindorf
Ph.D., Associate Professor in Medical Radiation Physics
Radiation Physics, Skåne University Hospital, Lund, Sweden
cecilia.hindorf@med.lu.se

Lena Jönsson
Senior Lecturer, Medical Radiation Physicist
Medical Radiation Physics, Department of Clinical Sciences, Lund University, Sweden
Lena_M.Jonsson@med.lu.se

Lena Jönsson, Ph.D., Medical Radiation Physicist, is employed as Senior Lecturer at the Department of Medical Radiation Physics, Lund University, Lund, Sweden. She is also currently working as a medical physicist and has more than 25 years of experience in diagnostic nuclear medicine and radiation protection. She is an acknowledged teacher at the programmes for medical physicists, radiographers, and biomedical engineers at Lund University and in continuous professional development (CPD) for different professions.

Raymond K. Wu
Chief of Physics
University of Arizona Cancer Center,
Arizona, USA
RayKWu@gmail.com

Raymond K. Wu earned his Ph.D. degree at Dartmouth College in 1974, and trained in Medical Physics at Thomas Jefferson University Hospital. He was Assistant Professor and then Associate Professor at Temple University Medical School. In 1985 he became Professor and Director of Medical Physics at Eastern Virginia Medical School serving in those positions until 2002. He has authored many publications including the NCRP Report 151, and the IAEA Safety Reports Series No. 47. He has served as the Chairman of the American College of Medical Physics. He is currently Chief of Medical Physics at the University of Arizona Cancer Center in Phoenix.

Taofeeq A. Ige
Secretary-General
Federation of African Medical Physics
Organizations (FAMPO)
igetaofeeq@yahoo.com

Taofeeq is a Chief Consultant Physicist at the National Hospital, Abuja. He is the immediate past Departmental Head of Medical Physics of the Hospital and also a Visiting Senior Lecturer to some universities in the country. He is a national, regional, and inter-regional project counterpart on IAEA medical-physics-related projects in the region and also a WHO consultant. He is an examiner in Physics to the Faculty of Radiology of both the West African College of Surgeons (WACS) and the National Post-Graduate Medical College of Nigeria (NPMCN). Taofeeq is also the pioneer Secretary-General of the Federation of African Medical Physics Organizations (FAMPO).

Kwan-Hoong Ng
Professor
Department of Biomedical Imaging,
University of Malaya, Kuala Lumpur,
Malaysia
ngkh@um.edu.my

Professor Ng's research contributions include radiation dosimetry, radiological protection, biophysical characterization of breast diseases, and medical imaging. He had served in key positions in IOMP and IUPESM. Dr. Ng has been serving as an IAEA consultant and expert for the last two decades, and was past-President of South East Asian Federation of Organizations for Medical Physics (SEAFOMP) and Asia-Oceania Federation of Organizations for Medical Physics (AFOMP). He has published over 200 papers in peer-reviewed journals, 12 book chapters, edited 8 books, and has written and presented over 400 scientific papers. He serves on the editorial boards of 12 journals. Recently he has contributed his expertise to the IAEA Fukushima Final Report.

Aik-Hao Ng
Physicist
Ministry of Health, Putrajaya,
Malaysia
hao06051982@yahoo.co.uk

Mr. Ng is a Medical Physicist and Assistant Director in Radiation Health and Safety Section, Ministry of Health (MOH), Malaysia. He graduated with a Master of Medical Physics (University of Malaya) and B.Sc. in Health Physics (University of Technology Malaysia). At MOH, his job focuses on controlling the use of ionizing radiation for medical purposes, particularly in developing regulations and guidelines as well as providing technical advice on radiation protection and hospital facilities design. He is pursuing a Ph.D. degree at the School of Medicine, University of Nottingham, United Kingdom.

Marina binti Mishar
Program Management Officer,
International Atomic Energy Agency,
Department of Technical Cooperation,
Division of Asia and the Pacific
m.b.mishar@iaea.org

Ms. Mishar is a program management officer in the Division of Technical Co-operation for Asia and the Pacific in the IAEA. She received her B.Sc and M.Sc. from the National University of Malaysia before working as a Scientific Officer in the Atomic Energy Licensing Board in Malaysia. After fifteen years of experience in the regulatory body, she accepted a tenured position at IAEA in the Technical Cooperation Division. Ms. Mishar is currently managing several radiation safety-related projects for IAEA Member States in Asia and the Pacific, including radiation protection for medical exposures.

Carmel J. Caruana
Head, Medical Physics Department
University of Malta
carmel.j.caruana@um.edu.mt

Professor Caruana specializes in diagnostic and interventional radiology, protection from ionizing radiation and other physical agents, and legislative (professional) Education and Training (E&T) issues in Medical Physics. He is past Chairperson of the E&T Committee of the European Federation of Organisations in Medical Physics (EFOMP) and main author of the role and E&T chapters of the EU-sponsored document, "European Guidelines on the Medical Physics Expert." He is also the main author of the first Eutempe-rx project (www.eutempe-rx.eu) module "Development of the Profession and the Challenges for the Medical Physics Expert (D&IR) in Europe" and an acknowledged expert in Radiation Protection legislation and E&T in Europe.

Ibrahim Duhaini
Chief Medical Physicist &
Radiation Safety Officer
Rafik Hariri University Hospital &
Nabih Berry Governmental University
Hospital
duhaini@yahoo.com

Academic Degrees: B.Sc., American University of Beirut; Teaching Diploma in Science, AUB; M.Sc. and Higher Studies in Medical Physics, Wayne State University, USA; Ph.D. in Nuclear Physics, Beirut Arab University, Lebanon. Professional Experience: Henry Ford, St. Mary Mercy, and Oakwood Hospitals (USA); Director of Radiation Safety at Hamad Medical Corporation; Consultant MP and RSO, Sidra Hospital (Qatar). Established and managed Radiotherapy Department at RHUH and NBGUH. Dr. Duhaini is a founder of LAMP and MEFOMP, and an active IOMP member. He established a Professional Consulting Service; "Radiation Expert Group" for E&T in Radiation.

Huda Al-Naemi
Executive Director,
Occupational Health And Safety
Department (OHS)
Hamad Medical Corporation
Halnaomi@hamad.qa

Dr. Al-Naemi started her career with Hamad Medical Corporation (HMC) as Medical Physicist, and obtained a Doctorate in Nuclear Physics from Cairo University. There she established an OHS Department and became the Executive Director. She represents Qatar in international events and works with global organizations such as IAEA, UNEP, WHO to implement projects at the national level. As a Consultant Physicist, she is a specialist lecturer in Medical Radiation. As the Chair of Radiation Professionals Licensing Committee, she runs the mandatory radiation protection training for all radiation workers as mandated by the Supreme Council of Health, Qatar.

Shada Wadi-Ramahi
Chief Medical Physicist
King Faisal Specialist Hospital
and Research Center
Shada_ramahi@yahoo.com

Dr. Wadi-Ramahi, a Fulbright Scholar, obtained her Ph.D. from Rush University Medical Center in Chicago in 2003, and became certified by the American Board of Radiology (ABR) in 2006. She worked for 11 years, February 2004 through November 2014, at the King Hussein Cancer Center as a clinical medical physicist and then became the head of the Medical Physics Section until November 2014, when she joined King Faisal Specialist Hospital and Research Center in Riyadh. Dr. Wadi-Ramahi has provided various consultation work concerning medical physics to IAEA. She is a co-author of the IAEA Human Health Series Report 25.

Nabaa Naji
Head of Department Assistant/
President of the Iraqi Medical Physics
Society (IMP)
Dept. of Physiology-Medical Physics
Unit/Mustansiriya Medical College/
Baghdad-Iraq
nabaaanaji@yahoo.com

Dr. Naji started her work in this field in 1996, and graduated with M.Sc. in 2002 and a Ph.D. in 2007. Currently she is working as a lecturer in medical physics in Mustansiria Medical College and a supervisor of the lab. She has published many papers, and participated in many events and trainings. In 2010, she was nominated to represent the Iraqi Medical Physicists in the AAPM Annual, accordingly Iraq received charter membership in the ICBMP. In 2013, she was elected President of IMPS. Dr. Naji helped in making Iraq a member of MEFOMP and IOMP, changing the curriculum of the M.Sc. to match the AAPM curriculum, and implementing the ISEP of medical physics in Iraq.

Hanan Awadah Al-Dousari
RSO and Physicist at Jaber Alahmad
Alsabah Center for Nuclear Medicine
and Molecular Imaging
Ministry of Health, Kuwait
Hanan3d@hotmail.com

Dr. Al-Dousari has a Ph.D. in Medical Physics from the University of Surrey, United Kingdom, and M.Sc. in Medical Imaging.

Riad Shweikani
Head, Nuclear Science and Technology
Training Center
Atomic Energy Commission of Syria
rshweikani@aec.org.sy

Since the beginning of 2015, Prof. Shweikani has been the head of the Nuclear Science and Technology Training Center at the Atomic Energy Commission of Syria. He was the head of the Radiological and Nuclear Regulatory Office from June 2010. He is involved widely in developing nuclear safety and security infrastructure in Syria. He worked in the radiological and nuclear regulatory office since 2002 and contributed to the process of producing the radiation law and general regulations in Syria. He was the director of the M.Sc. Program on Radiation protection and safety of radioactive sources at Damascus University, supported by IAEA (2006–2014).

Ali Al-Remeithi
Health Physicist
Federal Authority for Nuclear
Regulation (FANR)
ali.alremeithi@fanr.gov.ae

Dr. Al-Remeithi has been a Health Physicist in the Radiation Safety Department at FANR since 2013, mostly involved with licensing and inspecting medical and industrial facilities. He has a B.Sc. in Medical Imaging from Higher Colleges of Technology - UAE 2010 and M.Sc. in Radiation and Environmental Protection from the University of Surrey, UK, 2013. Guest lecturer at the Radiation Protection Educational Program 2014, organized by the Dubai Health Authority. He has been certified as a judicial officer by Ministry of Justice since 2014. Dr. Al-Remeithi contributed in organizing FANR's Meet Your Regulator Workshop in 2014, and has participated in national and regional IAEA workshops. He is a founding member of the Emirates Nuclear Society.

Ruth E. McBurney
Executive Director
Conference of Radiation Control
Program Directors, Inc.
rmcburney@crcpd.org

Ruth E. McBurney is the Executive Director of the Conference of Radiation Control Program Directors (CRCPD), where she manages and directs the administrative office for the organization. She also has nearly 30 years of experience in state radiation control programs. Ms. McBurney is currently serving on the Board of Directors of the National Council on Radiation Protection and Measurements, and is a past President of the Health Physics Society. Ms. McBurney holds an M.Sc. Degree in Radiation Sciences from the University of Arkansas for Medical Sciences, and is a Certified Health Physicist.

Simone Kodlulovich Renha
President
Asociación Latinoamericana de Física
Médica (ALFIM)
simone@cnen.gov.br

Graduated in Physics, M.Sc. and Ph.D. in medical physics. Researcher of the National Commission of Nuclear Energy. Lecturer of CNEN postgraduate program and proofreader of scientific journals. IAEA and PAHO expert. Member of the Radiological Protection Committee and Quality in Computed Tomography of the Brazilian College of Radiology and member of ABNT and IEC committees in diagnostic fields. Member of the International Medical Physics Certification Board and IOMP committees: E&T and Women subcommittee. Consultant of Latin-American Affairs Subcommittee of AAPM. President of ALFIM since 2010.

Lidia Vasconcellos de Sá
Researcher
National Nuclear Energy Commission,
Institute of Radiation Protection and
Dosimetry (IRD), Medical Physics
Department
lidia@ird.gov.br

Ph.D. in Nuclear Engineering, researcher at the National Nuclear Energy Commission in the IRD. Experience in Nuclear Engineering, Licensing, and Control of Radioactive Facilities, Radioisotope Applications, Diagnostic Radiology, Radiotherapy, and Nuclear Medicine. Member of: CNEN Advisory Committee for Standards and Regulations, RPCB, Brazilian Society of Nuclear Medicine. Coordinator of Medical Exposures in IRD/IAEA Specialization Course. IAEA Expert for NM, safety and security of sources, evaluation of regulatory bodies. Coordinator of Brazilian Standards Board in medical field. IEC TC/SC62C member. IOMP training group member. UNSCEAR contact point in Brazil.

Ileana Fleitas Estévez
National Consultant, PAHO-Cuba
Pan American Health Organization,
PAHO
fleitasi@paho.org

Cuban citizen Ileana Fleitas is a nuclear engineer with a Master's degree in Medical Physics and post-graduate training in Public Health. She has 19 years of experience as a medical physicist in the area of diagnostic imaging, radiation protection, and regulation for medical devices. She is author of scientific papers and chapters in the fields of quality control, quality assurance, radiation protection, and rational use of medical devices. Since 2003 she has worked for the Pan American Health Organization as part of the Radiological Health Program, participating in several courses and seminars in North and America.

Ahmed Meghzifene
Head, Dosimetry and Medical
Radiation Physics Section
International Atomic Energy Agency
a.meghzifene@iaea.org

Ahmed Meghzifene was awarded a research grant in France (1989) and one in Canada (1991), both in in medical dosimetry. After his return to Algeria, he dedicated his efforts to the establishment of dosimetry and medical physics infrastructure. In 1997, he joined the medical physics group at the IAEA as a medical radiation physicist. In 2007, he was appointed Section Head of IAEA Medical Physics Section. During the past 15 years, he has contributed to the promotion of the medical physics profession and supported education activities. He has published more than 25 papers and 3 book chapters, and has delivered numerous keynote talks at international conferences.

Ola Holmberg
Unit Head
Radiation Protection of Patients Unit;
Radiation Safety and Monitoring
Section; NSRW; International Atomic
Energy Agency; Vienna; Austria
o.holmberg@iaea.org

Dr. Ola Holmberg is a medical physicist and head of the Radiation Protection of Patients Unit at the International Atomic Energy Agency, Vienna. He has worked with radiotherapy at Skåne University Hospital, Malmö; St. Luke's Hospital, Dublin; and The Netherlands Cancer Institute, Amsterdam. Prior to working with the IAEA he was the Chief Physicist at Copenhagen University Hospital, Herlev, Denmark. He served as the Scientific Secretary for the International Conference on Radiation Protection in Medicine, Bonn, co-founded the Radiation Oncology Safety Information System (ROSIS), and participated in an ICRP Task Group on prevention of accidental exposures in radiotherapy.

Maria del Rosario Pérez
Scientist
World Health Organization,
Department of Public Health and
Environment
perezm@who.int

Physician (radiation oncologist) responsible for the technical coordination of the World Health Organization (WHO) Global Initiative on Radiation Safety in Health Care Settings, actively involved in the revision of the International Radiation Basic Safety Standards (BSS), and currently working to support its implementation. She represents WHO at the Inter-Agency Committee on Radiation Safety, the UNSCEAR, the IAEA Radiation Safety Standards Committee (RASSC), the Article 31 Euratom Group of Scientific Experts and its Working Party on Medical Exposures, and serves as WHO liaison technical officer for the ICRP.

Miriam Mikhail
Technical   Officer,   Diagnostic
Radiologist
World Health Organization
mikhailm@who.int

Dr. Miriam Mikhail, a diagnostic radiologist, is a technical officer at WHO Headquarters in Geneva where she partakes in multiple global radiation in healthcare initiatives. She completed medical school and radiology residency at The Mayo Clinic, as well as an intervening 1-year internship at McGill University, and a cardiac imaging fellowship in Florida. Then Dr. Mikhail worked 7 years as a staff hospital-based radiologist. She has international medical experience and, among professional acumen, previously served on the American College of Radiology (ACR) Guidelines and Standards Committee for Ultrasound as one of a team of authors of guidelines still in place.

Fridtjof Nüsslin
Professor of Biomedical Physics
Technische Universität
München, Germany
nuesslin@lrz.tum.de

Professional career: Studies of Physics and Physiology. Professor and Chair, Medical Physics, University of Tübingen. Since 2004: Professor of Biomedical Physics, Clinic for RadioOncology at Technische Universität München. Professional Positions: President DGMP, EFOMP and IOMP. Since 2012: Past President IOMP. Scientific Activities: Dosimetry and Treatment Planning, Conformal Radiotherapy, Image Guidance, Advanced Technologies in Radiooncology (hadrons, laser applications in imaging and particle beam therapy, MR-Linac), Biological and Molecular Imaging, Modeling in Tumorbiology. Training of Medical Physicists in Nuclear and Radiological Emergency.

Eliseo Vaño
Professor of Medical Physics
Radiology Department.
Faculty of Medicine. Complutense
University of Madrid and San Carlos
University Hospital
eliseov@med.ucm.es

Eliseo Vaño is a full Professor of Medical Physics and advisor to the Spanish Ministry of Health for Radiation Protection in Medical Exposures. He has published numerous articles about Radiation Protection dealing with Diagnostic and Interventional Radiology. He is a member of the ICRP and currently the Chairman of the Committee on Protection in Medicine. Prof. Vaño is a member of the Article 31 Group of Experts of the EURATOM Treaty and a member of the Spanish delegation in UNSCEAR. He is a consultant to the IAEA for topics concerning medical exposures.

Fred A. Mettler, Jr.
Professor Emeritus and Clinical
Professor
Department of Radiology at the
University of New Mexico School of
Medicine, USA
fmettler@salud.unm.edu

Dr. Fred Mettler received a B.A. from Columbia University, an M.D. from Thomas Jefferson University, and an M.Sc. in Public Health from Harvard University. He has authored over 360 scientific publications, including 21 textbooks. He is an Emeritus Commissioner of ICRP, an emeritus member of NCRP, and a member of the Nuclear and Radiation Studies Board of the U.S. National Academies. He was the U.S. Representative to UNSCEAR for 28 years, as well as the Health Effects Team Leader of the International Chernobyl Project and has served as an expert on radiology and radiation effects and accidents for the U.S. Department of HHS, Homeland Security, WHO and IAEA. He is currently a health advisor to the Japanese Cabinet for the Fukushima nuclear disaster.

Kevin Nelson
Medical Health Physicist and RSO
Mayo Clinic Arizona, USA
Nelson.Kevin2@mayo.edu

Dr. Nelson began his career as a medical health physicist in 1980 at the University of Minnesota. Shortly after obtaining PhD in Health Physics, he worked at 3M. He joined Mayo Clinic in Jacksonville, Florida in 1995 with work emphasis in diagnostic radiology and regulatory affairs. Dr. Nelson was elected President of the Health Physics Society in 2007 and was a Fellow of the Society. He received his certification from the American Board of Health Physics in 1992.

Kelly Classic
Health Physicist
Mayo Clinic
classic.kelly@mayo.edu

Ms. Classic has been with Mayo Clinic since receiving a Masters in Health Physics from Purdue University in 1984. She has authored or co-authored 40 papers and 5 book chapters, and has written a primer titled Radiation Answers for the Health Physics Society. She is an Assistant Professor of Radiologic Physics in Mayo Clinic's College of Medicine.

I

Introduction

1

Introduction

International Organization for Medical Physics Statement

Kin Yin Cheung

President of the International Organization for Medical Physics, 2012-2015
Hong Kong Sanatorium & Hospital, Hong Kong

Slavik Tabakov

President of the International Organization for Medical Physics, 2015-2018
Department of Medical Engineering and Physics, King's College London, UK

CONTENTS

MEDICAL EXPOSURE has been the largest radiation exposure to the human population from man-made radiation sources, per United Nations Scientific Committee on the Effects of Atomic Radiation's report (UNSCEAR 2008). Exposure of the population due to medical imaging procedures has grown very rapidly over the past few decades. For this reason, the contribution to population dose from medical imaging has increased significantly over the years. According to UNSCEAR (UNSCEAR 2008), more than 3,600,000,000 radiological procedures have been carried out every year around the world. The global annual effective dose per capita due to medical exposure has increased by about 100% to 0.64 mSv during the period 1993 to 2008. With increasing use of CT in diagnostic radiology, particularly in the developed countries, a higher rate of increase in annual effective dose per capita can be expected. In the United States, for instance, the annual effective dose per capita has increased from 0.54 mSv to 3.0 mSv during the period 1980 to 2006 (NCRP 2009). Increasing availability of radiation medicine to the global population due to expanding clinical service scope, improvement in service quality through advancing technologies and sophisticated clinical procedures, and changes in clinical diagnostic procedures to those more dependent on medical imaging, such as computed tomography (CT) in patient management in the clinics, might have contributed to this rapid increase in population dose. This trend of increasing population dose is likely to continue, perhaps at an increasing rate. The phenomenon has raised some concerns in the medical and radiation safety communities as well as in the public media on the potential cancer risks to patients, especially young and pregnant patients receiving diagnostic radiological procedures. There may be a need to review and strengthen the current practice of radiation protection in medicine, particularly in the management of patient dose. In addressing such issues, it is important to put the risks and benefits of medical exposure into the right perspective. For instance, reports on the hazards of medical exposure, if not addressed appropriately, could be misleading to members of the public. This in turn could raise unnecessary fears for patients and their relatives, especially pregnant patients and parents of young patients, and deter them from taking needed medical exposure. This in turn could put patients' lives at even bigger risk.

1.1 CANCER RISKS OF MEDICAL EXPOSURE

The cancer risks of ionizing radiation are well documented in the literature (e.g., international and national authoritative documents published by the International Commission on Radiological Protection (ICRP) (ICRP 2007), UNSCEAR (UNSCEAR 2012), and the Biological Effects of Ionizing Radiation (BEIR) (NRC 2006). It is important that all diagnostic and therapeutic medical exposures are justified and the radiation dose to patients arising from the radiological procedures be optimized according to the ICRP principles.

In the case of medical imaging, under normal operating conditions, the amount of radiation exposure to the patient under examination should be very small if appropriate protective and dose optimization measures are taken in the imaging procedure. Typical patient doses arising from commonly performed medical examination procedures are widely published. Such information can be found in the websites of the International Atomic Energy Agency (IAEA) and professional organizations such as the American College of Radiology

(ACR) (ACR 2015). It has been documented (ICRP 2007, UNSCEAR 2012) that epidemiological methods used for the estimation of cancer risks do not have the power to directly reveal cancer risks in the dose range up to around 100 mSv. The amount of radiation involved in a diagnostic imaging procedure is well below this dose level. The carcinogenic risks, if any, associated with medical imaging are likely to be very small with current imaging technologies. The risks and benefits of a radiological procedure in medicine should both be evaluated in the right perspective when justifying or giving advice to patients, particularly pregnant patients and parents of young patients with a medical exposure.

There are situations when abnormal occurrence appears in medical exposure and when that happens, the patient affected can be exposed to a higher level of ionizing radiation. Incidents of this sort could be serious in therapeutic radiology when a high level dose is involved. Incidents of abnormal occurrence can happen in the clinics due to various reasons. The key issues identified by IAEA (IAEA 2000) in their analysis of a large number of reported medical radiation incidents in radiotherapy were inadequacy in communication between staff and between staff and patient, and in verification or quality assurance, training, and documentation. Many of the reported radiation incidents in medicine could be prevented if a robust quality management system was in place, and if the staff was qualified and competent for the work they performed, with better awareness of safety issues and the need for taking preventive measures against human errors and miscommunication while performing their duties. A good sense of safety culture in the clinics can be effective in minimizing the possibility of radiation incidents.

As discussed above, prevention of abnormal occurrences, especially in radiation therapy, and reduction and optimization of patient dose and control of occupational exposure for members of staff are some of the key objectives in radiation protection in medicine.

1.2 THE ROLE OF MEDICAL PHYSICISTS IN RADIATION PROTECTION IN MEDICINE

Medical physicists in healthcare are key members of the radiology, nuclear medicine, and radiation oncology teams. They provide the required scientific support to all clinical departments in patient management involving the use of radiation. They play a key role in research, development, and implementation of improved or new imaging and radiation therapy modalities, techniques, and procedures. Apart from contributing to clinical services, medical physicists have responsibility for addressing radiation safety issues in the clinics. They support radiologists, radiation oncologists, and referring physicians by providing professional advice with supporting scientific data, if needed, on justification of radiological procedures, radiation dose optimization, and protection measures. They provide guidance to frontline staff such as radiation therapists and technologists on radiation safety measures and procedures for protection of patients and members of staff. They also provide counseling service to patients receiving medical exposures, to address any radiation safety concerns they may have. The professional competence and standard of practice of medical physicists is important in the effective discharge of such duties. The roles and responsibilities of medical physicists in medicine and their academic and professional qualifications are discussed in more detail in the section on the role of the International Organization for Medical Physics (IOMP) and medical physicists, below.

1.3 THE ROLE OF INTERNATIONAL PROFESSIONAL ORGANIZATIONS

International statutory organizations such as the European Commission (EC), the IAEA, the International Labor Organization (ILO), the Pan American Health Organization (PAHO), the UNSCEAR, and the World Health Organization (WHO), and international professional organizations such as the European Society for Radiotherapy and Oncology (ESTRO), the ICRP, the International Commission on Radiation Units and Measurements (ICRU), the International Radiation Protection Association (IRPA), IOMP, and the International Society of Radiology (ISR) have been playing different but important roles in promoting the global development of radiation medicine, particularly radiation safety in healthcare. They established, in consultation with all stakeholders, international standards and guidelines on radiation safety and protection in healthcare. These standards and guidelines are often adopted by state regulators in their legislative control of the safe use of radiation in their countries. They are also used as references or guiding principles by medical physicists and other healthcare professionals in diagnostic and therapeutic radiology, in establishing their service and safety standards and working protocols in the clinics. International organizations play a leading role in promoting scientific exchange amongst healthcare professionals, by supporting and working together with national professional organizations in organizing or hosting international scientific conferences and seminars. They also play a key role in the training of healthcare professionals, by setting guidelines and standards on education and training and organizing educational workshops and training courses for healthcare professionals and administrators practicing in radiation medicine. In some circumstances, they support countries to resolve their specific needs in the development or improvement of radiation medicine, particularly for medical physics and radiation safety.

1.4 INTERNATIONAL ORGANIZATION FOR MEDICAL PHYSICS

The international organization that practically represents all medical physicists in the world is the International Organization for Medical Physics (IOMP) (IOMP 2015). The IOMP was founded in 1963 and currently has

84 National Member Organizations (NMO) with more than 18,000 individual members around the world. The Organization has, together with the respective NMOs, formed six Regional Organizations (RO): European Federation of Organizations for Medical Physics (EFOMP), Asian-Oceania Federation of Organizations for Medical Physics (AFOMP), Latin American Medical Physics Association (ALFIM), Southeast Asian Federation for Medical Physics (SEAFOMP), Federation of African Medical Physics Organizations (FAMPO), and Middle East Federation of Organizations for Medical Physics (MEFOMP).

IOMP is charged with a mission to advance medical physics practice worldwide by disseminating scientific and technical information, fostering the educational and professional development of medical physics, and promoting the highest quality medical services for patients. Just as IOMP represents the medical physicists of the world, IRPA represents the health physicists (radiation protection specialists). Radiation protection specialists work in many sectors in which radiation is used directly, such as nuclear power generation. The IRPA president, Dr. Renate Czarwinski, discusses the role of IRPA in radiation protection in medicine in Chapter 2.

1.5 THE ROLE OF INDUSTRIAL ORGANIZATIONS

Protection of the patient, particularly dose reduction in radiation medicine, can be achieved mainly in two ways. One way is to optimize the examination techniques and procedures, machine performance, and exposure conditions, so as to acquire the needed diagnostic information with the minimum amount of patient dose. The amount of dose reduction through these procedures would depend on the knowledge and skills of the health professionals who perform the examination and related procedures. The other method is to use low radiation dose equipment technologies or even non-ionizing radiation modalities. In the case of medical imaging, while non-ionizing radiation imaging modalities cannot replace those using ionizing radiation, at least in the near future, low dose imaging technology is the only option in most clinical cases. Industrial companies have a key role to play in this aspect of radiation protection.

Achievement in modern medicine is made possible only because of the innovation and invention in research institutions and the equipment industry in producing the medical devices that serve clinical needs in diagnostic and therapeutic radiology. As in other branches of medical devices, radiological equipment technology has evolved from many generations of development ever since Roentgen discovered the X-ray in 1895. In diagnostic radiology, a much lower radiation dose is now needed to acquire the information required to make an accurate disease diagnosis than before. This is made possible through improvement in, among other items, image receptor technology and image reconstruction methodology and algorithms. In the case of CT examination using the latest generation of low-dose CT scanners, to acquire the image data with similar or even better diagnostic quality, the amount of patient dose needed can be two or more orders of magnitude lower than that given when acquired with conventional CT technology. The current generation of CT technology is capable of performing sophisticated imaging modalities such as CT angiography and brain perfusion with much lower exposure dose than those performed with previous technology. The typical effective dose a patient would receive during a CTA with low-dose CT is about 1 mSv, which is lower than that of a head scan performed a few years ago. Improvement in imaging equipment technology has helped reduce the amount of radiation dose to individual patients receiving radiological examination. However, it cannot help reduce the population dose because of increasing demand for more accurate diagnostic information to support clinical decisions in patient management.

Similar evolution has occurred in therapeutic radiology equipment technology. The current generation of radiotherapy equipment technologies is integrated with imaging systems for high-precision on-line localization of the target volume for treatment delivery. They are capable of delivering a therapeutic radiation dose that conforms to the size and shape of the disease volume in the patient, with better protection of adjacent normal tissue structures than ever before. The risk of normal tissue damage often limits the amount of therapeutic dose that can be given to the patient, and this in turn compromises the probability of local control and cure. Better protection of normal tissues will allow higher doses to be given to the treatment target volume to improve local control, and at the same time reduces the probability of developing radiation-induced side effects in the patient. Improvement in radiotherapy equipment technologies by manufacturers has benefited the patients in two ways, i.e., better quality treatment and better protection of the patient.

1.6   IOMP SUPPORT FOR VARIOUS RADIATION PROTECTION EDUCATION AND TRAINING ACTIVITIES

Contemporary medicine uses medical technology extensively. The rapid increase of development and implementation of this technology during the past decades naturally led to the increased need for specialists to assure its safe and effective use. This was the main reason for the expansion of the number of medical physicists around the word. While at the time of formation of the International Organization for Medical Physics (IOMP) in 1963 there had been around 6,000 medical physicists in the world, during 1994 this number was around 12,000, and during 2013 it was more than 18,000 — and further increases are expected. The significant increase of medical physics professionals in the past 20 years is also marked with the increase of the proportion of these specialists working in the field of medical imaging.

Against this background, it was natural to see the international recognition of medical physicists by the International Labor Organization (ILO, an Agency of the United Nations). The profession is now explicitly included (for the first time) in the latest International Standard Classification of Occupations (ISCO-08) under sub-major group 21, Science and Engineering Professionals (Unit Group 2111, "Physicists and Astronomers"). The inclusion of our profession in ISCO is further explained with a specific note "... medical physicists are considered to be an integral part of the health work force alongside those occupations classified in sub-major group 22, Health professionals ..." (Smith, Nüsslin 2013).

Most medical technology used in medical imaging and radiation oncology applies ionizing radiation, hence one of the main tasks of medical physicists is associated with radiation protection. IOMP and its members and Regional Organizations have developed and supported many activities aiming to increase the level of education and training of medical physicists in this area. Although these activities are based on the contribution of many highly specialized medical physicists, one has to mention the leading role of the largest national professional organizations — the American Association of Physicists in Medicine (AAPM) and the UK Institute of Physics and Engineering in Medicine (IPEM). The expertise and guidelines of these organisations were later used in the formation of various education and training programs, all of which include radiation protection.

One of the largest international institutions for professional education and training is the International College on Medical Physics, based at the International Center of Theoretical Physics (ICTP, operating under the aegis of UNESCO/IAEA) in Trieste, Italy. The first such college was founded in 1982, and since 1992 it has run on a regular basis (usually bi-annually).

Alongside the college (focussing on Medical Imaging and Radiation Protection), ICTP hosts many other medical physics workshops, courses, and conference, mainly related to IAEA activities. During 2013, ICTP started an international Master's program in Medical Physics and also a Radiotherapy School (both supported by the IOMP). During its 30 years of medical physics activities, ICTP has trained almost 2000 colleagues, mostly from the developing countries. The transfer of knowledge and experience to the developing countries is a major objective of the ICTP College of Medical Physics. Each participant receives a full set of lecturing materials, including Power Point slides, e-learning materials, access to websites, etc. These have triggered dozens of medical physics activities and courses in the developing countries (Bertocchi et al. 2014).

The spread of medical physics knowledge and building competencies was greatly accelerated by the development of e-learning. Medical physics was one of the first professions in the world to develop its own original e-learning materials. The projects European Medical Radiation Learning Development (EMERALD) and European Medical Imaging Technology Training (EMIT) (1994–2004) developed extensive e-learning materials — training tasks, image databases (including 4000+ images), simulations, and one of the first e-learning web sites (www.emerald2.eu) (Tabakov et al. 2011a, Tabakov et al. 2011b). Further, the project European Medical Imaging Technology e-Encyclopaedia for Lifelong Learning (EMITEL) (2005–2013) developed the first e-Encyclopaedia of Medical Physics (3100+ articles) with a Multilingual Dictionary of terms (translated into 29 languages) – both major reference materials for the profession (www.emitel2.eu) (Tabakov et al. 2011). These materials are used by thousands of colleagues all over the world. The success and impact of medical physics e-learning through these projects was recognised with the first educational award of the European Union — the Leonardo da Vinci Award.

The EMERALD and EMIT projects included 245 training tasks, building competencies in various fields, including quality assurance, but not directly in radiation protection (this being more difficult to train on a large international basis). However, later projects, such as European Training and Education for Medical Physics Experts in Radiology (EUTEMPE-RX), Medical Radiation Protection Education and Training (MEDRAPET), and European Diagnostic Reference Levels for Paediatric Imaging (PiDRL) developed radiation protection materials for the countries of the European Union (the European Federation EFOMP takes an active part in these projects) (Damilakis, Paulo, Christofides 2013). Additionally a number of websites actively provide such information (Smajo 2013, Sprawls, Duong 2013), the most frequently visited radiation protection website being the Radiation Protection of Patients (RPOP) of IAEA (https://rpop.iaea.org/RPoP/RPoP/Content/index.htm) (IAEA 2015, Rehani 2013). A recent IAEA Technical Meeting on Patient Safety in Radiotherapy was centered on the IAEA project SAFRON — an integrated voluntary reporting registry of radiation oncology incidents and near misses. This activity relies on global collaboration in safety reporting and prospective risk analysis implementation in radiotherapy practice. Using collaboration with the major international organizations associated with Radiation Safety, and based on the Bonn Call for Action, the project aims to form a basis for future developments in radiotherapy patient safety.

IOMP is also directly involved in co-organizing national and international training courses. In the past 20 years the IOMP Education and Training Committee has organized more than 70 workshops and seminar courses with attendees from 85 countries. About half of these events have been in collaboration with the American Association of Physicists in Medicine (AAPM) and IAEA (Niroomand-Rad et al. 2014). These activities greatly helped the professional development in Eastern Europe, Asia, and Latin America. Later a number of concerted activities were initiated to help professional development in Africa (Tabakov 2013). This IOMP assisting with larger-scale-capacity building projects is also related with the newly planned activities for accreditation of educational programs and certification of medical physicists — both due to begin very soon.

Finally, IOMP works closely with CRC Press/Taylor & Francis in developing and delivery of the Series in Medical Physics and Biomedical Engineering (Suh et al. 2014), in which a number of books, including this one, target Radiation Protection issues.

1.7   CONCLUSION

Increasing use of medical imaging in modern medicine has resulted in a rapid increase of population radiation dose in recent years. The media reports on potential risk of radiation dose to patients have aroused some concerns in their communities. Although cancer risks due to medical imaging is very low, there is a need to strengthen radiation safety, dose reduction in particular, in radiation medicine based on the ALARA principle. Practically everyone involved in the delivery of medical radiation exposure has a role to play in radiation safety in medicine. Medical physicists, who are responsible for radiation safety and protection in radiation medicine, should take the lead in strengthening radiation safety in the clinics, especially in patient dose reduction and optimization. In order to be able to perform their tasks effectively and appropriately, medical physicists must be fully qualified, with the level of education and clinical training as recommended by international organizations such as IOMP and IAEA. They should also be undergoing continual professional development, to enable them to face future clinical challenges. National and international professional and statutory organizations also play an important role in strengthening radiation safety in medicine, by providing guidance on such issues as safety and quality standards, operational procedures and work protocols, and educational and professional qualifications for the healthcare professionals.

Equipment manufacturers also play an important role in improving radiation safety in healthcare by producing equipment technologies that are safe and suitable for meeting specific clinical service needs. This would not be possible if manufacturers did not have a good knowledge of what these problems and needs are, along with the workflow and procedures involved in clinical practice. This essentially demands a close collaboration between manufacturers and members of the clinical teams in the research and development of such technologies. In radiation medicine, these should include the radiologists, radiation oncologists, medical physicists, and technologists who have a keen interest in research and development work. The recent development and launching of low-dose CT and other imaging equipment is a good demonstration of user–manufacturer collaboration. This has a positive impact on patient dose reduction and improvement in radiation safety in radiology.

To better achieve the goal of strengthening radiation safety in healthcare and better protection of the patients from excessive or unnecessary radiation exposure, a closer collaboration between all the role players including medical physicists, radiologists, radiation oncologists, referring practitioners, technologists, regulators and equipment manufacturers is essential.

1.8   REFERENCES

American College of Radiology (2015). *Website on radiology information resources for patients: www.radiologyinfo.org* (Accessed January 25, 2015).

Bertocchi L, Benini A, Milano F, Padovani R, Sprawls P, Tabakov S (2014). *50 Years ICTP and its Activities in the Field of Medical Physics.* J. Medical Physics International, Vol.2 No.2, pp. 410–415, www.mpijournal.org/MPI-v02i02.aspx (Accessed January 25, 2015).

Damilakis J, Paulo G, Christofides S (2013). *A Study with European Professional Societies on Medical Radiation Protection Education and Training.* J. Medical Physics International, Vol.1 No.2, pp. 123–128, www.mpijournal.org/MPI-v01i01.aspx (Accessed January 25, 2015).

IAEA (2000). *Safety Reports Series No. 17: Lessons Learned from accidental exposures in radiotherapy.* International Atomic Energy Agency, 2000.

IAEA (2015). *Website on radiation protection of patient.* rpop.iaea.org/RPOP/RPoP/Content/index.htm (Accessed January 25, 2015).

International Organization for Medical Physics (2015). *IOMP Website.* www.iomp.org (Accessed January 25, 2015).

ICRP (2007). *The 2007 recommendations of the International Commission on Radiological Protection. ICRP publication 103.* Ann ICRP 2007; 37 (2-4) 1-332.

National Council on Radiation Protection and Measurements (NCRP) (2009). *Ionizing radiation exposure of the population of the United States (NCRP Report No. 160).* Bethesda, MD: NCRP.

National Research Council (2006). *Health risks from exposure to low levels of ionizing radiation: BEIR VII - Phase 2. Committee to Assess Health Risks from Exposure to Low Levels of Ionizing Radiation.* Washington, DC. National Academies Press.

Niroomand-Rad A, Orton C, Smith P H S, and Tabakov S (2014). *A History of the International Organization for Medical Physics - 50 Years Anniversary part II.* J. Medical Physics International, Vol.2 No.1, pp. 7–17, www.mpijournal.org/MPI-v02i01.aspx (Accessed January 25, 2015).

Rehani M (2013). *Radiation Protection of Patients Website of the IAEA as a Major Resource for Medical Physicists.* J. Medical Physics International, Vol.1 No.1, pp. 38–41, www.mpijournal.org/MPI-v01i01.aspx (Accessed January 25, 2015).

Smajo M E (2013). *Free Educational Resource: Medical Physics Clinical Skills Workbook for Therapy Physics.* J. Medical Physics International, Vol.1 No.1, pp. 42–45, www.mpijournal.org/MPI-v01i01.aspx (Accessed January 25, 2015).

Smith P H S, Nüsslin F (2013). *Benefits to Medical Physics from the Recent Inclusion of Medical Physicists in the International Classification of Standard Occupations.* J. Medical Physics International, Vol.1 No.1, pp. 11–15, www.mpijournal.org/MPI-v01i01.aspx (Accessed January 25, 2015).

Sprawls P, Duong P-A T (2013). *Effective Physics Education for Optimizing CT Image Quality and Dose Management with Open Access Resources.* J. Medical Physics International, Vol.1 No.1, pp. 45–49, www.mpijournal.org/MPI-v01i01.aspx (Accessed January 25, 2015).

Suh T S, McGowan F, Ng K H, Ritenour R, Tabakov S, Webster J G (2014). *IOMP Collaboration with CRC PRESS/TAYLOR & FRANCIS.* J. Medical Physics International, Vol.2 No.2, pp. 403–406, www.mpijournal.org/MPI-v02i02.aspx (Accessed January 25, 2015).

Tabakov S (2013). *IOMP Project for Medical Physics Development in Africa.* J. Medical Physics World, Vol.4 No.2, pp. 19–20, www.iomp.org/sites/default/files/empw-2013-02.pdf (Accessed January 25, 2015).

Tabakov S, Roberts C, Jonsson B, Ljungberg M, Lewis C, Strand S, Lamm I, Milano F, Wirestam R, Simmons A, Deane C, Goss D, Giraud J (2011a). *EMERALD and EMIT e-Learning Materials for Medical Physics Training, in Medical Physics and Engineering Education and Training - part I.* ISBN 92-95003-44-6, ICTP, Trieste, Italy, pp. 212–215, www.emerald2.eu/mep/e-book11/ETC_BOOK_2011_ebook_s.pdf (Accessed January 25, 2015).

Tabakov S, Smith P, Milano F, Strand S-E, Lamm I-L, Lewis C, Stoeva M, Cvetkov A, Tabakova V (2011b). *EMITEL e-Encyclopaedia of Medical Physics with Multilingual Dictionary, in Medical Physics and Engineering Education and Training - part I.* ISBN 92-95003-44-6, ICTP, Trieste, Italy, pp. 234–240, www.emerald2.eu/mep/e-book11/ETC_BOOK_2011_ebook_s.pdf (Accessed January 25, 2015).

United Nations Scientific Committee on the Effects of Atomic Radiation (UNSCEAR) (2008). *Report to the General Assembly: volume I, Annex A- Medical Radiation Exposures.* New York: United Nations.

United Nations Scientific Committee on the Effects of Atomic Radiation (UNSCEAR) (2012). *Report of the United Nations Scientific Committee on the Effects of Atomic Radiation.* 59th session, May 21–25, 2012. General Assembly Official Records. 67th session, Supplement No. 46.

International Radiation Protection Association Statement

Renate Czarwinski

President of the International Radiation Protection Association
Head of Section on Safety of Radiation Sources, Radiation Incidents and Type Approvals
Federal Office for Radiation Protection, Berlin, Germany

CONTENTS

T HE USE OF IONIZING RADIATION in healthcare is by far the largest contributor to the radiation exposure of the general population from artificial sources (UNSCEAR 2010). Safety and quality assurance in the use of radiation in medicine aims to reduce unnecessary radiation risks while maximizing the benefits. Improvements in quality and safety in radiation medicine require a strong radiation safety culture. Every healthcare practitioner involved in practices that utilize ionizing radiation has a role in assuring that benefits are maximized and risks are minimized.

2.1 THE INTERNATIONAL RADIATION PROTECTION ASSOCIATION

The International Radiation Protection Association (IRPA) is an international nongovernmental organization (NGO) whose primary purpose is to serve as a means whereby those engaged in radiation protection activities worldwide may communicate more readily with each other, and through this process, advance radiation protection throughout the world (IRPA 2014). IRPA was established in 1964 when the Health Physics Society in the United States formed an international section representing health physicists from 46 countries. The first General Assembly took place later that year in Paris. The meeting was hosted by what was then called the French Section of the Health Physics Society. A constitution was adopted with emphasis on providing protection of people and the environment from the hazards caused by ionizing radiation. Today IRPA is an independent NGO with approximately 18,000 members from 50 associate societies representing 63 countries from around the world (IRPA 2014).

The vision of IRPA today is to be recognized by its members, stakeholders, and the public as the international voice of the radiation protection profession in the enhancement of radiation protection culture and practice worldwide. The expertise of its members includes such branches of knowledge as science, medicine, engineering, technology, and law, to provide for the protection of people and their environment from hazards caused by ionizing radiation, and thereby to facilitate the safe use of medical, scientific, and industrial radiation practices for the benefit of mankind. In the field of medical care, radiation protection activities are often conducted under the leadership of medical physicists or medical health physicists, i.e., those who are specialized in the practice of radiation protection in medicine. For purposes of this discussion, medical physics practices are assumed to include radiation protection from sources of radiation used in medicine.

The basic objectives of IRPA are to:

- encourage the establishment of radiation protection societies throughout the world as a means of achieving international cooperation and expanding radiation protection efforts,

- provide for and support international meetings for the discussion of all aspects of radiation protection,

- encourage international publications dedicated to radiation protection,

- encourage research and educational opportunities in those scientific and related disciplines that support radiation protection,

- encourage the establishment and continuous review of universally acceptable radiation protection standards or recommendations through the international bodies concerned.

IRPA invests considerable effort to provide for and support regional and international meetings for the discussion of radiation protection matters. Through its congresses, IRPA assists the development of protection policies, criteria, methods, and radiation protection culture. The International Congresses of IRPA itself are the most important of these meetings. These have been held about every four years since 1966. For all Associate Societies of IRPA and individual members, it is an important objective to attend regional and international IRPA Congresses. The next International Congress will be held in Cape Town, South Africa, in 2016, when IRPA celebrates its 50th Anniversary. It is the 14th Congress in this series. For many related professions it is an excellent opportunity to communicate on the achievements, scientific knowledge, and operational experience in radiation protection. An emerging concern in radiation protection is the clear understanding of the radiation risks in different sectors, including the medical applications of ionizing radiation. Public understanding of risks and risk communication are therefore essential topics in IRPA's activities. Appropriate communication has to be extended beyond traditional stakeholders in radiation safety. Currently, for instance, IRPA is jointly working with the World Health Organization (WHO) and the International Organization of Medical Physics (IOMP) on the development of new guiding principles focused on safety culture in the medical sector, as part of the IRPA series of guiding principles. In addition, IRPA is involved in the WHO program on global safety in health care settings.

A further essential activity of IRPA is the continuation of the discussion of the revised dose limits for the lens of the eyes and the implications for dosimetry and methods of protection, in particular in the medical application of ionizing radiation.

IRPA promotes excellence in the practice of radiation protection through national and regional Associate Societies for radiation protection professionals. Education and training is a key factor in establishing effective national radiation protection programs. The IRPA Education and Training Plan has three objectives:

- cooperation with international and regional organizations dealing with education and training in radiation protection,

- stimulation of education and training by organizing discussion forums during IRPA Congresses,

- stimulation and support of education and training activities either by IRPA or by its Associate Societies.

IRPA encourages its members to undertake cooperative education and training activities by two or more Associate Societies; to promote education and training networks between Associate Societies sharing a common language or having regional proximity; and to activate the emergence of activities to attract young generations to the profession.

2.2 FOSTERING MEDICAL PHYSICS IN DEVELOPING COUNTRIES

In low- and medium-income countries where there is a need for medical physics and radiation protection services, IRPA encourages increased awareness of medical physics (including medical health physics) by governments and health authorities. This awareness is particularly important following the advancement of medical imaging and the increasing significance of cancer.

By virtue of their education and professional training, medical physicists are key players in radiation medicine, including diagnostic imaging by X-ray and nuclear medicine technology, radiation oncology, radiation protection, and related activities. The highest relevance of radiation protection in developing countries relates to the medical application of ionizing radiation to cancer management. All healthcare facilities that utilize ionizing radiation require the services of medical physicists who, with appropriate training and experience in radiation safety matters, can undertake all regulated activities of the radiation protection officer. Large healthcare facilities are encouraged to employ medical health physicists who can dedicate the time necessary to assure quality and the safe use of ionizing radiation.

IRPA and IOMP work together with other relevant organizations to:

- issue joint guidance and recommendations on the implementation of International Atomic Energy Agency (IAEA) Basic Safety Standards (BSS) in healthcare, including the roles of medical physicists and radiation protection officers,

- pursue the implementation of plans for education, professional training, certification, and registration of medical physicists who undertake the roles required by the BSS,

- encourage joint training courses and programs in radiation protection in healthcare to ensure that medical physicists involved in radiation protection activities have appropriate training and experience,

- promote the roles of medical physicists and radiation protection officers to governments, health authorities, and hospitals, to ensure that all healthcare facilities have adequate access to medical physicists and radiation protection officers.

2.3 COLLABORATION BETWEEN IOMP AND IRPA ON THE USE OF IONIZING RADIATION IN HEALTHCARE

In 2010, IOMP and IRPA signed a "Statement of Collaboration on the use of ionizing radiation in healthcare" as a platform for collaboration on activities to improve the safe use of radiation in healthcare without compromising quality of care. Specific objectives were to develop guidance for fostering and enhancing radiation protection culture in healthcare and to foster medical physics in developing countries (IRPA 2010). These objectives are based on the terminology and responsibilities of healthcare personnel as describe in the IAEA "Radiation Protection and Safety of Radiation Sources: International Basic Safety Standards," widely referred to as the BSS.

Under this statement, IOMP and IRPA work together to

- collaborate with WHO and IAEA to produce guidance for radiation protection and medical physics professionals;

- conduct workshops as needed to determine the elements for a definition of radiation protection culture, elements, or traits of such a culture, the criteria for assessing the success of the effort, the assessment tools to be used, methods for engaging stakeholders, and the role of radiation protection and medical physics professionals;

- draft a discussion document on radiation protection culture and make the document available to the IRPA and IOMP membership for comment and discussion and feedback. A draft document was presented for discussion at the IRPA13 Congress in Glasgow, May 2012, and approved/published during the European IRPA Congress in Geneva in June 2014.

Partnerships with international organizations as well as with professional bodies and other nongovernmental organizations are essential to achieve an understandable and sustainable radiation protection regime in medicine.

2.4   REFERENCES

IRPA (2010). *Statement of Collaboration between IRPA and IOMP to Develop Guidance for Fostering and Enhancing Radiation Protection Culture in Health Care.* http://irpa.net/members/54511/%7B86A1C65B-ED4E-4D73-868C-58E875B8D42F%7D/EC61-12 IRPA-IOMP Statement of Collab-GuidRPCHealthCare.docx, (Accessed December 29, 2014).

International Radiation Protection Association (2014). *www.irpa.net.* (Accessed December 29, 2014).

United Nations Scientific Committee on the Effects of Atomic Radiation (UN-SCEAR) (2010). *2008 report to the General Assembly: volume I, Annex A- Medical Radiation Exposures.* New York: United Nations.

II

The System of Radiation Protection

CHAPTER 3

The System of Radiation Protection

Christopher H. Clement
Scientific Secretary, International Commission on Radiological Protection

CONTENTS

THE INTERNATIONAL COMMISSION ON RADIOLOGICAL PROTECTION (ICRP) is an independent international organization that advances for the public benefit the science of radiological protection, in particular by providing recommendations and guidance on all aspects of protection against ionizing radiation. It is a Registered Charity (a not-for-profit organization) in the United Kingdom, and has a Scientific Secretariat in Ottawa, Canada.

ICRP consists of eminent scientists and policy makers in the field of radiological protection. The nearly 250 members of the Main Commission, Committees, and Task Groups are volunteers, most of whose employers pay for their time and travel expenses to work with ICRP. Some volunteer their time outside of regular work or after retirement. Members are invited to serve with ICRP based on the skills and knowledge they bring to the work, and as such do not represent their countries or employers when working with ICRP.

ICRP recommendations form the basis of radiological protection standards, legislation, programs, and practice worldwide. In preparing its recommendations, ICRP considers the fundamental principles and quantitative bases upon which appropriate radiation protection measures can be established, while leaving to the various national protection bodies the responsibility of formulating the specific advice, codes of practice, or regulations that are best suited to the needs of their individual countries.

3.1 HISTORICAL CONTEXT

Just months after the discovery of X-rays in November 1895 (Röntgen 1895), radiation damage to the skin was already being observed in early experimental investigators, who developed conditions including erythema, dermatitis, and ulceration (Grubbé 1933) (Leppin 1896). Nonetheless, X-rays, with their ability to help see inside the human body, were used almost immediately in the medical field, including in military field hospitals as early as 1897 (Churchill, 1898). Therapeutic uses were also tested as early as 1896 (Belot 1905). With

Curie's discovery of radium in 1898 (Curie 1898), use of radiation in medicine continued to increase, as did reports of radiation-induced damage in practitioners and patients.

3.2   FIRST RADIOLOGICAL PROTECTION RECOMMENDATIONS

On December 12, 1896, just one year after the discovery of X-rays, the first radiological protection recommendations were published in the Western Electrician by Wolfram Fuchs (Fuchs 1896). He reported having "applied the X-ray to all parts of the body" in 1,400 cases over nine months, and "but four instances of the slow healing burns which have lately attracted considerable attention through the columns of the press." Fuchs noted that "the injury may be regarded as slight in comparison with the benefits resulting from this wonderful discovery," adding, in the next sentence: "however, it is desirable, of course, to prevent the inconvenience and pain of these 'sunburns.' " Perhaps it is a bit of a stretch, but this evaluation of a net benefit might be considered the first published relatively general justification of the use of ionizing radiation in medicine.

Fuchs offered three specific suggestions:

- "First — Shorten the exposure to a minimum."

- "Second — Place the tube not nearer than 12 inches from the body."

- "Third — Rub vaseline well into the skin and leave a coating on the part of the body to be exposed; also, use glass plates for the protection of the parts not to be exposed."

These three suggestions look very much like the basic concepts of practical protection from external radiation still used today: minimize time, maximize distance, and use shielding.

3.3   FORMATION OF THE INTERNATIONAL COMMISSION ON RADIOLOGICAL PROTECTION

In 1925, the first International Congress of Radiology (ICR) was held in London. Here, the International X-ray Unit Committee, now the International Commission on Radiation Units and Measurements, was formed. Discussions were also held regarding creation of an international radiological protection committee. This occurred at the second ICR in Stockholm in 1928, when the International X-ray and Radium Protection Committee (IXRPC) was created.

Arising from the 1928 ICR were the first "International Recommendations for X-Ray and Radium Protection" (ICR 1929). Weighing in at just three and a half pages, and published together in English, German, and French, the focus of these recommendations was squarely aimed at protection of "X-ray and radium workers" in medical facilities. They included advice on maximum working hours and minimum holiday allotments ("not less than one month's holiday a year"), sunshine and fresh air in the rooms of the X-ray department, precautions against electrical hazards of X-ray equipment, and even decorating suggestions ("All rooms should preferably be decorated in light colours."). These are not features found in modern radiological protection recommendations! However, also included was advice to which a modern practitioner of radiological protection might better relate. For example, it included admonitions for operators to avoid unnecessary exposures to a direct beam of X-rays, staying "as remote as practicable from the X-ray tube," and using shielding around the X-ray tube.

After the ICR in Stockholm, the IXRPC met next during the ICRs held in Paris in 1931, and then in Zurich in 1934. During the latter the IXRPC was faced with the hosts insisting on having four (of 11) participants, and the German authorities replacing the Jewish German member with another person (Lindell 1996). New rules were established to maintain full control over future membership, and the "C" in IXRPC changed from Committee to Commission, perhaps to emphasize its more independent nature.

The IXRPC met again in Chicago at the 1937 ICR, and then not until after World War II in London in 1950 at which time it was renamed the International Commission on Radiological Protection (ICRP).

So, just two decades after the discovery of X-rays, ICRP (as the IXRPC) was born when the great benefits of the use of ionizing radiation in medicine were first being realized, while at the same time concerns within the radiology community were increasing about skin burns and other negative effects being seen in practitioners, researchers, and patients. The first business of the organization was radiological protection in medicine. This remains an important part of the business of ICRP, which now encompasses all aspects of protection of people and the environment from the effects of ionizing radiation, whatever the source and whatever the conditions of exposure.

To this day ICRP continues to be the leading international organization with respect to the development and maintenance of the system of radiological protection. It has published nearly 150 reports on all aspects of protection from ionizing radiation, since 1977 in its dedicated journal The Annals of the ICRP. The recommendations of ICRP form the basis of radiological protection standards, legislation, guidance, and practice worldwide.

3.4   THE SYSTEM OF RADIOLOGICAL PROTECTION

ICRP Publication 103, The 2007 Recommendations of the International Commission on Radiological Protection (ICRP 2007b) is the most recent set of "fundamental" recommendations. This publication describes the entire system of radiological protection, and thus is heavily referenced in this chapter.

3.5 SCOPE

The scope of the system of radiological protection includes all exposures to ionizing radiation from any source, regardless of its size and origin. This includes both natural and man-made sources of radiation. However, the system can only apply fully when the source of exposure or the pathways leading to the exposure of individuals can reasonably be controlled.

A single set of fundamental principles of protection is applied in all circumstances, although how they are applied in each instance may differ. However, this broad scope does not mean equal treatment in terms of protection in all circumstances. In general, a graded burden of obligation is expected; for example, through regulatory controls or procedural arrangements, depending on the amenability of control and the level of risk. Two concepts help delineate the domain of radiological protection control: exclusion and exemption.

Certain situations can be excluded on the basis that they are not amenable to control, either because control is not possible under any conceivable circumstance (consider exposure to potassium-40 that occurs naturally in the human body) or where control is obviously impractical (consider exposure to cosmic radiation at ground level).

In addition, exemption from regulatory requirements is sensible for situations where such controls are regarded as unwarranted, i.e., where the effort to control is excessive compared to the associated risk.

In either case, it is normally up to national regulators to decide what is impractical to control (and thus excluded), and under what circumstances control is unwarranted (and thus excluded).

3.6 AIMS

3.6.1 Primary Aim

The primary aim of the system of radiological protection is "to contribute to an appropriate level of protection for people and the environment against the detrimental effects of radiation exposure without unduly limiting the desirable human actions that may be associated with such exposure." (ICRP 2007b)

3.6.2 Human Health Objectives

The human health objectives of the system of radiological protection "are relatively straightforward: to manage and control exposures to ionizing radiation so that deterministic effects are prevented, and the risks of stochastic effects are reduced to the extent reasonably achievable." (ICRP 2007b)

Deterministic effects are, in principle, preventable by keeping doses below a threshold, while conversely for stochastic effects the assumption is that even a small dose might give rise to a correspondingly small risk. So, in short, the system aims to prevent harm that is preventable, and manage what cannot be prevented.

3.7 BASIS

The system of radiological protection is based on scientific knowledge, ethical values, and experience.

Science can describe what is. In radiological protection, science can give us information about, for example, how radioactive materials move through the environment and through human bodies once inhaled or ingested, how radiation interacts with matter, and about effects of radiation on human health and the environment. All of these aspects are important for radiological protection, but for the purposes of this chapter it is the effects of radiation on humans that is central.

Ethics can help us decide what should be. In radiological protection, value judgments are necessary, for example, to decide between what might be tolerable or intolerable, in terms of radiation exposure, in a particular circumstance. Ethical principles explicitly relied upon in the system of radiological protection include prudence (precaution and wisdom) and justice (equity). Beneficence (doing good) is among the key ethical principles clearly embodied in the system, but not explicitly mentioned in the 2007 Recommendations.

Experience helps us decide what is practicable. It helps, for example, in deciding on the appropriate balance between scientific precision and rigor, and simplifications that ease understanding and implementation of the system of protection.

3.7.1 Biological Basis

For the purposes of radiological protection, adverse health effects of radiation exposure are divided into two general categories: stochastic effects and deterministic effects.

Stochastic effects are defined in ICRP Publication 103 as "Malignant disease and heritable effects for which the probability of an effect occurring, but not its severity, is regarded as a function of dose without threshold." Stochastic effects include cancer and heritable effects.

Deterministic effects have been described in detail in ICRP Publication 118 (ICRP 2012a). In the glossary of this publication a deterministic effect is described as "Injury in populations of cells, characterized by a threshold dose and an increase in the severity of the reaction as the dose is increased further. Also termed 'tissue reaction.' "

The simplistic characteristics of these two types of effect, and the categorization of adverse health effects into these two categories, are simplifications for the purposes of radiological protection. These simplifications are among the many needed to make the system of radiological protection practicable. They are useful for protection purposes, but should not be confused with biological or clinical descriptions, which may need to retain more complexity.

3.7.2 Stochastic Effects

The definition of a stochastic effect says that probability is "a function of dose without threshold" but does not specify how probability varies with dose. Nonetheless, the shape of this dose-response relationship is a central feature of the system of radiological protection. This subject is treated in detail in ICRP Publication 99, Low-dose Extrapolation of Radiation-related Cancer Risk (ICRP 2005c).

The system of radiological protection uses the Linear No-Threshold (LNT) model as a prudent basis. This model assumes that any dose, no matter how small, will produce a proportionate increment in probability of incurring cancer or heritable effects. LNT is a scientifically plausible model, but is not universally accepted as the biological truth. Nonetheless, use of the LNT model is considered to be prudent for public policy aimed at avoiding unnecessary risk from exposure to ionizing radiation. The LNT model also has practical benefits; for example, it makes the addition of doses sensible, even if received at different times, under different circumstances, and through different routes.

3.7.3 Deterministic Effects (Tissue Reactions)

Deterministic effects (tissue reactions) were treated most recently in detail in ICRP Publication 118 (ICRP 2012a). For these types of effects, the key parameter, in terms of radiological protection, is the threshold. Recalling that one of the two primary aims for protection of human health is to prevent deterministic effects, the threshold provides the dose that one must stay below to achieve this aim.

ICRP Publication 118 focused on estimating dose thresholds, defined as the dose resulting in 1% of individuals exhibiting a specified effect. Although this "is not a 'true' threshold in the sense of the effect not occurring at all, it is used here in a practical sense for protection purposes. The use of a smaller level ... would entail a greater extrapolation of response frequencies to even lower doses, with concomitant greater uncertainties attached to the value. The use of a higher level would have less uncertainties in the value, and this may be acceptable in practical situations for some endpoints but not others. However, it would be even further from the 'true' threshold." (ICRP 2012a) Thus, defining a "threshold" "does not imply that no biological effects occur at lower doses; it merely defines the dose above which a specified effect becomes clinically apparent in a small percentage of individuals" (idem).

3.8 DOSE AND RISK

A central feature of the system of radiological protection is calculation of the dose of radiation received. Three dose quantities are used: the physical quantity of absorbed dose (which can relate to any mass of matter), the protection quantities of equivalent dose (which relates to specific organs and tissues), and the effective dose (which relates to the whole body).

3.8.1 Absorbed Dose

The starting point is the physical quantity absorbed dose, D, defined as the mean energy imparted to matter by ionizing radiation divided by the mass of the matter. The SI unit of absorbed dose is joules per kilogram (J kg^{-1}), and is given the special name of gray (Gy). As an example, if ionizing radiation imparts 12 J to 3 kg of water, the absorbed dose to the water is 12 J / 3 kg = 4 Gy. Absorbed dose is a measurable quantity, unlike the protection quantities.

In practical applications of radiological protection, absorbed dose is averaged over tissue volumes, often entire organs. A basic premise is that, at relatively low doses, the absorbed dose averaged over a specific organ or tissue is correlated with radiation detriment for stochastic effects with a sufficient accuracy for protection purposes. However, there are cases in which the highly heterogeneous deposition of energy is important. "Specific dosimetric models have been developed to take account of such heterogeneity in the distribution and retention of activity and of sensitive regions in these particular cases" (ICRP 2007b).

3.8.2 Equivalent Dose

Different ionizing radiations (alpha, beta, gamma, neutron, and various charged particles) have differing levels of effectiveness in causing biological effects. Equivalent dose is based on absorbed dose, but takes into account the relative biological effectiveness (RBE) of various radiation types.

RBE is the absorbed dose of the radiation of interest divided by the same absorbed dose of a reference radiation (typically relatively low energy photons) producing the same level of biological effect. RBE values depend on a wide variety of factors such as the tissue or cell type used, the dose, the dose rate, any dose fractionation, the biological effect, and the specific characteristics of the reference radiation. As a result, there are a wide variety of RBE values for any given radiation type and energy.

Based on experimental results of RBE, and applying judgment to ensure a practicable system, a set of tissue weighting factors, w_R, has been established to represent their relative effectiveness in producing stochastic effects (see Table 3.1).

Table 3.1: Recommended radiation weighting factors

Radiation Type	w_R
Photons	1
Electrons and muons	1
Protons and charged pions	2
Alpha particles, fission fragments, heavy ions	20
Neutrons	Continuous function of energy, ranging from 2.5 to 20*

* w_R for neutrons is calculated using the following formula, where En is the neutron energy:

$$w_R = \begin{cases} 2.5 + 18.2e^{-[\ln (E_n)]^2/6}, & E_n < 1 \text{ MeV} \\ 5.0 + 17.0e^{-[\ln (2E_n)]^2/6}, & 1 \text{ MeV} \leqslant E_n \leqslant 50 \text{ MeV} \\ 2.5 + 3.25e^{-[\ln (0.04E_n)]^2/6}, & E_n > 50 \text{ MeV} \end{cases}$$

(Adapted with permission from ICRP 2007b, Table 3.2)

Table 3.2: Recommended tissue weighing factors

Tissue	w_T
Red bone-marrow, Colon, Lung, Stomach, Breast, Remainder tissues*	0.12
Gonads	0.08
Bladder, Oesophagus, Liver, Thyroid	0.04
Bone surface, Brain, Salivary glands, Skin	0.01
Total	1

* The w_T for the remainder tissues applies to the mean dose of the following 13 organs and tissues for each sex: Adrenals, Extrathoracic region, Gall bladder, Heart, Kidneys, Lymphatic nodes, Muscle, Oral mucosa, Pancreas, Prostate (male), Small intestine, Spleen, Thymus, Uterus/cervix (female).
(Adapted with permission from ICRP 2007b, Table 3)

Table 3.3: Detriment-adjusted nominal risk coefficients (10-2 Sv-1, or percent per sievert) for stochastic effects after exposure to radiation at low doses

Exposed Population	Cancer	Heritable Effects	Total
Whole	5.5	0.2	5.7
Adult	4.1	0.1	4.2

(Adapted with permission from ICRP 2007b, Table 1)

Table 3.4: Recommended dose limits in planned exposure situations

Type of limit	Occupational	Public
Effective dose	20 mSv per year, averaged over defined periods of 5 years, and no more than 50 mSv in any single year	1 mSv in a year
Equivalent dose to lens of the eye	20 mSv per year, averaged over defined periods of 5 years, and no more than 50 mSv in any single year	15 mSv in a year
Equivalent dose averaged over 1 cm^2 area of skin	500 mSv in a year	50 mSv in a year
Equivalent dose to hands and feet	500 mSv in a year	-

(Adapted with permission from Table 6 of ICRP 2007b, as modified by ICRP 2012a)

Equivalent dose to a tissue is the sum over all radiation types of the absorbed dose of each radiation type multiplied by the relevant radiation weighting factors. The SI unit of equivalent dose is joules per kilogram (J kg^{-1}), and is given the special name of sievert (Sv).

As an example, if the liver is subjected to an absorbed dose of 1 mGy from photons ($w_R = 1$), the equivalent dose to the liver is 1 mSv. However, if the liver is subjected to an absorbed dose of 1 mGy from alpha particles ($w_R = 20$), the equivalent dose to the liver is 20 mSv, reflecting the substantially higher effectiveness of alpha particles relative to photons in causing biological effects.

When more than one type of radiation is involved, a summation over all radiation types is required. For example, if the liver is subjected to an absorbed dose of 0.5 mGy from photons, and 0.1 mGy of alpha particles, the equivalent dose to the liver is 0.5 mGy × 1 + 0.1 mGy × 20 = 2.5 mSv.

3.8.3   Effective Dose

Different organs and tissues have different sensitivities to ionizing radiation. Building on equivalent dose, which already takes into account the differing effectiveness of radiation types, effective dose goes one step further and takes into account the differing sensitivities of organs and tissues. This is done by assigning tissue weighing factors (w_T) (see Table 3.2) to the organs and tissues of the body, based on epidemiological studies of cancer induction and risk estimates for heritable effects.

Effective dose (H) is the sum over all tissues of the equivalent dose in each tissue (D_T)- multiplied by the relevant for each tissue; thus, $H = \Sigma_T\ w_T\ D_T$. The SI unit of effective dose is joules per kilogram (J kg^{-1}), and is given the special name of sievert (Sv). This reflects the overall detriment, and therefore effective dose is sometimes referred to informally as the "whole-body dose." As an example, if the equivalent dose to the stomach is 0.5 mSv, and the effective dose to the gonads is 10 mSv, and there is no other dose to any other tissue, then the equivalent dose is 0.5 mSv × 0.12 + 10 mSv × 0.08 = 0.86 mSv.

3.8.4   Practical Dose Calculation

Calculation of equivalent and effective doses in most cases has been simplified by providing tabulated results; for example, those found in ICRP Publication 119, Compendium of Dose Coefficients, based on ICRP Publication 60 (ICRP 2012b). Note that at the time of publication of this text ICRP is developing a new set of coefficients based on the 2007 Recommendations (ICRP 2007). The first of these are found in ICRP Publication 116, Conversion Coefficients for Radiological Protection Quantities for External Radiation Exposures (ICRP 2010a).

3.8.5   Risk and Limitations of Equivalent and Effective Dose

Equivalent and effective dose, and their predecessors, have proven useful for the purpose for which they were designed: radiological protection. Based on risk, these quantities use age- and sex-averaged weighting factors selected using epidemiological results from diverse populations, in addition to many other factors generalized at a population level, such as the size and placement of organs and biokinetic models.

ICRP has calculated detriment-adjusted nominal risk coefficients for stochastic effects after exposure to radiation at low doses (see Table 3.3).

The figures in Table 3.3, and the oft-quoted total risk of "5% per Sv," can be considered reasonable estimates for moderate effective doses (from around 100 Sv to a few Sv) for a large and diverse population. Below around 100 mSv these estimates remain scientifically plausible, but become increasingly uncertain, although in this region it is known that the risk per unit dose is not enormously higher as this would be detected in epidemiological studies. Above a few sieverts of effective dose, deterministic effects begin to predominate, and therefore the applicability of effective dose (which is based on stochastic effects) starts to break down.

As a result, the figures in Table 3.3 are not useful for calculating the risk to an individual with reasonable certainty, and this is doubly so at lower doses (below around 100 mSv). Retrospective evaluation of radiation, related risks is better accomplished using age- and sex-specific data, as well as any other relevant factors relating to the exposure(s) and the individual(s) in question. Even then, significant differences in individual radiosensitivity have been observed, the reasons for which are not all well known, further complicating attempts to calculate risks to individuals.

3.9   FUNDAMENTAL PRINCIPLES OF PROTECTION

If all radiation effects were deterministic in nature, with relatively high thresholds, then the system of radiological protection could be quite straightforward. In this case, it might be enough to simply avoid exposures that would exceed these thresholds, thereby avoiding any negative effects of radiation. Exposures below these thresholds would be considered "safe" and those above potentially "dangerous."

Unfortunately, life is not this simple. The probabilistic nature of stochastic effects, as embodied by the LNT model that posits small risks even for the lowest doses, means there is no clear distinction between levels of exposure that are safe and dangerous. The appropriate level of protection (referred to in the primary aim of the system) is based on what is deemed acceptable in a given circumstance. Here, ethical considerations such as justice (equity) and beneficence (doing good) play a key role, leading to the three fundamental principles of the ICRP system of protection:

- Justification,

- Optimization, and

- Limitation.

3.9.1   Justification

"The principle of justification: Any decision that alters the radiation exposure situation should do more good than harm." (ICRP 2007b)

In other words, in introducing a new source of radiation, or reducing or removing a risk of potential exposure from an existing one, the benefits should outweigh the detriments. This principle applies under all circumstances.

3.9.2   Optimization

"The principle of optimisation of protection: the likelihood of incurring exposures, the number of people exposed, and the magnitude of their individual doses should all be kept as low as reasonably achievable, taking into account economic and societal factors." (ICRP 2007b)

Going beyond the principle of justification, the principle of optimization demands that the margin of benefits over detriments (the net benefit) to society and individuals should be maximized. This concept defines the best level of protection under the prevailing circumstances. This principle applies under all circumstances.

However, to address considerations of justice (fairness) that could be compromised by severely inequitable outcomes of optimization, restrictions on doses or risk to individuals from a particular source are introduced. These include "soft" restrictions such as dose and risk constraints (discussed later), as well as the "hard" restrictions imposed by dose limits.

3.9.3 Dose Limitation

"The principle of application of dose limits: The total dose to any individual from regulated sources in planned exposure situations other than medical exposure of patients should not exceed the appropriate limits recommended by the Commission." (ICRP 2007b)

Dose limits are regarded as "hard" restrictions on doses to individuals as they are typically introduced into radiological protection legislation, with penalties associated with causing an individual to exceed these limits. Dose limits do not apply in all circumstances (only in planned exposure situations, as described later). In other circumstances reliance on the "softer" restrictions noted above is more appropriate.

3.10   STRUCTURE AND KEY FEATURES

The system of radiological protection organizes exposures to individuals in two different ways: by the exposure situation (planned, existing, or emergency), and by the category of exposure (occupational, medical, and public). This organization allows the fundamental principles to be applied in the most sensible way given the characteristics of the exposure situation and category of exposure in question.

3.10.1   Exposure Situations

The three exposure situations (planned, existing, and emergency) address all conceivable circumstances.

"Planned exposure situations are where radiological protection can be planned in advance, before exposures occur, and where the magnitude and extent of the exposures can be reasonably predicted." (ICRP 2007b)

"Existing exposure situations are those that already exist when a decision on control has to be taken. There are many types of existing exposure situations that may cause exposures high enough to warrant radiological protective actions, or at least their consideration. Radon in dwellings or the workplace, and naturally occurring radioactive material (NORM) are well-known examples." (ICRP 2007b)

"Emergency exposure situations are unexpected situations that may require urgent protective actions, and perhaps also longer-term protective actions, to be implemented" (ICRP 2007b)

3.10.2   Categories of Exposure

An important simplification in the system of radiological protection is that "individuals are subject to several categories of exposure, which can be dealt with separately ... . For example, most workers who are exposed to radiation sources as part of their work are also exposed to environmental sources as members of the public, and to medical exposure as patients" (ICRP 2007b). The control of exposures in one category need not be influenced by the exposures from the others.

"Occupational exposure is ... all radiation exposure of workers incurred as a result of their work ... that can reasonably be regarded as being the responsibility of the operating management." (ICRP 2007b).

Occupational exposure can occur in planned, existing, and emergency exposure situations. "Medical exposure refers to ... the exposure of individuals for diagnostic, interventional, and therapeutic purposes" (ICRP 2007b). It can also refer to exposures of non-occupational comforters and carers (e.g., family and friends), and volunteers in biomedical research. All medical exposures are in planned exposure situations. Even "emergency" medical procedures, including protection measures, are planned in advance.

"Public exposure encompasses all exposures of the public other than occupational exposures and medical exposures" (ICRP 2007b). Public exposures are possible in planned, existing, and emergency exposure situations.

3.11   RADIOLOGICAL PROTECTION IN MEDICINE

The objective of radiological protection in medicine is to provide optimal protection to staff, patients, and members of the public, when radiation is used for medical purposes. This includes the use of radiation in diagnosis (generally imaging) and treatment, and also in medical procedures that do not fit neatly into just one of these categories, such as fluoroscopically guided procedures. ICRP Publication 105, Radiological Protection in Medicine (ICRP 2007c) is highly recommended reading for those with a keen interest in this subject, as it treats it much more thoroughly than is possible here. Another useful reference is ICRP Supporting Guidance 2, Radiation and your patient: A guide for medical practitioners (ICRP 2001). More specific radiological protection advice related to medicine is provided in other ICRP publications, including:

- ICRP Publication 97, Prevention of High-dose-rate Brachytherapy Accidents (ICRP 2005a)

- ICRP Publication 98, Radiation Safety Aspects of Brachytherapy for Prostate Cancer Using Permanently Implanted Sources (ICRP 2005b)

- ICRP Publication 102, Managing patient dose in Multi-Detector Computed Tomography (MDCT) (ICRP 2007a)

- ICRP Publication 112, Preventing Accidental Exposures from New External Beam Radiation Therapy Technologies (ICRP 2009b)

- ICRP Publication 113, Education and Training in Radiological Protection for Diagnostic and Interventional Procedures (ICRP 2009c)

- ICRP Publication 117, Radiological Protection in Fluoroscopically Guided Procedures Performed Outside the Imaging Department (ICRP 2010b)

- ICRP Publication 120, Radiological Protection in Cardiology (ICRP 2013)

3.12 PROTECTION OF WORKERS AND THE PUBLIC

Radiological protection of workers (receiving occupational exposures) and members of the public (receiving public exposures) in medicine is not fundamentally different from protection in other industries. Dose limits must be respected (see Table 3.4) for planned exposures, and protection must be optimized. Optimization is generally reducing all doses to "levels that are as low as reasonably achievable, economic and societal factors being taken into account". (ICRP 2007b)

3.13 PROTECTION OF COMFORTERS AND CARERS, AND VOLUNTEERS IN BIOMEDICAL RESEARCH

The vast majority of medical exposures are received by patients. Nonetheless, the definition of medical exposures also includes those received by comforters and carers, and volunteers in biomedical research. In these latter cases, the use of dose constraints is central. Here, a dose constraint is an input in the planning phase that restricts the dose that would be received by a comforter, carer, or volunteer in biomedical research. It provides a basic level of protection without imposing a rigorous limit.

Comforters and carers are friends and relations voluntarily helping in the support and comfort of a patient. Dose constraints should be used in defining protection policies for visitors to patients, and families at home caring for discharged nuclear medicine patients. "A value of 5 mSv per episode for an adult (i.e., for the duration of a given release of a patient after therapy) is reasonable ... [but] higher doses may well be appropriate for [for example] the parents of very sick children." However, "young children, infants, and visitors not engaged in direct comforting or care should be treated as members of the public (subject to the public dose limit of 1 mSv/year)." (ICRP 2007c)

For the protection of volunteers in biomedical research, the "key aspects include the need to guarantee a free and informed choice by the volunteers, the adoption of dose constraints linked to the societal worth of the studies, and the use of an ethics committee that can influence the design and conduct of the studies." (ICRP 2007c)

3.14 PROTECTION OF PATIENTS

Patients may be exposed to ionizing radiation as a result of diagnostic, interventional, or therapeutic procedures. Radiological protection of patients is different in a number of respects from most other exposures, and therefore requires a slightly different approach.

The exposure of patients is deliberate, it is voluntary in nature, and there is an expectation of a direct health benefit for the patient. There is some degree of informed consent, which implies that the patient is aware of the expected benefit of the procedure to be undertaken as well as the risks involved, including the risks of radiation exposure. The degree of informed consent, and thus information provided, can be minimal for very low-risk procedures such as a chest X-ray, or quite extensive for higher-risk procedures such as for radiation therapy.

3.14.1 Dose Limits

Given these features, this is the one case where, for planned exposure situations, the fundamental principle of dose limitation does not apply, and dose constraints are not recommended for individual patients. Dose limits and dose constraints are tools designed to ensure an equitable distribution of risks among individuals when protection is optimized. Since essentially all of the dose, and all of the benefit, apply to a single individual — the patient — there is no dose distribution per se, and therefore ensuring an equitable distribution is irrelevant. Looking at it another way, the use of dose limits or dose constraints may unnecessarily restrict the dose that can be delivered, reducing the effectiveness of diagnosis or treatment, and in the end doing more harm than good, or at least making optimized treatment impossible.

Nonetheless, the other fundamental principles of radiological protection, justification and optimization, continue to apply.

3.14.2 Justification

Justification is about ensuring that more good than harm is done. "Most of the assessments needed for the justification of a radiological practice in medicine are made on the basis of experience, professional judgement, and common sense." (ICRP 2007c). However, it is important to note that in any circumstance there may be a wide variety of justified options (i.e., procedures that would result in more good than harm). Selecting the best procedure from among them goes beyond the question of justification.

In radiological protection of patients, justification is undertaken at three levels:

Level 1: At the most general level, the question is whether the proper use of radiation in medicine is justified. That the proper use of radiation in medicine does more good than harm, and is therefore justified, "is now taken for granted" (ICRP 2007c).

Level 2: At this level, the question is whether a specified procedure with a specified objective is justified. The aim is to judge whether the improvement in diagnosis or treatment is greater than the risk of exposure to radiation. This level of justification is normally done by national and international professional bodies and/or health or radiological protection authorities. The benefits to be considered go beyond the direct health benefits to the patients to include benefits to their families and society. Similarly, although the patients receive the main exposures, exposures to staff and the public should also be considered, as should the risk of accidental exposures. Decisions on justification at this level may change from place to place depending on, e.g., socioeconomic conditions, or simply the availability of equipment and trained personnel. They may also change over time as new techniques and technologies arise, and more information becomes available regarding risks and benefits.

Level 3: Beyond justifying a procedure in general, the application of the procedure to an individual patient should be justified, i.e., judged to do more good than harm to the specific patient under the specific circumstances. This would include, for example, checking that the necessary diagnostic information is not already available from a recent procedure. Beyond this, no specific justification is normally needed for simple diagnostic procedures where the procedure has already been justified at the general level (level 2). For more complex procedures, individual justifications may be made more efficiently by defining patient categories and referral criteria in advance.

3.14.3 Optimization

The basic aim of optimization of protection is to maximize the margin of benefits over detriments (maximize the net benefit).

Optimization of protection in medicine is best described as management of the radiation dose to the patient commensurate with the medical purpose. First, the procedure must gather the necessary diagnostic information or achieve the desired treatment outcome. Restricting exposures to such an extent that this is not achieved is not productive, and contrary to optimization. Thus, optimization of doses to patients does not necessarily mean reduction of doses.

"For example, diagnostic radiographic equipment often uses antiscatter grids to improve the image quality, yet removing the grid would allow a reduction in dose by a factor of 2-4. For radiography of the abdomen of adults, where the scattered radiation is important, the net benefit would be reduced by removing the grid because the benefit of the dose reduction would be more than offset by the loss of quality of the image. The optimisation of protection would not call for the removal of the grid. In the radiography of small children, however, the amount of scattered radiation is less and the benefit of the dose reduction resulting from the removal of the grid is not fully offset by the small deterioration of the image. The optimisation of protection then calls for the reduction in dose allowed by the removal of the grid." (ICRP 2007c)

In radiological protection of patients, optimization is usually applied at two levels: first, in the design, selection, and construction of equipment and installations, and second in the day-to-day working procedures.

For optimization of patient doses in radiation therapy it is necessary to differentiate between the dose to the target tissue and the dose to the rest of the patient. The dose to the target tissue must be sufficiently high to ensure effective treatment. However, this being achieved, optimizing the protection of tissues outside the target volume involves keeping doses to these tissues "as low as reasonably achievable" — to minimize the probability of unacceptable complications in normal tissue. This is an integral part of dose planning.

Optimization of patient doses in diagnostic and interventional procedures can often be aided by the use of diagnostic reference levels (DRLs). These help evaluate whether doses to patients (in general, not individual patients) are unusually high or low for a specific imaging procedure.

3.14.4   Diagnostic Reference Levels

The purpose of a DRL is to help in the optimization of protection of patients by avoiding radiation dose that does not contribute to the clinical purpose of the imaging procedure. "A diagnostic reference level can be used:

- to improve a regional, national, or local distribution of observed results for a general medical imaging task, by reducing the frequency of unjustified high or low values;

- to promote attainment of a narrower range of values that represent good practice for a more specific medical imaging task; or

- to promote attainment of an optimum range of values for a specified medical imaging protocol."

(ICRP 2007c) This is accomplished by comparing the value of the DRL to the distribution of values observed in practice for a given medical imaging task. To ensure that this comparison is meaningful it is important that the reference group of patients, usually defined as being within a certain range of physical parameters such as height and weight, is appropriate for the selected DRL.

If this comparison reveals that procedures consistently cause the DRL to be exceeded, then a local review of procedures and equipment is in order. This may reveal opportunities to improve optimization.

The selection of DRL values should be made by professional medical bodies in conjunction with national health and radiological protection authorities. The values should be based on relevant regional, national, or local data.

DRLs should apply to easily measured quantities that are reasonable relative indicators of patient dose. Depending on the procedure, these might include the entrance surface air kerma (in mGy), dose length product (DLP), dose area product (DAP), milliampere seconds (mAs), or administered activity (in MBq).

3.15   ACKNOWLEDGEMENTS

This chapter is based primarily on ICRP Publication 103, The 2007 Recommendations of the International Commission on Radiological Protection (ICRP 2007b), and ICRP Publication 105, Radiological Protection in Medicine (ICRP 2007c). As such, due credit is owed to the many contributors to those publications.

The description of the historical context draws extensively on The History of ICRP and the Evolution of its Policies by R.H. Clarke and J. Valentin, published as part of ICRP Publication 109 (ICRP 2009a).

3.16   REFERENCES

Belot, J., 1905. Radiotherapy in skin disease. Rebman.

Churchill, W.S., 1898. The Story of the Malakind Field Force. Longman's Green & Co., London.

Curie, M., 1898. Rayons emis par les composes de l'uranium et du thorium. C. R. Acad. Sci. Paris 126, 1101.

Fuchs, W.C., 1896. Effect of the Röntgen Rays on the Skin. Western Electrician, December 1896, 291.

Grubbé, E.H., 1933. Priority in the therapeutic use of X-rays. Radiology XXI, 156–162.

ICR, 1929. International Recommendations for X-ray and Radium Protection. A Report of the Second International Congress of Radiology. P.A. Nordstedt & Söner, Stockholm, pp. 62–73.

ICRP, 2001. Radiation and your patient: a guide for medical practitioners. Also includes: Diagnostic reference levels in medical imaging - review and additional advice. ICRP Supporting Guidance 2. Ann. ICRP 31(4).

ICRP, 2005a. Prevention of High-dose-rate Brachytherapy Accidents. ICRP Publication 97. Ann. ICRP 35(2).

ICRP, 2005b. Radiation Safety Aspects of Brachytherapy for Prostate Cancer Using Permanently Implanted Sources. ICRP Publication 98. Ann ICRP 35(3).

ICRP, 2005c. Low-dose Extrapolation of Radiation-related Cancer Risk. ICRP Publication 99. Ann. ICRP 35(4).

ICRP, 2007a. Managing patient dose in Multi-Detector Computed Tomography (MDCT). ICRP Publication 102. Ann. ICRP 37(1).

ICRP, 2007b. The 2007 Recommendations of the International Commission on Radiological Protection. ICRP Publication 103. Ann. ICRP 37 (2-4).

ICRP, 2007c. Radiological Protection in Medicine. ICRP Publication 105. Ann. ICRP 37 (6).

ICRP, 2009a. Application of the Commission's Recommendations for the Protection of People in Emergency Exposure Situations. ICRP Publication 109. Ann. ICRP 39 (1).

ICRP, 2009b. Preventing Accidental Exposures from New External Beam Radiation Therapy Technologies. ICRP Publication 112. Ann ICRP 39(4).

ICRP, 2009c. Education and Training in Radiological Protection for Diagnostic and Interventional Procedures. ICRP Publication 113, Ann. ICRP 39 (5).

ICRP, 2010a. Conversion Coefficients for Radiological Protection Quantities for External Radiation Exposures. ICRP Publication 116, Ann. ICRP 40(2-5).

ICRP, 2010b. Radiological Protection in Fluoroscopically Guided Procedures Performed Outside the Imaging Department. ICRP Publication 117. Ann. ICRP 40(6).

ICRP, 2012a. ICRP Statement on Tissue Reactions/Early and Late Effects of Radiation in Normal Tissues and Organs - Threshold Doses for Tissue Reactions in a Radiation Protection Context. ICRP Publication 118. Ann. ICRP 41(1/2).

ICRP, 2012b. Compendium of Dose Coefficients based on ICRP Publication 60. ICRP Publication 119. Ann. ICRP 41(Suppl.).

ICRP, 2013. Radiological protection in cardiology. ICRP Publication 120. Ann. ICRP 42(1).

Lindell, B., 1996. The history of radiation protection. Rad. Prot. Dosim. 68, 83–95.

Leppin, O., 1896. Aus kleine Mitteilungen. Wirkung der Röntgenstrahlen auf die Haut. Dtsch. Med. Wschr. 28, 454.

Röntgen, W.C., 1895. Über eine neue Art von Strahlen. Sitzungsberichte d. Phys. Mediz. Ges. Würzburg 9, 132.

III

Overview of Medical Health Physics (Radiation Protection in Medicine)

III

CHAPTER 4

Radiation Protection in Medicine (Medical Health Physics)

Richard J. Vetter

Mayo College of Medicine, Mayo Clinic, Rochester, MN USA

Magdalena S. Stoeva

Publication Committee, International Organization for Medical Physics (IOMP)

CONTENTS

M EDICAL PHYSICS is the science associated with the accuracy, safety, and quality of the use of radiation in medical procedures including medical imaging and radiation therapy. Various chapters in this book focus on the safe use of radiation in specific areas of medicine. This chapter is intended to review the scientific literature of radiation protection in medicine, to provide readers with a brief background in the development of safety practices and the science of radiation protection in medicine. This review is not intended to be exhaustive because such a review would occupy several book volumes. Rather, this chapter, based on an earlier review by Vetter (2005), is intended to provide readers with an appreciation for the worldwide effort of scientists and practitioners to protect healthcare providers, members of the public, and patients from unwarranted radiation exposure in medicine, a science often referred to as medical health physics. Readers are referred to subsequent chapters in this book for specific guidance on safe use of radiation in medicine.

4.1 INTRODUCTION

Practitioners in radiation medicine and medical health physicists must be knowledgeable in the principles of radiation safety and in the applications of radiation in medicine. Advances in radiation medicine have resulted in new modalities and procedures, some of which have significant potential to cause serious harm; for example, radiologic procedures that require very long fluoroscopy times. Early users of radiation developed an interest in protection against radiation hazards long before the profession of medical health physics. It is well known that the enthusiastic use of X-rays and radioactive materials shortly after their discovery resulted in a number of deleterious biological effects. Kathren and Ziemer (1980) provide a succinct summary of many of these events, the first of which were reported in Nature (Edison 1896; Morton 1896). Protection against hazards of X-rays was advocated by the American Roentgen Ray Society, which was formed in 1900 (Christie 1956). Recommendations for radiation protection from the early years of radiation use in medicine have been reviewed by Kathren (1962).

Over the years, many equipment improvements, interventions, and procedures have been developed to reduce radiation exposure to patients, workers, and members of the public while improving the effectiveness of ionizing radiation in the diagnosis and treatment of disease. Ironically, some of these advances have increased radiation doses to patients from select procedures, and some have resulted in deleterious biological effects (Shope 1996). Continuous attention must be paid to the safe use of radiation and to find ways to reduce radiation doses without adversely affecting quality.

4.2   RADIATION PROTECTION IN MEDICINE

The philosophy and objectives of radiation protection described by the International Commission on Radiological Protection (ICRP 1991) apply to the protection of healthcare personnel, patients, and members of the public. Both technical and administrative controls are used to prevent injury and reduce risk. Controls include engineering, e.g., shielding built into walls, personal protective equipment (PPE) such as lead aprons, and administrative controls, such as monitoring a patient at the completion of a procedure. Each type of radiation and its application generates different radiation protection challenges. Consequently, actions taken to keep radiation exposure as low as reasonably achievable (ALARA) below regulatory limits differ among the major disciplines of radiology, nuclear medicine, and radiation oncology. These areas are treated separately below and in more detail in subsequent chapters.

4.3   MEASUREMENTS AND DOSIMETRY

Measurements of radiation fields and occupational dose (external and internal) in medical environments are done the same way as in other environments (St. Germain 1995) Typically, medical workers who receive the highest average annual radiation dose are workers in cardiac catheterization laboratories including those that conduct ablation procedures, cyclotron and radiopharmaceutical production, and nuclear medicine (Al-Haj and Lagarde 2002).

The development of internal dosimetry in medicine was driven by the need to know the radiation dose to specific organs of patients (Stelson et al. 1995), but the same fundamentals apply to internal dosimetry of workers. A specific group of workers in whom application of internal dosimetry is particularly important is women who work with radiopharmaceuticals and who could become pregnant. Russell et al. (1997a; 1997b) identified radiopharmaceuticals that could cross the placenta or be taken up unintentionally by fertile women who work with these materials. These authors also provided absorbed dose estimates to the fetus from both maternal and fetal self-dose contributions for various gestation periods. Not surprisingly, the highest estimated fetal doses resulted from administration of therapeutic radioiodine. The authors advise that their dose estimates should be used with caution because many factors may cause doses to vary among individual cases. Fortunately, we have learned to apply methods that minimize the risk of uptake of these radiopharmaceuticals by medical radiation workers.

4.4   RADIOLOGY

Medical radiology involves the exposure of a patient to radiation for the purpose of producing an image on film or another receptor. Radiation sources include ultrasound, magnetic and electric fields (magnetic resonance imaging), and ionizing radiation (X-rays). Webster (1995) reviewed the evolution of medical X-ray technology from Roentgen's discovery to modern-day computed tomography (CT). Only protection against ionizing radiation will be addressed in this chapter.

Numerous advances have been made in technology and techniques for producing quality images (Seibert 1995). Radiation protection in radiology is a mature field, but introduction of new modalities, procedures and techniques has stimulated additional research. Equipment design has been modified to reduce both patient and worker dose. For example, in therapeutic interventional cardiology where skin burns in patients have been observed (Shope 1996), progressive scanning video systems in catheterization laboratories reduce doses to patients, physicians, and technical personnel (Holmes et al. 1990). Studies have shown that pulsed fluoroscopy significantly reduces patient dose from long fluoroscopy times in electrophysiology procedures designed to locate and ablate cells in the heart responsible for fibrillation and related disorders (Scanavacca et al. 1998). In addition to changes in X-ray systems, patient dose has been reduced through improvement in training of cardiologists, which includes operational radiation protection, and routine patient dose measurements (Vaño et al. 1998).

In most countries it is common practice for personnel to wear a protective lead apron whenever they could be exposed to scatter from an X-ray fluoroscope. In some busy patient-care environments, personnel wear lead aprons for several hours each day, e.g., during catheter ablation procedures, which could result in back strain. The dose under the lead apron for radiofrequency catheter ablation of atrial fibrillation is approximately 2 μSv per hour of fluoroscopy (Macle et al. 2003). A composite material with approximately 30% reduced weight has been developed, which provides similar attenuation to that of the conventional lead apron (Yaffe et al. 1991), and Vaño et al. (1995) showed that tungsten gloves offer the same protection as lead gloves but also allow better tactile perception.

CT fluoroscopy has been used increasingly in interventional radiologic procedures. Radiation doses to medical staff from CT fluoroscopy can be significant with doses to radiologists' hands of 0.6 to 1.5 mGy min-1 during biopsy procedures (Nickoloff et al. 2000). Thin leaded gloves reduced the scattered radiation to the hands by only 11% to as much as 44% depending on the kVp and type of glove. Floor-mounted shields reduced the scattered radiation levels to the body by 94% to 99%. Carlson et al. (2001) demonstrated that use of an intermittent mode of image acquisition resulted in personnel radiation dosimeter readings that were below measurable levels. When it is necessary to use the continuous mode of acquisition, a needle-holder can be used to keep the hands of the radiologist out of the direct primary beam during needle placement and advancement.

Medical personnel and the public are shielded from the source of X-rays by the use of shielding placed in the walls of the radiographic room (Archer 1995). After NCRP (1993) recommended an annual average effective dose limit of 1 mSv to members of the public, Metzger et al. (1993) used a Monte Carlo code to demonstrate that most shielding in radiographic facilities was adequate even though it may have been designed for the previous public dose limit of 5 mSv. They pointed out that physicists frequently used several conservative assumptions when designing barriers, e.g., neglecting the attenuation provided by the gypsum wallboard.

The primary and secondary barriers for radiographic rooms are often designed with the use of lead to protect people in areas outside the rooms. Christensen and Sayeg (1979), Glaze et al (1979), and Simpkin (1987, 1995) showed that gypsum wallboard works effectively for low-energy applications such as mammography. Simpkin (1988) showed that gypsum wallboard is effective for both low-energy narrow and broad-beam transmissions. Simpkin (1996) also measured scatter from mammography units over a range of scattering angles and provided a quick method for estimating the unshielded dose to occupied areas. Archer et al. (1994) provided broad-beam transmission data for a variety of diagnostic X-ray shielding materials and developed a unified database for single- and three-phase attenuation measurements. While mammography units can be shielded with gypsum wallboard, the evolution of CT has created the opposite challenge. Faster machines and greater throughput require that care be given to designing a CT room barrier to assure adequate shielding for the floor near the head of the machine (Langer and Gray 1998).

An emerging issue in radiology, especially in special-procedures laboratories including cardiac labs, is the association of tissue dose with development of cataracts and cardiovascular disease. In 2011, the International Commission on Radiological Protection issued a statement on tissue reactions (formerly termed non-stochastic or deterministic effects) to recommend lowering the threshold for cataracts and the occupational equivalent dose limit for the crystalline lens of the eye (ICRP 2011). Even though additional research is probably necessary to understand the mechanistic basis for radiation cataractogenesis and to better quantify the risk of low-dose, protracted radiation exposures, the U.K Health Protection Agency endorsed the conclusion reached by the ICRP in their 2011 statement that the equivalent dose limit for the lens of the eye should be reduced from 150 to 20 mSv per year, averaged over a five year period, with no year's dose exceeding 50 mSv (Bouffler, et al. 2012). Some countries have not been quick to adopt the ICRP recommendation, which is discussed further in ICRP Report No. 118 (ICRP 2012). Following a study of the status of eye lens radiation dose monitoring in European hospitals, Carinou et al. (2014) concluded that the proposed eye lens dose limit can be exceeded in interventional radiology procedures and that personnel should be properly trained in how to use protective equipment in order to keep eye lens doses as low as reasonably achievable. They also highlighted a need to improve the design of eye dosimeters to ensure satisfactory use by workers. In a retrospective assessment of the cumulative eye lens doses of interventional cardiologists, Jacob et al. (2013) estimated cumulative eye lens dose ranged from 25 mSv to more than 1600 mSv and concluded that without eye protection, interventional cardiologists may exceed the new ICRP lifetime eye dose threshold of 500 mSv. McVey et al. (2013) proposed the use of a general dose reduction factor of 5 when using eyewear with a lead equivalence of $0.5 - 0.75$ mm. They also concluded that the forehead of the wearer provides the most robust position to site a dosimeter used to estimate the dose to both eyes as part of a personal monitoring regime. Using an algorithm based on patient kerma-area product and other factors, Antic et al. (2013) used patient dose to estimate staff eye dose during interventional cardiology procedures. Although their results are based on a local practice, they may provide useful reference for other cardiology practices in assessing the eye dose using patient dose values. Even with these enlightening recent studies, more research is needed to develop acceptable methods for quantitating eye dose in special procedures and cardiology laboratories.

4.5   NUCLEAR MEDICINE

The origins of nuclear medicine are in therapeutic applications of radionuclides, but in 1924, George de Hevesy recognized that radionuclides could be used as "tracers" for medical purposes (Early 1995). Nuclear medicine has evolved through a number of growth phases and now includes subspecialties including nuclear cardiology (Williams 2003), applications of positron emitters (Lowe and Wiseman 2002), and radioactive monoclonal antibody therapy (Wiseman et al. 2003).

4.5.1   Diagnostic Nuclear Medicine

Because nuclear medicine technologists handle large quantities of radioactive materials during elution of ^{99m}Tc generators and preparation and administration of radiopharmaceuticals, their potential radiation exposure is significant. Typically, local shielding is used to reduce whole-body and hand doses. McElroy (1981) demonstrated that several different syringe shields reduced index finger exposure at least 20-fold but that reduction of exposure to middle and ring fingers varied from 0 to 250-fold. A lead-sleeve-type syringe shield provided maximum reduction, with or without glass. He also reported that technologist technique influences the effectiveness of the shield.

The handling and processing of radiopharmaceuticals presents the potential for radioactive contamination of the patient care environment. Aerosol inhalation procedures, for example, might cause contamination due to leakage around the mouth of the patient. Braga et al. (1998) demonstrated that an adherent mask reduced contamination events from 70% of procedures without the mask to 5% of procedures with the mask. Another application where contamination is not uncommon is nuclear cardiology, where radiopharmaceuticals are injected while a patient is on a treadmill. Mosman et al. (1999) reported a 4-part plan to reduce contamination events: a training program for new resident physicians, a procedure to closely inspect the intravenous apparatus, a mobile radioactive waste container, and a clear designation of duties for personnel to be included in the exercise procedure protocol.

Patients who receive diagnostic levels of radiopharmaceuticals are often free to move about the medical center both before and after their scans. Also, these patients may be scheduled to undergo additional exams, tests, or procedures subsequent to the nuclear medicine exam. A procedure of particular concern is hemodialysis where the equipment comes in direct contact with the blood of the patient. Serrano et al. (1991) demonstrated that hemodialysis of patients who have been administered a Tl-201 stress test resulted in no significant contamination of dialysis equipment. Dose rate from the patient was less than 3 μGy h^{-1} at 10 cm. Average activity concentration in the effluent was 44.4 Bq mL^{-1}, and total activity eliminated by the patient during the dialysis cycle was approximately 6 × 103 Bq. Dialysis tubing was stored for decay prior to disposal.

Since patients are allowed to move about the hospital, they expose members of the public and create the potential for contamination outside nuclear medicine. Benedetto et al. (1989) taped TLD chips to the abdomen of patients who had received a variety of common diagnostic radiopharmaceuticals. TLD readings correlated well to ion chamber survey meter measurements. These authors postulated scenarios for exposure of co-workers and family members and predicted doses between 7 μSv and 20 μSv per procedure. Ho and Shearer (1992) evaluated radioactive contamination in patient restrooms within the nuclear medicine area and several restrooms in public and staff areas and found the highest level of removable contamination in a men's restroom located across from the main public cafeteria of a 700-bed hospital. They concluded that the skin dose to someone who used a contaminated toilet seat would be very low even if all the radioactive contamination were to transfer to the skin. These studies demonstrate that patients who receive diagnostic radiopharmaceuticals and who move about the hospital unrestricted do not cause others to receive significant radiation doses.

4.5.2 Therapeutic Nuclear Medicine

Radioiodine has been used for decades to treat benign and malignant thyroid disease (Vetter 1997). Over this period of time, medical centers have learned to protect personnel from internal contamination and external exposure associated with handling GBq quantities of ^{131}I and other radionuclides. Two specific issues of current interest are disposal of therapeutic quantities of radioactive materials and limitation of public exposure.

In the United States, excreta from nuclear medicine patients are exempt from regulations that address disposal of radioactivity into sanitary sewerage (USNRC 1995). Fenner and Martin (1997) studied the behavior of ^{131}I compounds in municipal sewerage and discovered that transport time of radioiodine was much longer than that of normal sewage, perhaps due to absorption of radioiodine by organic materials on the walls of the sewer lines. Some countries limit the amount of radioactivity from patients that is disposed of in the sewer. Leung and Nikolic (1998) describe a method for collecting toilet discharge in a holding talk for physical decay to allow more patients to be treated without exceeding permissible concentrations of radioiodine in sewerage. Goddard (1999) described a similar system and determined that the maximum effective dose to a member of the public due to the holding tanks was 40 μSv per year.

In some cases, excreta may be discarded as solid waste by a patient who is incontinent. Evdokimoff et al. (1994) evaluated this potential out of concern that such waste might be rejected by a landfill where the waste was surveyed for radioactivity. They installed radiation detectors to survey trash and medical waste to ensure that no radioactive waste would be sent to the landfill.

In 1997, the U.S. Nuclear Regulatory Commission relaxed the regulations that had limited the release of patients who had received therapeutic quantities of radiopharmaceuticals or permanent brachytherapy implants (USNRC 2014). Consequently, the opportunity for radioactive material to show up in household waste increased. Some municipalities have installed radiation monitors at solid waste landfills to prevent the disposal of detectable radioactivity. To reduce detection of radioactive material that is generated from authorized activities, Siegel and Sparks (2002) recommended standardization of the design and installation of landfill radiation monitors, use of portable spectrometers to identify the offending radionuclide, and establishment of procedures to allow short-lived radionuclides to be disposed of immediately at the landfill since disposal poses no danger to the facility staff, the public, or the environment.

In the United States, patients who contain therapeutic quantities of radioactive materials or permanent brachytherapy implants may be released from the hospital if the dose to a member of the public is predicted not to exceed 5 mSv, which is higher than most countries allow (USNRC 2014). Numerous investigations have shown that the current practice of releasing such patients does not result in members of the public exceeding that limit. By placing TLDs on the chest of 35 partners and 38 children of patients treated with ^{131}I for hyperthyroidism and cancer, Mathieu et al. (1999) showed that doses to all family members were lower than 0.5 mSv. Barrington et al. (1999) conducted a similar study on hyperthyroid patients and found that 97% of adult family members and 89% of children received doses less than 5 mSv. The dose received by adults during travel was a small fraction of the total dose (median 0.03 mSv). They concluded that, with appropriate radiation protection advice, these patients could continue to be treated as outpatients. Grigsby et al. (2000) conducted a similar study on patients who were treated for thyroid carcinoma. The dose to 65 household members ranged from 0.01 mSv to 1.09 mSv (mean 0.24 mSv). Leslie et al. (2002) measured dose rates using adult and infant phantoms and concluded that current restrictions could be made less stringent and still keep doses to members of the public within dose limits. Zanzonico (1997) measured thyroid activity and external absorbed dose and estimated the total thyroid dose to adult family members of immediately released ^{131}I-treated hyperthyroid patients to be approximately 2.7×10^{-3} mGy per MBq administered, with an effective dose of approximately 5×10^{-3} mSv per MBq administered to a hyperthyroid patient. Ryan et al. (2000) and Rutar et al. (2001a,b) showed that patients treated with ^{131}I-labeled antibodies can also be released without exceeding the dose limit to members of the public. In their study, Rutar et al. (2001a) predicted doses to family members would range from 0.95 to 4.23 mSv. Observed doses were 0.01 to 4.09 mSv. Taken together, these studies suggest that the NRC guidance on release of patients treated with therapeutic radiopharmaceuticals is appropriate and protects the public to the recommended dose limit of 5 mSv (USNRC 2014). Additional precautions on care of radionuclide therapy patients is provided in ICRP Publication 94 (2004).

Studies of the fate of ^{131}I in patients on hemodialysis who have received radioiodine for hyperthyroidism or thyroid carcinoma show that the effective half-life varies from approximately 1 day to more than 7 days (Homer and Smith 2002, Magne et al. 2002, Toubert et al. 2001). This variability is affected by the removal rate of textsuperscript131I from the blood during hemodialysis and the ability of the kidneys to filter textsuperscript131I. Measured dose rates 30 cm away from the patient were highest at 42 h after treatment (0.064 mSv h^{-1}). Staff time near the patient was minimal. The maximum time near the patient was 15 min during connection of the patient to the hemodialysis machine. Thus, radiation hazard was negligible to staff during hemodialysis of patients who had been treated with ^{131}I for thyroid disease. When the predicted doses to members of the public exceed 5 mSv, the patient must be hospitalized for radiation protection purposes. NCRP Report No. 155 (2006) and Thompson (2001) provide precautions in the management of these patients. Achey et al. (2001) describe their experiences with 50 such patients and provide considerable data on external exposure rates, release activities, half-lives, and post-discharge room contamination levels. By following precautions, including appropriate training of healthcare staff, occupational doses can be kept well under the maximum permissible dose of 50 mSv. As new therapeutic agents are developed, changes in safety techniques may be required. Espenan et al. (1999) evaluated the safety of handling somatostatin analogues labeled with a variety of radionuclides and learned that personnel protection, contamination control, and other safety techniques required significant modification from their procedures for handling high dosages of radioiodine. Radioimmunotherapy with ^{90}Y-labeled Zevalin is a new and promising treatment for non-Hodgkin's lymphoma (Wiseman et al. 2003). Since ^{90}Y is primarily a beta emitter, external exposure control is not a serious matter during administration of the radiolabeled monoclonal antibodies. Most of the Zevalin is retained in the body, but approximately 7% is eliminated in the urine during the first week (Zhu 2003). Standard universal precautions are sufficient to prevent contamination during administration. Patients should receive basic instruction on the use of standard precautions to prevent contamination of family members.

As more experience is gained in treating radiopharmaceutical therapy and permanent-implant brachytherapy patients as outpatients, procedures will become more standardized. For example, an algorithm has been developed to determine time of release and duration of post treatment precautions (Zanzonico et al. 2000), and an interactive software program has been written to automate generation and printing of radiation safety recommendations specific to clinical, dosage, and social considerations (Friedman and Ghesani 2002).

Medical emergencies following administration of therapeutic quantities of radioiodine are not common, and death of a patient soon after administration of radioiodine is rare. Guidance on handling such emergencies is provided in NCRP Report No. 155 (2006). A number of authors have shared their experiences in handling these emergencies; two case reports are cited here (Griffiths et al. 2000, Greaves and Tindale 2001). In both cases the patients required considerable care and died soon after ablation therapy. These authors stressed the importance of communicating specific radiation protection guidelines to clinical staff and pathologists to minimize radiation dose and contamination.

4.5.3   Positron Emission Tomography (PET)

Positron-emitting radiopharmaceuticals present a greater risk of occupational radiation exposure than radiopharmaceuticals labeled with ^{99m}Tc and similar radionuclides because positron annihilation results in the generation of 511 keV photons. Patients, unit dosages, the pneumatic transport system, and gas lines could be significant sources of radiation. Kearfott et al. (1992) discussed methods for reducing exposures from these sources. For example, doses from gas lines may be kept ALARA by careful planning of the routing of the line and by using high flow rates and small-bore tubing. Doses from pneumatic transport systems can be kept small by designing the system to limit the number of failures of the system. Use of shielding in surrounding walls may be necessary to reduce radiation dose from patients who have received dosages of PET radiopharmaceuticals. However, room shielding can be costly, especially when simplifying assumptions result in significant overestimate of the shielding requirements. Methe (2003) describes an approach using resources readily available to medical health physicists. In the scenario he described, 1.6 to 22.2 mm of lead was required to protect various dose points of interest from patients who were being imaged or who had just received a dosage.

Production of PET radiopharmaceuticals results in the release of gaseous effluent containing the PET radionuclide (Kleck et al. 1991). Public exposure depends on a number of characteristics of the facility including stack height and location relative to members of the public. In the facility described by Kleck et al., the effluent concentration limited production to four runs per week.

4.6 RADIATION ONCOLOGY

After Roentgen's discovery of X-rays, therapeutic applications expanded rapidly. Some treatments were successful while others resulted in severe normal tissue damage. Orton (1995) reviewed the development of external beam radiation therapy from early applications to the use of high-energy linear accelerators. In today's modern radiation oncology environment, considerable efforts are made to protect medical personnel and the public from the high-energy radiation and high-activity sources used to treat patients.

4.6.1 External Beam Shielding

McGinley and Miner (1995) published a history of shielding for radiation treatment rooms from the time of the discovery of X-rays to modern day. Design of a radiation barrier for external beam radiation is addressed in NCRP Report No. 51 (1977), and an update of this report is in the draft stage.

Linear accelerators used for treatment of cancer are now capable of generating photons with energies that exceed the threshold for generation of neutrons. The higher photon energies require thicker radiation barriers, and the neutron component requires consideration of materials appropriate for shielding neutrons. Thus, high-energy accelerators require a more complicated barrier design than low-energy accelerators. Barrier design is described in NCRP Report No. 151 (2005).

Cost of construction increases with photon energy due to the additional volume of concrete that is needed for the wall and the additional space taken up by the thicker wall. This is particularly problematic for medical facilities that are upgrading radiation therapy capabilities by replacing lower-energy machines with high-energy accelerators. Barish (1990, 1993) describes the use of metals in conjunction with a matrix of Portland cement, which permits the dimensions of the radiation therapy room walls to be reduced by a factor of approximately 2. Caution must be used in the application of this material depending on the neutron component of the beam. McGinley (1992) showed that a sizable neutron field might be present outside a room where lead or steel is used with concrete. He found that the photoneutron dose was minimized when the metal part of the shield was positioned inside the treatment room in front of the concrete and when steel was used in place of lead. Kase et al. (2003) reported that non-uniform distribution of the iron could affect photon and neutron transmission. They also warn that significant error, as high as a factor of ten, can be made in estimating transmission of neutrons through high-density concrete if the density is not scaled properly from normal concrete. Finally, since modern medical linear accelerators can be operated at more than a single energy, use factor has become a significant consideration in barrier design. Kron et al. (1995) reported that 80% of the total dose delivered at isocenter in their clinic was delivered by 6-MV X-rays and that less than 25% of their clinical treatment fields extended beyond 200 cm^2. Physicists might want to consider more than a single use factor, especially in academic medical centers where treatment protocols may be driven by advances in research. However, LaRiviere (1984) pointed out that estimation of workload on the high side, assumption that the machine leakage is at applicable regulatory limit, and use of the primary beam TVL for leakage radiation yields progressively more conservative results with increasing energy above 6 MeV.

4.6.2 Brachytherapy

Brachytherapy is the use of radioactive sealed sources in the treatment of benign and malignant disease. Brachytherapy applications use either temporary or permanent implants. Use of radiation sources for brachytherapy creates the potential for serious personal injury to the patient and to anyone who handles the sources. The objectives of a brachytherapy radiation safety program are to protect patients from adverse medical events involving the radiation source, members of the public from exposure to the sources, and medical personnel who handle the sources or care for the patient during the brachytherapy treatment period.

Brachytherapy safety includes source control, dose limitation, reduction of doses ALARA, and regulatory compliance. An effective brachytherapy safety program includes a commitment to the ALARA principle, adequate resources for the program, and periodic review of the effectiveness of the program. Responsibility for implementation of the program rests with a team that includes the Radiation Safety Officer (RSO), medical physics staff, and radiation oncologists.

4.6.3 Low-Dose-Rate Brachytherapy

Regardless of the technique used or whether the implant is temporary or permanent, low-dose-rate brachytherapy will result in occupational radiation dose to members of the therapy team and any healthcare personnel who care for the patient. During temporary implants the patient must be hospitalized in a manner that prevents radiation dose to healthcare staff and members of the public from exceeding appropriate regulatory limits. Some hospitals meet the public dose limits by placing the patient in a room at the end of a hospital corridor and leaving the adjacent room vacant. Others have added shielding to the walls of a room that is used any time a patient is hospitalized following application of brachytherapy sources or in some cases administration of therapeutic quantities of radiopharmaceuticals. Retrofitting existing rooms can be expensive (Gitterman and Webster 1984), but leaving adjacent rooms vacant results in an opportunity cost to the hospital. Thomadsen et al (1983) discuss a variety of techniques that can be used to find the optimum solution to this problem, which include leaving adjacent rooms empty, use of portable shields, addition of shielding material to the walls of a designated room, or changing the radionuclides that are used.

Portable bedside shields can be used to reduce the radiation dose in areas adjacent to the brachytherapy treatment room as well as to the healthcare staff. Exposure rates to staff are variable depending on the type of treatment, the sources used, and the activity in which staff are engaged (Glasgow et al, 1985; Smith et al. 1998). The optimal placement of lateral bedside shields to reduce staff exposures from gynecologic implants logically depends on the need to shield adjacent areas as well as protecting staff, and may require placement of more than one bedside shield (Papin et al. 1990).

4.6.4 High-Dose-Rate Brachytherapy

The application of high-dose-rate (HDR) brachytherapy usually requires the use of specialized equipment that moves a high-activity radiation source from a shield into a catheter that was previously placed in the tumor bed of the patient. Treatment times are relatively short, on the order of minutes; thus, there is little room for error. The American Association of Physicists in Medicine (AAPM) developed recommendations for an HDR brachytherapy program, which include quality assurance and emergency procedures (Kubo et al. 1998). Radiation exposure is often not an issue for radiation oncology staff during treatments because the HDR unit is usually located in a shielded room equipped with a door interlock to interrupt treatment if someone were to open the door. Dose rates in adjacent areas are kept within regulatory limits by using existing shielding, such as an external beam treatment room, or by adding shielding to the wall as necessary. Occasionally, the HDR may be moved to another location such as an intraoperative radiation treatment room that does not contain additional shielding in the walls. If exposure rates in adjacent areas exceed limits, mobile shielding has been shown to provide adequate shielding (Sephton et al. 1999).

4.7 RADIATION ACCIDENTS IN MEDICINE

Even though medical professionals are careful to avoid accidents associated with high-activity radioactive sources used in medicine, a number of severe accidents have occurred. The lack of attention to detail can have devastating consequences. The journal Health Physics devoted an entire issue to the Goiania radiation accident that caused several deaths, numerous radiation injuries, and millions of dollars in clean-up costs as a result of a carelessly abandoned teletherapy unit (Maletskos and Lipstein 1991). Lushbaugh et al. (1986) reviewed a number of accidents that resulted in high skin doses. Nenot (1998) reviewed a number of radiation accidents and concluded that human factors, such as lack of elementary safety rules and inadequate training, play a major role in most of the accidents occurring in industry and in the medical field. A report of the International Commission on Radiological Protection focused on prevention of accidental exposures to therapy patients (ICRP 2000). This report suggested that at least some of the accidents described in the report would not have occurred if operators had reached a level of safety awareness in which they could actively participate in advancing safety objectives. The report also concluded that the prevention of accidental exposures requires the systematic application of quality assurance. These reports all point out the importance of attention to detail when working with radiation sources that are capable of delivering high radiation doses.

4.8   THE ROLE OF RECOMMENDATIONS AND REGULATIONS

Throughout this review, reference has been made to recommendations and regulations as they pertain to maximum permissible doses to various populations, and other matters of interest to medical health physicists. Those who serve in a radiation safety capacity, e.g., medical health physicists at medical facilities, are intimately familiar with the regulation of the medical use of radioactive materials and sources of radiation. In a review of the historical development of radiation standards, Kocher (1991) stressed the increasing importance of ALARA in reducing radiation exposures to workers and the public. In a review of the history and trends of radiation protection standards, Hendee (1993) discussed the trend towards more rigorous limits that require increased commitments of personnel and resources. The U.S. Nuclear Regulatory Commission (USNRC 2000) revised its 1979 policy statement on the medical use of byproduct material to focus its regulations on those medical procedures that pose the highest risk and to structure its regulations to be risk-informed and more performance-based. Thus, recent focus has shifted away from check-list type programs such as intensive review of records during inspections toward performance issues such as prevention of errors that result in incorrect radiation doses to target or normal tissue. Ostrom et al. (1996) conducted detailed investigations and analyses of seven such errors (termed misadministrations at that time). Results of these investigations indicated that the institutional traditions of some licensees contributed to the potential for error. The lack of procedures, or procedures that were not clearly written, strongly contributed to many of these errors. Some of the licensees had not effectively implemented quality management programs at the time. Finally, limited involvement of the radiation safety officer and authorized users contributed to the potential for errors.

4.9   CONCLUSION

Awareness of the need for emphasis on radiation protection contributes significantly to the safety of healthcare providers, patients, and the public. Contributions are most evident in facility design where they are involved in design of radiation protection barriers, in monitoring of personnel and the patient care environment through radiation surveys and personnel dosimetry, in development of procedures that emphasize careful handling of radioactive sources and use of personal protective equipment, in development of practices to limit radiation exposure of the public to acceptable levels, and in proper disposal of radioactive sources. Medical health physicists are often challenged to maximize protection of personnel while minimizing the cost of resources necessary to keep radiation doses ALARA. They are also challenged to keep pace with medical colleagues who continue to develop new modalities for diagnosis and treatment of benign conditions and cancer and who expand uses of current technology. Advances in medical health physics will continue to be based on evidence gathered through basic and applied research. Periodic review of the evidence will help medical health physicists to focus on the issues and to advance the science.

4.10 REFERENCES

Achey B, Miller KL, Erdman M, King S. 2001. Some experiences with treating thyroid cancer patients. Health Phys 80: S62–6.

Al-Haj AN, Lagarde CS. 2002. Statistical analysis of historical occupational dose records at a large medical center. Health Phys 83: 854-60.

Antic v, Ciraj-Bjelac O, Rehani M, Aleksandric S, Arandjic D, Ostojic M. 2012. Eye Lens Dosimetry In Interventional Cardiology: Results Of Staff Dose Measurements And Link To Patient Dose Levels. Radiol Prot. Ncs 236. http://rpd.oxfordjournals.org/content/early/2012/11/14/rpd.ncs236.short (accessed August 23, 2015).

Archer BR. 1995. A history of shielding of diagnostic x-ray facilities. Health Phys 69: 458.

Archer BR, Fewell TR, Conway BJ, Quinn PW. 1994. Attenuation properties of diagnostic x-ray shielding materials. Med Phys. 21: 1499–507.

Barish RJ. 1990. Practical high-density shielding materials for medical linear accelerator rooms. Health Phys 58:37–39.

Barish RJ. 1993. Evaluation of a new high-density shielding material. Health Phys 64: 412–416.

Barrington SF, O'Doherty MJ, Kettle AG, Thomson WH, Mountford PJ, Burrell DN, Farrell RJ, Batchelor S, Seed P, Harding LK. 1999. Radiation exposure of the families of outpatients treated with radioiodine (iodine-131) for hyperthyroidism. Eur J Nucl Med 26: 686–92.

Benedetto AR, Dziuk TW, Nusynowitz ML. 1989. Population exposure from nuclear medicine procedures: measurement data. Health Phys 57: 725–31.

Bouffler S, Ainsbury E, Gilvin P, Harrison J. 2012. Radiation-induced cataracts: the Health Protection Agency's response to the ICRP statement on tissue reactions and recommendation on the dose limit for eye lens. J. Radiol. Prot. 32: 479–488.

Braga FJ, Souza JF, Trad CS, Santos AC, Ghillardi Netto T, Elias J Jr, Hindie E, Iazigi N. 1998. An improved mouth-piece to prevent environmental contamination during radioaerosol inhalation procedures. Health Phys 75: 424–7.

Carinou E, Ginjaume M, O'Connor U, Kopec R, Sans Merce M. 2014. Status of eye lens radiation dose monitoring in European hospitals. J. Radiol. Prot. 34: 729–739.

Carlson SK, Bender CE, Classic KL, Zink FE, Quam JP, Ward EM, Oberg AL. 2001. Benefits and safety of CT fluoroscopy in interventional radiologic procedures. Radiology 219: 515–20.

Christensen RC, Sayeg JA. 1979. Attenuation characteristics of gypsum wallboard. Health Phys 36: 595–600.

Christie AC. 1956. The American Roentgen Ray Society. Amer. J. Roentgenol 76: 1.

Early PJ. 1995. Use of diagnostic radionuclides in medicine. Health Phys 69: 649–661.

Edison TA. 1896. Notes. Nature 53: 421.

Espenan GD, Nelson JA, Fisher DR, Diaco DS, McCarthy KE, Anthony LB, Maloney TJ, Woltering EA. 1999. Experiences with high dose radiopetide therapy: the health physics perspective. Health Phys 76: 225–235.

Evdokimoff V, Cash C, Buckley K, Cardenas A. 1994. Potential for radioactive patient excreta in hospital trash and medical waste. Health Phys 66: 209–211.

Fenner FD, Martin JE. 1997. Behavior of Na^{131}I and meta(^{131}I) iodobenzyl-guanidine (MIBG) in municipal sewerage. Health Phys 73: 333–9.

Friedman MI, Ghesani M. 2002. Interactive software automates personalized radiation safety plans for Na^{131}I therapy. Health Phys 83: S71–84.

Gitterman M, Webster EW. 1984. Shielding hospital rooms for brachytherapy patients: design, regulatory and cost/benefit factors. Health Phys 46: 617–625.

Glasgow GP, Walker S, Williams JD. 1985. Radiation exposure rates near brachytherpy patients containing 137Cs Sources. Health Phys 48: 97–104.

Glaze SA, Scheiders NJ, Bushong SC. 1979. Use of gypsum wallboard for diagnostic x-ray protective barriers. Health Phys 36: 587–593.

Goddard C. 1999. The use of delay tanks in the management of radioactive waste from thyroid therapy. Nucl Med Commun 20:85–94.

Greaves C, Tindale 2001. W. Radioiodine therapy: care of the helpless patient and handling of the radioactive corpse. J Radiol Prot. 21: 381–92.

Griffiths PA, Jones GP, Marshall C, Powley SK. 2000. Radiation protection consequences of the care of a terminally ill patient having received a thyroid ablation dose of ^{131}I-sodium iodide. Br J Radiol. 73: 1209–12.

Grigsby PW, Siegel BA, Baker S, Eichling JO. 2000. Radiation exposure from outpatient radioactive iodine (^{131}I) therapy for thyroid carcinoma. JAMA 283: 2272–4.

Hendee WR. 1993. History, current status, and trends of radiation protection standards. Med Phys. 20:1303–14.

Ho SY, Shearer DR. 1992. Radioactive contamination in hospitals from nuclear medicine patients. Health Phys 62: 462–466.

Holmes DR Jr, Wondrow MA, Gray JE, Vetter RJ, Fellows JL, Julsrud PR. 1990. Effect of pulsed progressive fluoroscopy on reduction of radiation dose in the cardiac catheterization laboratory. J Am Coll Cardiol 15: 159–162.

Homer L, Smith AH. 2002. Radiation protection issues of treating hyperthyroidism with 131 I in patients on haemodialysis. Nucl Med Commun. 23: 261–4.

International Commission of Radiological Protection. 1991. 1990 Recommendations of the International Commission on Radiological Protection. ICRP Publication 60. Oxford: ICRP.

International Commission of Radiological Protection. 2000. Prevention of accidental exposures to patients undergoing radiation therapy. ICRP Publication 60. Oxford: ICRP.

International Commission on Radiological Protection. 2004. Release of Patients after Therapy with Unsealed Radionuclides, ICRP Publication 94. Oxford: ICRP.

International Commission on Radiological Protection. 2011. ICRP statement on tissue reactions. Ottawa: ICRP. http://www.icrp.org/docs/ICRP statement on Tissue Reactions.pdf. (accessed October 4, 2014).

International Commission on Radiological Protection. 2012. ICRP statement on tissue reactions and early and late effects of radiation in normal tissues and organs – threshold doses for tissue reactions in a radiation protection context. ICRP Publication 118. Ottawa: ICRP.

Jacob S1, Donadille L, Maccia C, Bar O, Boveda S, Laurier D, Bernier MO. 2013. Eye lens radiation exposure to interventional cardiologists: a retrospective assessment of cumulative doses. Radiat Prot Dosimetry 153:282–293.

Kase KR, Nelson WR, Fasso A, Liu JC, Mao X, Jenkins TM, Kleck JH. 2003. Measurements of accelerator-produced leakage neutron and photon transmission through concrete. Health Phys 84: 180–187.

Kathren RL. 1962. Early x-ray protection in the United States. Health Phys 8: 503–511.

Kathren RL, Ziemer PL. 1980. Health physics: A backward glance. New York: Pergamon Press.

Kearfott KJ, Carey JE, Lemenshaw MN, Faulkner DB. 1992. Radiation protection design for a clinical positron emission tomography imaging suite. Health Phys 63: 581–589.

Kleck JH, Benedict SH, Cook JS, Birdsall RL, Satyamurthy N. 1991. Assessment of 18F gaseous releases during the production of 18F-fluorodeoxyglucose. Health Phys 60: 657–60.

Kocher DC. 1991. Perspective on the historical development of radiation standards. Health Phys. 61: 519–27.

Kron T, Aldrich B, Javanovic K, Howlett, S, Hamilton C. 1995. Workload and use factor of medical linear accelerators in radiotherapy. Health Phys 69: 971–975.

Kubo HD, Glasgow GP, Pethel TD, Thomadsen BR, Williamson JF. 1998. High dose-rate brachytherapy treatment delivery: report of the AAPM Radiation Therapy Committee Task Group No. 59. Med Phys 25: 375–403.

Langer SG, Gray JE. 1998. Radiation shielding implications of computed tomography scatter exposure to the floor. Health Phys 75: 193–196.

LaRiviere PD. 1984. Transmission in concrete of primary and leakage x rays from a 24-MV medical linear accelerator. Health Phys. 47:819–27.

Leslie WD, Havelock J, Palser R, Abrams DN. 2002. Large-body radiation doses following radioiodine therapy. Nucl Med Commun 23: 1091–7.

Leung PM, Nikolic M. 1998. Disposal of therapeutic ^{131}I waste using a multiple holding tank system. Health Phys 75: 315–321.

Lowe VJ, Wiseman GA. 2002. Assessment of lymphoma therapy using 18F-FDG PET. N Nucl Med 43: 1028–1030.

Lushbaugh CC, Fry SA, Ricks RC, Hubner KF, Burr WW. 1986. Historical update of past and recent skin damage radiation accidents. Br J Radiol Suppl. 19: 7–12.

Macle L, Weerasooriya R, Jais P, Scavee C, Raybaud F, Choi KJ, Hocini M, Clementy J, Haissaguerre M. 2003. Radiation exposure during radiofrequency catheter ablation for atrial fibrillation. Pacing Clin Electrophysiol 26: 288–91.

Magne N, Magne J, Bracco J, Bussiere F. 2002. Disposition of radioiodine (131)I therapy for thyroid carcinoma in a patient with severely impaired renal function on chronic dialysis: a case report. Jpn J Clin Oncol. 32: 202–5.

Maletskos CJ, Lipstein; ed. 1991. The Goiania radiation accident. Health Phys. 60: 1–113.

Mathieu I, Caussin J, Smeesters P, Wambersie A, Beckers C. 1999. Recommended restrictions after ^{131}I therapy: measured doses in family members. Health Phys 76: 129–36.

McElroy NL. 1981. Efficacy of various syringe shields for ^{99m}Tc. Health Phys 41: 535–542.

McGinley PH. 1992. Photoneutron production in the primary barriers of medical accelerator rooms. Health Phys 62: 359–362.

McGinley PH, Miner MS. 1995. A history of radiation shielding of x-ray therapy rooms. Health Phys 69: 759–765.

McVey S1, Sandison A, Sutton DG. 2013. An assessment of lead eyewear in interventional radiology. J Radiol Prot. 33:647–659.

Methe BM. 2003. Shielding design for a PET imaging suite: a case study. Health Phys 84: S83–S88.

Metzger R, Richardson R, Van Riper KA. 1993. A Monte Carlo model for retrospective analysis of shield design in a diagnostic x-ray room. Health Phys 65: 164–71.

Morton WJ. 1896. Notes. Nature 53: 421.

Mosman EA, Peterson LJ, Hung JC, Gibbons RJ. 1999. Practical methods for reducing radioactive contamination incidents in the nuclear cardiology laboratory. J Nucl Med Technol 27: 287–9.

National Council on Radiation Protection and Measurements. 1977. Radiation protection design guidelines for 0.1-100 MeV particle accelerator facilities. NCRP Report No. 51. Bethesda: NCRP.

National Council on Radiation Protection and Measurements. 1993. Limitation of exposure to ionizing radiation. NCRP Report No. 116. Bethesda: NCRP.

National Council on Radiation Protection and Measurements. 2005. Structural Shielding Design and Evaluation for Megavoltage X- and Gamma-Ray Radiotherapy Facilities. NCRP Report No. 151. Bethesda: NCRP.

National Council on Radiation Protection and Measurements. 2006. Management of Radionuclide Therapy Patients. NCRP Report No. 155. Bethesda: NCRP.

Nenot JC. 1998. Radiation accidents: lessons learnt for future radiological protection. Int J Radiat Biol. 73: 435–42.

Nickoloff EL, Khandji A, Dutta A. 2000. Radiation doses during CT fluoroscopy. Health Phys. 79:675–81.

Orton CG. 1995. Uses of therapeutic x rays in medicine. Health Phys. 69: 662–76.

Ostrom LT, Rathbun P, Cumberlin R, Horton J, Gastorf R, Leahy TJ. 1996. Lessons learned from investigations of therapy misadministration events. Int J Radiat Oncol Biol Phys. 34: 227–34.

Papin PJ, Ramsey MJ, LaFontaine RL, LePage RP. 1990. Effect of bedside shielding on air-kerma rates around gynecologic intracavitary brachytherapy patients containing 226Ra or 137Cs. Health Phys 58: 405–410.

Russell JR, Stabin MG, Sparks RB. 1997a. Placental transfer of radiopharmaceuticals and dosimetry in pregnancy. Health Phys 73: 747–55.

Russell JR, Stabin MG, Sparks RB, Watson E. 1997b. Radiation absorbed dose to the embryo/fetus from radiopharmaceuticals. Health Phys 73: 756–69.

Rutar FJ, Augustine SC, Colcher D, Siegel JA, Jacobson DA, Tempero MA, Dukat VJ, Hohenstein MA, Gobar LS, Vose JM. 2001a. Outpatient treatment with (131)I-anti-B1 antibody: radiation exposure to family members. J Nucl Med 42: 907–15.

Rutar FJ, Augustine SC, Kaminski MS, Wahl RL, Siegel JA, Colcher D. 2001b. Feasibility and safety of outpatient Bexxar therapy (tositumomab and iodine I 131 tositumomab) for non-Hodgkin's lymphoma based on radiation doses to family members. Clin Lymphoma 2: 164–72.

Ryan MT, Spicer KM, Frei-Lahr D, Samei E, Frey GD, Hargrove H, Bloodworth G. 2000. Health physics consequences of out-patient treatment of non-Hodgkin's lymphoma with ^{131}I-radiolabeled anti-B1 antibody. Health Phys 79: S52–5.

Scanavacca M, d'Avila A, Velade JL, Reolao JB, Sosa E. 1998. Reduction of radiation exposure time during catheter ablation with the use of pulsed fluoroscopy. In J Cardiol 63: 71–74.

Seibert JA. 1995. One hundred years of medical diagnostic imaging technology. Health Phys 69: 695–720.

Sephton R, Das KR, Coles J, Toye W. Pinder P. 1999. Local shielding of high dose rate barachytherapy in an operating theatre. Australas Phys Eng Scie Med 22: 113–117.

Serrano M, Olson A, Man C, Galonsky R, Stein R. 1991. Contamination and radiation exposure from 201Tl in patients undergoing dialysis after a nuclear medicine study. Health Phys 60: 365–366.

Shope TB. 1996. Radiation-induced skin injuries from fluoroscopy. Radiographics 16: 1195–1199.

Siegel JA, Sparks RB. 2002. Radioactivity appearing at landfills in household trash of nuclear medicine patients: much ado about nothing? Health Phys 82:367–72.

Simpkin DJ. 1987. Shielding requirements for mammography. Health Phys 53: 267–279.

Simpkin DJ. 1988. Photon transmission of monoenergetic broad beams through gypsum wallboard. Health Phys 54: 561–563.

Simpkin DJ. 1995. Transmission data for shielding diagnostic x-ray facilities. Health Phys 68: 704–709.

Simpkin DJ. 1996. Scatter radiation intensities about mammography units. Health Phys 70: 238–244.

Smith CL, Chu WK, Granville D, Nabity L. 1998. An examination of radiation exposure to clinical staff from patient implanted with 137Cs and 192Ir for the treatment of gynecologic malignancies. Health Phys 74: 301–308.

St. Germain, J. 1995. Radiation monitoring with reference to the medical environment. Health Phys 69: 728–749.

Stelson AT, Watson EE, Cloutier RJ. 1995. A history of medical internal dosimetry. Health Phys 69: 766–782.

Thomadsen B, Van De Geijn J, Buchler D, Paliwal, B. 1983. Fortification of existing rooms used for brachytherapy patients. Health Phys. 45:607–615.

Thompson MA. 2001. Radiation safety precautions in the management of the hospitalized ^{131}I therapy patient. J Nucl Med Technol. 29: 61–6.

Toubert ME, Michel C, Metivier F, Peker MC, Rain JD. 2001. Iodine-131 ablation therapy for a patient receiving peritoneal dialysis. Clin Nucl Med. 26: 302–5.

U.S. Nuclear Regulatory Commission. 1995. Title 10, Chapter 1, Code of Federal Regulations - Energy, Part 20, Standards for protection against radiation, subpart K, waste disposal. http://www.nrc.gov/reading-rm/doc-collections/cfr/part020/full-text.html. (accessed October 3, 2014).

U.S. Nuclear Regulatory Commission. 2000. Medical use of byproduct material; policy statement, revision. Fed Regist. 65: 47654–47660.

U.S. Nuclear Regulatory Commission. 2014. Title 10, Part 35.75. Release of individuals containing unsealed byproduct material or implants containing byproduct material. http://www.nrc.gov/reading-rm/doc-collections/cfr/part035/part035-0075.html (accessed October 3, 2014).

Vaño E, Fernandez JM, Delgado V, Gonzalez L. 1995. Evaluation of tungsten and lead surgical gloves for radiation protection. Health Phys 68: 855–8.

Vaño E, Arranz L, Sastre JM, Moro C, Ledo A, Garate MT, Minguez I. 1998. Dosimetric and radiation protection considerations based on some cases of patient skin injuries in interventional cardiology. Br J Radiol 71: 510–516.

Vetter RJ. 1997. Regulations for radioiodine therapy in the United States: current status and the process of change. Thyroid. 7:209–11.

Vetter RJ. 2005. Medical health physics: a review. Health Phys. 88: 653–664.

Webster EW. 1995. X rays in diagnostic radiology. Health Phys 69: 610–635.

Williams KA. 2003. Measurement of ventricular function with radionuclide techniques. Curr Cardiolo Rep 5: 45–51.

Wiseman GA, Kornmehl E, Leigh B, Erwin WD, Podoloff DA, Spies S, Sparks RB, Stavin MG, Witzig T, White CA. 2003. Radiation dosimetry results and safety correlations from ^{90}Y-ibritumomab tiuxetan radioimmunotherapy for relapsed or refractory non-Hodgkin's lymphoma: combined data from 4 clinical trials. J Nucl Med 44: 465–474.

Yaffe MJ, Mawdsley GE, Lilley M, Servant R, Reh G. 1991.Composite materials for x-ray protection. Health Phys 60: 661–4.

Zanzonico PB. 1997. Radiation dose to patients and relatives incident to ^{131}I therapy. Thyroid 7: 199–204.

Zanzonico PB, Siegel JA, St Germain J. 2000. A generalized algorithm for determining the time of release and the duration of post-release radiation precautions following radionuclide therapy. Health Phys 78: 648–59.

Zhu X. 2003. Radiation safety consideration with therapeutic ^{90}Y Zevalin. Health Phys 85: S31–S35.

IV

Radiation Protection in Diagnostic Radiology

CHAPTER 5

Radiation Protection in Diagnostic Radiology

Elizabeth Benson

King's College Hospital, London, UK

Cornelius Lewis

King's College Hospital, London, UK

CONTENTS

THE INTERNATIONAL RECOMMENDATIONS for radiation protection are set out by the International Commission on Radiological Protection (ICRP). A series of publications, Annals of the ICRP, are produced periodically to communicate these recommendations. In response to the ICRP, other bodies, including the International Atomic Energy Agency (IAEA), publish Basic Safety Standards (BSS) as an international guide for implementation of the recommendations. The BSS are accompanied by various IAEA Safety Report Series documents providing additional and more in-depth guidance.

Regional and national governments and political organizations (for example, the European Union) interpret the Basic Safety Standards as Directives or other statutory instruments to introduce ICRP recommendations into legislative frameworks. This book is concerned with the principles of radiation protection rather than specific legal interpretation and so the discussion is based on international recommendations and standards.

5.1 PRINCIPLES OF RADIATION PROTECTION IN DIAGNOSTIC RADIOLOGY

Both the ICRP (ICRP 2008) and the IAEA (IAEA 2014) discuss the requirement for a radiological protection system. This system must consider the types of radiation exposure that will take place and those individuals who may be exposed. Following from this, the system must provide for assessment of these exposures and application of the principles of radiation protection to ensure the safety of those individuals.

In Publication 103 (ICRP 2008), the ICRP defines three types of exposure situations: planned, emergency, and existing. Use of X-rays in diagnostic radiology is a planned exposure situation, i.e., a radiation source is deliberately introduced and operated to give rise to an expected exposure. Conversely, emergency exposure situations result from any unexpected radiation exposure and actions will be required to reduce the adverse consequences of such an exposure. Emergency exposure situations may arise from planned exposure situations; for example, the malfunction of a Computed Tomography (CT) unit during use, requiring a technologist to enter the room while the CT is in operation. Existing exposure situations are those that are already present, such as exposure to natural background radiation or exposure to residual radiation following an emergency exposure situation.

The ICRP (ICRP 2008) further defines planned exposure situations as normal and potential. Normal exposures are those that are expected to occur; for example, the routine operation of a diagnostic radiology department. Potential exposures are those that are not expected to occur; these may be the result of deviations from planned procedures or accidents. Both normal and potential exposures contribute to the three categories of exposure described in Publication 103: occupational exposures, public exposures, and medical exposures of patients.

IRCP defines three fundamental principles of radiological protection: justification, optimization, and limitation (ICRP 2008). Chapter 3 of this book provides an in-depth discussion of these principles. Considering their application to the three categories of exposure, all apply to occupational and public exposures (Section 5.2). However, the principal of limitation is not applied to the medical exposure of patients, only justification and optimization (Section 5.3).

5.2 PROTECTION OF EMPLOYEES AND THE PUBLIC

The implementation of a radiological protection and safety system is fundamental to meeting both the recommendations of the ICRP and requirements of the IAEA set out in the BSS. IAEA Safety Series Report No. 39 (IAEA 2006) gives further guidance on implementation of the Basic Safety Standards. One of the key requirements of the BSS is that the relevant regulatory body (sometimes referred to as the Competent Authority) must be notified of the intention to conduct planned exposure situations. In the case of medical exposure to patients it is also necessary to apply to the regulatory body for authorization, resulting in registration or licensing.

The licensee or registrant may also be the employer and has a number of responsibilities in relation to planned exposure situations (IAEA 2014). These include:

- establishment and periodic review of a radiological protection and safety program;

- ensuring that safety assessments are carried out prior to the commencement of work;

- ensuring that shielding is in place where there is a potential for public exposure;

- ensuring that radiation equipment is safely maintained;

- implementation of procedures for protection and safety;

- ensuring that occupational exposure does not exceed relevant dose limits;

- implementation of procedures for the reporting of incidents.

A full description of the responsibilities of the licensee, registrant, or employer can be found in the BSS. Routinely the licensee or registrant will delegate these responsibilities to a competent individual with advanced knowledge of diagnostic radiology facilities and equipment. The BSS title for such an individual is the radiation protection officer (RPO) although national regulatory systems may define alternative titles. The following sections discuss how the responsibilities of the RPO are met practically within the discipline of diagnostic radiology.

5.2.1 Safety Assessment

A safety assessment must be submitted to the regulatory body as part of the application for authorization. The safety assessment for a facility, or a practice carried out within a facility, must provide a critical analysis of the following in relation to protection and safety (IAEA 2014):

- likelihood and consequence of structural, radiation equipment, and software failure;

- potential procedural failures and their consequences;

- external factors that may affect safety;

- the workload associated with the facility.

This analysis must be carried out prior to commencing the use of radiation at the facility. Furthermore, the introduction of any new practice involving radiation at a facility must be subject to a prior safety assessment. Following from this, it will be necessary to carry out safety assessments at different stages throughout the lifetime of a facility or practice as circumstances and technologies change.

Safety assessment may also be referred to as risk or hazard assessment and must be documented and reviewed regularly. A key issue in conducting a safety assessment is to establish the areas where a radiological protection and safety program requires improvement. If a safety assessment indicates that actions are required to improve protection and safety, the implications should be considered and changes made where necessary.

5.2.2 Facility Design and Shielding

When designing a facility where there are planned exposures, consideration must be given to the safety and protection of workers and the public. For most diagnostic radiology equipment, some degree of shielding will be required to constrain radiation levels outside the facility and ensure that limits, as set out by the ICRP, are not exceeded (ICRP 2008). Shielding should be determined at the design stage to avoid problems later in the process. This requires cooperation between all those involved in the facility design and construction.

The type of diagnostic radiology equipment used in a facility and its layout have major design implications. These issues are discussed in some detail in a publication from the Radiological Protection Institute of Ireland (RPII) (RPII 2009).

Consideration must be given to both primary and secondary sources of radiation. Primary radiation describes the radiation beam emitted from the X-ray tube prior to any interaction and also the attenuated X-ray beam post interaction with the patient, detector, couch assembly, etc. Secondary radiation arises from scatter and leakage. Manufacturers of radiation-generating equipment are required by the BSS to provide information on dose-rate distributions during normal operation, which should be used when assessing shielding requirements.

The purpose of shielding is to reduce the radiation dose rate in surrounding areas to an acceptable level within set constraints that may be set at a National level (Table 5.1 lists design dose constraints for some European countries). Design dose constraints can be modified if the area under consideration is not continuously occupied. For example, an office would be assumed to be continuously (100%) occupied and so the dose level outside the barrier would be set at the design constraint. However, it would be reasonable to assume that a corridor or washroom would not be occupied continuously, so the design dose level can be adjusted and increased according to occupancy factors. Table 5.2 gives examples of occupancy factors recommended by the British Institute of Radiology (BIR) and National Council for Radiation Protection and Measurements (NCRP) for use with their shielding design methodologies (discussed below).

Table 5.1: Recommended dose constraints

Country	Value (mSv/yr)		Comments
Sweden	0.1		RT shielding, external beam, site limit (nuclear)
Belgium	0.02 (0.5)	mSv/week	Shielding. Outside room (patients with sources)
Ireland	0.3		General
Germany	0.3		Site limit (nuclear)
UK	0.3		Site limit (nuclear), DR room design
Finland	0.1		Site limit (nuclear)

Table 5.2: Occupancy factors

	BIR,%	NCRP
Control rooms, nurses' stations, reception areas, offices, children's indoor play areas	100	1
Shops, living quarters, occupied space in nearby buildings	100	-
Laboratories, pharmacies, other work areas fully occupied by an individual, attended waiting rooms, adjacent X-ray rooms	-	1
Reporting areas	20-50	1
Staff rooms	20-50	1/5
Patient rooms	-	1/5
Adjacent wards, clinic rooms	20-50	1/2
Corridors	5-12.5	1/5
Corridor doors	-	1/8
Public toilets, bathrooms, store rooms, unattended waiting rooms,	5-12.5	1/20
Changing rooms	5-12.5	-
Stairways	5-12.5	1/40
Unattended car parks	5-12.5	1/40
Unattended vending areas, outdoor areas with seating, patient holding areas	-	1/20
Outdoor areas with transient traffic, unattended elevators	-	1/40

The most important and often most difficult factor to determine when undertaking shielding calculations is workload. Ideally, colleagues in the Radiology Department will be able to give good estimates for workload from which the relevant information required for shielding calculations can be extracted. If the new facility is merely replacing old equipment, historical data will be available. Where no workload information exists, for example with new developments, BIR and NCRP provide examples for typical installations and there are other sources in the literature; for example, from the reports of the UK Health Protection Agency (HPA) (NCRP 2004). If any uncertainty exists it is prudent to make a generous assumption. The cost of planning for an extra thickness of lead or concrete is much lower than adding further shielding after equipment is installed.

The two accepted approaches to shielding calculation, from The British Institute of Radiology (BIR) (Sutton et al. 2012) and the National Council on Radiation Protection and Measurements (NCRP), Report No. 147 (NCRP 2004) are summarized below.

5.2.2.1 BIR Shielding Method

The BIR method considers primary and secondary radiation separately. The unattenuated primary beam usually requires most consideration when orientated directly at a wall, e.g., in chest radiography. The air kerma at the wall may be calculated in one of two ways using either entrance surface dose (ESD) or detector dose (see Figure 5.1).

The entrance surface dose method sums the patient ESDs and the inverse square law corrects to give ESD immediately outside the barrier. A backscatter factor must also be applied. The detector dose method applies where the beam is completely intercepted by the patient, detector, and detector assembly. The air kerma at the detector for the exposure (DAK) is determined by the workload and inverse square law corrected to the outside of the barrier. A guideline DAK of 10 μGy is given for this method along with factors for the attenuation resulting from various detector set-ups.

There are three sources of secondary radiation; tube leakage, wall scatter, and patient scatter. As patient scatter is by far the largest component in diagnostic installations, the other two may effectively be ignored. Patient scatter can be estimated using a scatter factor, S, defined as:

$$S = \frac{K_S}{KAP} \tag{5.1}$$

where

K_S = scatter air kerma at 1 m

KAP = product of entrance air kerma and X-ray beam area.

In this methodology, it is argued that as in almost all cases the direction of the incident X-ray beam will be parallel to the closest barrier, the increase in scatter factor as the scatter angle increases is balanced by the increase in distance between the scatter source and the barrier. This empirical assump-

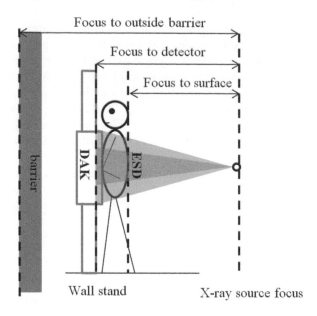

Figure 5.1 ESD and detector dose methods for primary shielding

tion removes the dependence of scatter with angle, allowing scatter to be determined using the formula:

$$S_{MAX} = [(0.031 \times kV) + 2.5]\mu Gy(Gy\ cm^2)^{-1} \qquad (5.2)$$

where
S_{MAX} is the maximum scatter factor at 1m
kV is the tube potential.

To apply this equation practically, maximum scatter factors are determined for the range of procedures to be used and then summed. This total S_{MAX} may then be adjusted using the inverse square law to determine scatter kerma at the barrier distance.

When determining the shielding for a CT scanner room, only scattered radiation need be considered. CT scatter air kerma (KCT) at 1 m from the isocenter is determined from the number of head and body scans taken per week (Nb and Nh) and the average Dose Length Products (DLP) for these scans (DLPb and DLPh). Scatter factors (SCTb and SCTh) are given in the BIR publication (Sutton et al. 2012).

$$K_{CT} = (N_b \times 50 \times DLP_b \times S_{CTb}) + (N_h \times 50 \times DLP_h \times S_{CTh})\ (5.3)$$

Where high dose equipment with heavy workloads is used, tertiary scatter from ceilings and walls should also be considered. The degree to which this scatter must be taken into account depends on shielding height. The minimum recommended height for a shielding barrier is 2 m. A method to derive tertiary scatter is given in the BIR publication but will not be described here.

Once both the primary and scatter components have been assessed for each barrier, the transmission required to reduce the incident kerma to the design constraint may be calculated using the formula

$$B = \frac{dose\ constraint}{annual\ incident\ air\ kerma\ \times\ occupancy\ factor} \qquad (5.4)$$

where
B is the Broad beam transmission factor.

The thickness of barrier shielding required to provide transmission (B) is dependent on tube potential (kVp) and can either be determined by reading directly from graphs of transmission data or from the following empirical formula:

$$x = \frac{1}{\alpha\gamma} ln\left[\frac{B^{-\gamma} + (\beta/\alpha)}{1 + \beta/\alpha}\right] \qquad (5.5)$$

where
x is the thickness of shielding material
B is the transmission factor
α, β and γ are fitting parameters dependent on tube potential and material (Sutton et al. 2012).

5.2.2.2 NCRP Method

In contrast to the BIR method, the NCRP method specifies design constraints based on the designation of areas, i.e., controlled or uncontrolled. The constraint for controlled areas is recommended as 0.1 mGy week^{-1} or 5mGy y^{-1} (based on a 50-week year) and the constraint for uncontrolled areas is recommended as 0.02 mGy week^{-1} or 1 mGy y^{-1}. These constraints are derived from the limits for pregnant workers and members of the public, respectively.

Several parameters, similar to those defined in the BIR report, must be determined to utilize the NCRP method:

- distance (d) from the source to the radiosensitive organs of the nearest person in the occupied area;

- occupancy factor (T) as discussed above.

- Workload (W) is the tube current-time product over a specified period usually expressed in terms of mA-minutes. This may be defined per patient and multiplied by the number of patients treated in a set time period to give the total workload. A table of reference workloads for different rooms and different operating potentials is given.

- The use factor (U) is the fraction of primary beam workload directed towards a given primary barrier (one that will attenuate the primary beam). Typical values are tabulated.

- Unshielded primary air kerma Kp(0) at distance dp from the X-ray tube focal spot is calculated using the unshielded primary air kerma per patient at 1 m (K_p^1), the use factor, and the number of patients per week (N):

$$K_p(0) = \frac{K_p^1 U N}{d_p^2}.$$ (5.6)

- Pre-shielding factors, take into account the attenuation of the detector assembly, and a table of equivalent thicknesses is defined, although no information is available for digital detector assemblies.

- Leakage and scattered radiation is taken into account and a table of indicated values is given. These must be combined, corrected for the number of patients, and corrected for distance to give the total air kerma from unshielded secondary radiation.

These parameters are then used to calculate the broad beam transmission factor $B(x)$ and the thickness of shielding material (x) required using the formula

$$B(x) = \left(\frac{P}{T}\right)\frac{d^2}{K^1 N}$$ (5.7)

where
P is the weekly shielding design goal (0.02 mGy week^{-1}) and
K^1 is the average unshielded air kerma per patient at 1 m from the source

$$x = \frac{1}{\alpha\gamma}ln\left[\frac{\left(\frac{NKT^1}{Pd^2}\right)^\gamma + \frac{\beta}{\alpha}}{1 + \frac{(\beta)}{\alpha}}\right]$$ (5.8)

where α, β and γ are fitting parameters with tabulated values provided, as for the BIR method.

5.2.2.3 Building Materials

There is a wide range of building materials that may be used to provide shielding. Important considerations are that the material should be fit for purpose and as straightforward as possible to install. The cost of the shielding material is also a significant consideration although this may be only a fraction of the cost of the work required to install it. It is advisable to employ a specialist shielding installer as they will be aware of the considerations that must be taken when shielding is installed. Some of the common types of shielding and considerations to be taken when using them are shown in Table 5.3:

Table 5.3: Considerations when using common shielding materials

Material	Considerations
Lead sheet	• normally bonded to board to maintain uniform thickness • sold in standard thicknesses, code 3 (1.32 mm) to code 8 (3.55 mm) • milled or rolled lead is manufactured to a European standard, machine cast lead is not – it is prudent to specify milled or rolled • weight of individual sheets
Brick	• inexpensive • may be used in conjunction with lead sheet • pre-existing brick walls may be used but must be checked carefully • take brick density into account in shielding calculations • cavities must be filled with mortar
Gypsum wallboard	• useful for lower-energy installations, e.g., mammography, dental
Concrete	• ensure density is specified – lightweight concrete is available • confirm details of floor construction – may contain hollowed-out sections • ensure cavities within blocks and poured concrete are filled

Continued on next page

Table 5.3 – *Continued from previous page*

Material	Considerations
Barium plaster	<ul><li>lead equivalence varies with tube potential</li><li>shelf life of three months</li><li>must be applied evenly and be crack-free when dry</li></ul>
Lead glass	<ul><li>used widely in control areas, vision panels, safety glasses, etc.</li></ul>

5.2.2.4   Shielding Assessment

Following shielding installation, it is important to carry out an inspection to ensure the correct level of protection is provided and that there are no breaches. If possible, shielding should be visually inspected immediately after installation and before wall finishes have been applied. At this time any obvious external breaches will easily be observed. However, on a rapidly moving construction schedule this may not be possible. In addition to visual inspection, assessment of transmission and internal breaches must be undertaken.

The assessment of transmission requires the use of a radiation source. This may be in the form of a radioactive isotope that emits at an energy similar to the diagnostic radiation that will be in use, for example, ^{241}Am (60 keV) or ^{99m}Tc (141 keV). It is essential to be aware of the properties of the shielding material that is to be tested, and to ensure the correct radioactive source is used. For example, ^{99m}Tc cannot be used to test the transmission of barium plaster due to its misleadingly low attenuation in the material. If it is not possible to use a radioactive source, a mobile X-ray system could be used. Alternatively, the shielding could be tested with the installed equipment, which has the advantage that the emissions used for testing will be those that the shielding has been designed for.

A safety assessment should always be carried out prior to the use of a radiation source for shielding assessment. This should be used to define a procedure based on local legislative considerations. Care should be taken not to cause any contamination when using radioactive materials. The safety of the workers carrying out the assessment must be considered, and time, distance, and shielding should be employed to ensure doses are kept as low as reasonably achievable (ALARA). The source must be controlled at all times and it may be necessary to designate areas temporarily during the course of the testing.

To assess shielding using a radiation source, a radiation detector such as a scintillation counter will be required. It is important that the detector has the correct properties to detect the emissions from the source in use. For transmission using X-ray equipment, it will be necessary to carry out several exposures while dose measurements are taken using an ionization chamber at each barrier. In order to calculate wall transmission, the following measurements will be required:

- distance from source to barrier;

- thickness of the barrier;

- background count rate, where relevant;

- unattenuated count rate reading taken at a known distance from the source or unattenuated dose reading at 1m;

- attenuated count rate or dose reading on the opposite site of the barrier from the source.

The unattenuated count rate or dose reading should be corrected to the position where the attenuated measurement was taken for each barrier, using the inverse square law. The transmission may then be calculated by comparing the two measurements at the same position:

$$Transmission, T = \frac{unattenuated\ count\ rate\ or\ dose\ reading}{attenuated\ count\ rate\ or\ dose\ reading} \times 100\% \quad (5.9)$$

The wall thickness, x, may then be calculated using the formula:

$$x = \frac{lnT}{-\mu}. \quad (5.10)$$

where μ is the linear attenuation coefficient of the radionuclide or X-ray beam used for testing, in the material of which the wall is constructed. The Handbook of Health Physics and Radiological Health (Shleien et al. 1998) provides data on values of μ for different beams and materials.

A radioactive source and suitable detector is usually the most effective means of checking for problems, as this should more accurately locate the breach. The source should be placed in the center of the room and the barriers thoroughly and slowly surveyed with the detector. Any breaches will lead to an increased count rate. The position of a breach should be noted and transmission calculated at that point to inform further actions.

When inspecting a new installation, there are certain areas that are susceptible to breaches. These include plug sockets, pipework, and any other area where lead sheet may be removed to provide a route through the barrier. Gaps around doors, vision panels, and windows should also be thoroughly checked. It is almost inevitable that minor breaches, for example, through door furniture, will be found. These need to be assessed by a competent person (Radiation Protection Officer) to assess the impact of the breach. Where breaches significantly impact on the design dose constraint, remedial work will be required.

5.2.3 Equipment

Successful equipment management is an important component of the quality assurance (QA) program that should be in place in any diagnostic radiology facility as a part of the protection and safety program. Quality assurance is a process of continual assessment and improvement of all the protection and safety measures within a facility. Throughout the lifetime of each piece of diagnostic equipment, there will be different requirements relating to the QA system. Below, the lifetime of diagnostic radiology equipment is considered in four stages: design and construction; operation; modifications; and decommissioning (IAEA 2006).

5.2.3.1 Design and Construction

When supplying equipment, manufacturers are required to demonstrate compliance with the relevant International Electrotechnical Committee (IEC) standards. Compliance with the IEC standards will usually ensure that additional local and/or regional medical device standards are met, for example, medical devices for use in Europe must have obtained the CE mark as described in the European Medical Devices Directive (EEC 1993).

Critical safety features expected to be included in the design and construction of diagnostic radiology equipment include (EEC 1993):

- minimization of the radiation exposure level to ALARA while still carrying out its required function;

- exposure to unintended, stray, or scatter radiation reduced as far as possible;

- visible and/or audible warning signal to indicate the emission of radiation;

- means of terminating exposure in case of emergency;

- emissions and geometry of the field controlled by user;

- appropriate image quality with lowest radiation exposure;

- electrically safe.

Once a manufacturer has supplied a piece of diagnostic radiology equipment it must be assessed under the quality assurance system to ensure it delivers all that is expected. This is the progression to the operation stage and acceptance testing.

5.2.3.2 Operation

Operation includes acceptance testing, commissioning, clinical use, and maintenance. Acceptance testing must be performed post installation, usually by a medical physicist to ensure that the equipment fulfils the requirements set out in the relevant legislation and guidance. Appendix II of BSS Report No. 39 provides a comprehensive discussion of diagnostic radiology equipment's radiation protection features. Any protocol for acceptance testing should take these requirements into account.

Commissioning, clinical use, and maintenance form a significant component of the facility quality assurance program. Commissioning tests must be carried out following acceptance of the equipment, and prior to clinical use, to provide baselines for future testing. Throughout the clinical use of the equipment it should be subject to periodic testing or quality control (QC) according to the facility quality assurance program. The results of routine QC tests should be compared to the baselines acquired at commissioning, to indicate any changes in the performance of the equipment.

All diagnostic radiology departments must have an ongoing QA program in place. This program includes all the routine actions necessary to ensure that medical exposure to ionizing radiation within the department is optimized and carried out safely, in accordance with relevant legislation. A QA system runs continuously and includes actions such as regular audit against legislation, reject analysis, quality control testing, and routine patient dose audits. As part of a successful QA program, maintenance of equipment must be carried out at regular intervals. A schedule of maintenance may be agreed upon with the manufacturer or other authorized agent to ensure that faults which arise and issues identified by QC testing can be rectified. When it is necessary to carry out extensive maintenance such as the replacement of an X-ray tube, or any other component affecting radiation protection and safety, it may be necessary to re-commission the equipment.

5.2.3.3 Modifications

In addition to routine maintenance, technological developments may necessitate hardware and software upgrades during the lifetime of the equipment. Depending on the nature of such modifications, it may be necessary to perform acceptance tests, commissioning tests, or indeed routine QC tests to ensure optimal equipment performance. The testing that is required should be defined under the QA program according to the modes of operation of the equipment.

5.2.3.4 Decommissioning

When diagnostic radiology equipment reaches the end of its life, it is important to quantify its performance. Given that a suitable and effective QA program is in place, performance data should be available over the lifetime of the equipment. However, it is prudent to carry out further tests as close as possible to the removal of the equipment. This ensures that, should the equipment be put to use in an alternative facility, evidence of its condition prior to decommissioning will be available. Decommissioning itself must be carried out by an authorized individual.

5.2.4 Policies and Procedures

As part of the facility protection and safety program, it is the responsibility of the licensee or registrant to implement policies and procedures to ensure compliance with the BSS. This will require a safety policy stating the designation of responsibilities for protection and safety within the facility. This policy should discuss the protection and safety program and the separate procedures that exist to ensure that the facility is compliant with the relevant recommendations. This policy should be distributed to and understood by all personnel working within the facility.

Supplementary to the safety policy, separate procedures should be defined for those activities that are described in the BSS, and National Regulations may require additional procedures to be defined. Examples of such procedures include those for personal dose monitoring, quality control of diagnostic radiology equipment, training, and the use of personal protective equipment. These procedures should be distributed and made available to all those who are expected to follow them.

5.2.5 Designation of Areas

The ICRP define two types of designated area, controlled and supervised. The BSS expands on these definitions. The purpose of designation is to ensure that occupational and public dose limits are adhered to and to ensure exposures are optimized.

5.2.5.1 Controlled Areas

A controlled area is defined in the BSS as any area in which specific protection and safety measures are or may be required to control exposures. Designation also takes into account the magnitude of exposures that are likely to occur in the area. National legislation and guidance will recommend the magnitude of exposure determining a controlled area. Diagnostic radiology rooms are generally designated as controlled areas.

It is necessary to restrict access to and demarcate controlled areas. This is usually achieved using physical barriers such as walls and doors. Signage is

Figure 5.2 UK designated area warning sign and warning light

also required for demarcation and should display the recommended International Organization for Standardization symbol, the trefoil (ISO 1975). Where mobile X-ray units are used, it may be necessary to demarcate a temporary controlled area.

X-ray room warning signs should be posted at all entrances to the room. For diagnostic X-ray rooms these will usually indicate that the area should not be entered when a warning light is illuminated. Warning lights should be placed at eye level (1,500 mm) at either side of double doors or at the opening side of a single door. Alternatively, if there is no space to affix the warning light by the side of the door, it can be positioned above the door. Figure 5.2 shows examples of warning signs and warning lights.

Appropriate instruction and training must be provided for those working within controlled areas, and written Local rules must be established. The Local Rules must include (IAEA 2014):

- details of the radiation protection officer;

- procedures for normal operation within the controlled area;

- procedures for unusual events;

- relevant investigation levels for personnel monitoring;

- procedures to be followed should an investigation level be exceeded.

Further additions to the Local Rules may be required by National regulatory bodies. All employees working in the controlled area should be aware of the Local Rules and have "read and understood" them.

5.2.5.2 Supervised Areas

Supervised areas are those areas in which occupational exposures may reach the levels that would require the area to be controlled. The designation of these areas is kept under review and may change with alterations to the use of the area. Supervised areas must be appropriately delineated with approved signage.

As part of the quality assurance program, the status of all designated areas should be reviewed on a regular basis. Changes in procedures, practices, and workload may require changes to the designation of some areas. An alteration to the use of a diagnostic radiology room should always indicate the requirement for a revised safety assessment to inform potential re-designation.

5.2.6   Personal Protective Equipment

In accordance with the BSS, the protection and safety program for any facility should follow the recommended hierarchy:

- engineering controls;

- administrative controls;

- personal protective equipment.

That is, all possible other controls must have been applied to any planned exposure situation before the use of personal protective equipment (PPE) to reduce exposure is considered.

Due to the nature of some procedures in diagnostic radiology, personal protective equipment may be necessary in some situations; for example, the wearing of lead aprons in interventional radiology theatres. These procedures must be thoroughly safety assessed and reviewed on a regular basis and specifically if usage changes. When carrying out a safety assessment to determine whether personal protective equipment is required, consideration should be given to the increased time of exposure and complexity of tasks that may result from the wearing of PPE.

Different types of PPE are available for different applications. Examples include lead aprons, thyroid shields, leaded glasses, and lead gloves. It is the duty of the employer to provide adequate and well-fitting personal protective equipment that complies with the relevant standards and specifications. In addition to these measures, rooms and equipment should be fitted with appropriate protection measures such as ceiling-suspended lead shields, portable lead shields, or lead curtains attached to the image intensifier or patient couch.

Workers should receive training on the correct use of PPE and room protection features and it is the responsibility of the workers to ensure that they comply with the training and instructions given.

5.2.7 Dose Assessment

The recommendations of the ICRP are that dose assessment is carried out for occupational, public, and patient exposure. Patient dose assessment is considered in Section 5.3.2, while the methods by which dose assessment is carried out for occupational and public dose are discussed below. The IAEA Safety Standards Series No. RS-G-1.3 provides a comprehensive guide to the assessment of occupational exposure (IAEA 1999a).

A dose assessment program enables optimization and limitation of doses to workers and the public. Doses measured in either case should not exceed the ICRP limits (Table 5.4). Doses to members of the public are limited through the design of facilities and safety assessments, discussed above. Doses to staff may be assessed through programs of personnel monitoring.

Table 5.4: Current recommended dose limits

Exposure category	ICRP recommendation
Occupational exposure	20 mSv/year average over 5-year period
Lens of the eye	20 mSv/year*
Skin	500 mSv/year
Hands and feet	500 mSv/year
Pregnant workers	1 mSv/year to the embryo/fetus
Public exposure	1 mSv/year
Lens of the eye	15 mSv/year
Skin	50 mSv/year
	Revised from 150 mSv/year in ICRP 103 to 20 mSv/year in ICRP 118 (2011)

5.2.7.1 Occupational Dose Assessment

Monitoring of individual doses is usually carried out using personal dosimeters worn on the body. The quantity used for individual monitoring is the personal dose equivalent or Hp(d), as defined by the International Commission on Radiation Units and Measurements (ICRU 2011):

Hp(d) – personal dose equivalent in soft tissue at an appropriate depth, d, below a specified point on the human body.

The depth, *d*, at which dose is assessed is related to the penetration of the radiation. Personal equivalent dose received as a result of exposure to strongly penetrating radiation is assessed at a depth of 10 mm, Hp(10). Personal equivalent dose received as a result of weakly penetrating radiation is assessed at 0.07 mm, Hp(0.07).

Hp(10) is used to provide an estimate of the effective dose and this is usually sufficient to assess occupational exposure. A dosimeter is placed on the torso as this area is more likely to be effected by strongly penetrating radiation. As the sensitive skin cells are around 0.05-0.1 mm below the skin surface, Hp(0.07) is used to give an estimation of skin dose and, in particular, extremity dose. The ratio of Hp(0.07) to Hp(10) will depend on the radiation type.

It is the responsibility of the employer or license holder to define a program for occupational dose monitoring. This will require contracting with an individual monitoring service that has been approved by the relevant regulatory authority. To be approved by the regulatory authority, a service is required to provide dosimeters with the required accuracy to measure Hp(10) and Hp(0.07) for the radiation in use, to process the dosimeters, and to employ suitably qualified personnel. Workers who are contracted to multiple employers must be monitored separately by each employer and it is a requirement that this information should be shared to ensure that the overall radiation dose to the individual does not exceed any limit. An example of this would be radiologists who work in a number of different hospitals (IAEA 1999b).

The requirement for monitoring of individuals should be based on the findings of a safety assessment, taking into account exposure during normal working practices. The frequency with which monitoring occurs should also be determined by taking into account the safety assessment and the magnitude of exposure the worker may be expected to receive. Typical monitoring periods are one to three months.

In order to ensure that occupational doses are optimized and that dose limits are not exceeded, it is necessary to work to a dose constraint, as described by the ICRP. Dose constraints set in occupational dose monitoring indicate a level of dose above which optimal procedures are not being followed. Whenever this constraint is exceeded, the cause must be investigated. In some scenarios the cause of increased dose received by the employee may be acceptable. If this is found to be the case, the employee may then be subject to additional monitoring precautions including registration as a "classified worker" under some National legislative systems.

There are no internationally prescribed values for dose constraints, although they may be set out within National regulatory requirements. Personnel dose constraints may also be set over a time period to coincide with monitoring, rather than over a year, to provide early warning if an employee is on a trajectory that may exceed the statutory annual limit.

There are several different types of personal dosimeters used in diagnostic radiology: thermoluminescent dosimeters (TLDs), optically stimulated luminescence dosimeters (OSL), photographic film and electronic dosimeters; and a selection of these are shown in Figure 5.3. The properties of these dosimeters must be carefully considered to establish which one is most suitable. Properties include accuracy, sensitivity, range, cost, durability, fading, effects from exposure to light, effects from different energies of ionizing radiation that are not intended to be measured, and ease of wearing.

Dosimeters for whole body monitoring should be worn on the trunk. If an individual is required to wear a lead apron, for example when involved

a) thermoluminescent dosimeters

b) optically stimulated luminescence dosimeter detector

c) photographic film detector

Figure 5.3   Different types of personal dosimeter

in interventional radiology, the dosimeter would usually be placed under the lead apron. Sometimes an additional dosimeter is provided to be worn outside the lead apron to capture further information about the dose to other areas of the body, for example the thyroid or eyes. It may be argued that if only one dosimeter is available it should be worn outside the apron and algorithms should be used to correct for doses elsewhere (NCRP 1995). This is a possibility; but, particularly for a classified worker, the employer must be confident that the algorithm provides an accurate dose estimate. Employers should provide very clear procedures on how, when, and where dosimeters should be worn.

The accuracy of the dose indicated is heavily influenced by the position in which the dosimeter is worn under any exposure circumstance. For example, a dosimeter worn at the front of the body will give an accurate indication of effective dose in exposure situations where the worker is exposed from the front and laterally. However, in a situation where the worker is exposed from the back, the effective dose reading will not provide an accurate indication of exposure.

Employers should set procedures in writing for an occupational dosimetry system. This should include information on exchange of dosimeters, monitoring periods, constraints applied, how to use dosimeters, and any training required. Employees are responsible for following those procedures, including wearing the personal dosimeters correctly and taking the required care of them.

Occupational eye exposure

Based on a reassessment of the threshold dose at which deterministic effects to the lens of the eye may occur, the ICRP has recommended a reduction in occupational exposure limits. These limits were first recommended in ICRP Publication No. 118 (ICRP 2012) approved in October 2011. Previously the occupational limit for exposure to the lens of the eye was 150 mSv per year. The ICRP recommended that this limit be reduced to 20 mSv per year, averaged over five years, with no more than 50 mSv in any one year.

The personal dose equivalent for the lens of the eye is specified at a depth of 3 mm, Hp(3). Hp(10) and Hp(0.07) may be used to assess Hp(3) with reasonable accuracy. If both Hp(10) and Hp(0.07) are below the limit for the lens it may reasonably be assumed that Hp(3) is also below this limit. This is adequate for assessment of occupational eye dose for the majority of diagnostic radiology activities.

It may be necessary to monitor eye doses more closely for employees involved in fluoroscopy and interventional procedures. This can be achieved using personal dosimeters attached to a headband or glasses. A careful safety assessment should be carried out, taking into account existing dose information, to establish which workers in these areas should be monitored.

5.2.7.2 Public Dose Assessment

The BSS define the responsibilities of regulatory bodies, employers, and licensees with regard to public dose assessment. It is the responsibility of the regulatory authority to review public exposure and to report on their findings. It is also the responsibility of the employer or licensee to ensure that public doses within their facilities are assessed and reported to the regulatory authority, as required.

The purpose of public dose monitoring, as for occupational dose monitoring, is to ensure that the exposure limits set out by the ICRP are not exceeded. As it is impractical for members of the public to be monitored individually, areas with public access may be monitored for a set period, e.g., two months, and average exposure levels determined. Such monitoring can alert the employer to any areas with potential problems.

The ICRU-defined quantities applied for area monitoring are the ambient dose equivalent H*(d) and the directional dose equivalent H'(d,Ω) (ICRU 1993). As for the personal dose equivalent, the ambient and directional dose equivalent are based on whether the radiation is strongly (d = 10 mm) or weakly (d = 0.07 mm) penetrating. The directional dose equivalent also takes into account the direction of the radiation field, Ω. H*(10) is used to obtain an approximate value for effective dose in an area (IAEA 1999b).

In common with personnel monitoring, care should be taken to ensure that area monitors are appropriate for measuring the required parameter. The type of radiation, the sensitivity and range of the instrument, the dose unit that is measured, battery life, and directionality are among the characteristics that must be considered. For the purposes of public exposure, area monitoring may be carried out using TLDs to give an integral exposure measurement over a set time frame. If a time-related output is required, for example, to monitor peak dose rates in an area, a monitor with a time base will be required. Careful consideration should also be given to the areas that are to be monitored to ensure that measurements are made where exposures would typically occur.

5.2.8   Pregnant Staff

ICRP Publication 103 states that protection of the fetus should be considered following notification of the pregnancy to the employer. Once notification has been given, the fetus should receive the same consideration as a member of the public. Therefore, a dose limit of 1 mSv over the remainder of the pregnancy must be applied. This dose limit may also require the frequency of monitoring to be reviewed, to ensure that adequate warning is available if the limit is likely to be exceeded.

It is the duty of the employer to carry out a detailed safety assessment of the role of the pregnant employee. This safety assessment will indicate any duties that may need to be modified or stopped for the duration of the pregnancy. In addition, radiation workers should be provided with information on the minimization of radiation dose during pregnancy. As part of the protection and safety management program, there should be a defined procedure for pregnant employees. All radiation workers within the facility should be aware of this policy.

5.2.9   Staff Supporting Patients

The safety and protection program of the facility should have in place procedures for employees required to support patients. This procedure should include provision of measures to minimize the possibility of holding, for example, sedation and mechanical supports. Those who are allowed to support patients such as friends and family, provided they are not pregnant, should be described in the procedure, along with those who are not allowed to support. Methods for positioning to ensure that doses are ALARA and instructions for the use of personal protective equipment should also be included (IAEA 2002).

Employees should only hold patients under exceptional circumstances. These circumstances should be defined in the relevant procedure. If it is necessary for an employee to support a patient during a diagnostic radiology procedure, the recommended PPE should be worn and the facility procedures for positioning to reduce exposure should be followed closely.

5.3 PROTECTION OF PATIENTS

This section will consider the medical exposure of patients. As discussed briefly in Section 5.1, when considering the use of ionizing radiation for diagnostic purposes, only the principles of justification and optimization are applied, as there can be no statutory limit to the dose a patient may receive. The patient may receive any exposure that is justified in order to provide diagnostic information or to deliver treatment.

5.3.1 Responsibilities

The key roles and responsibilities for the medical exposure of patients are illustrated in Figure 5.4. The facility protection and safety program should have sufficient policies and procedures in place to ensure that no individual undergoes a medical exposure unless the following conditions are met (IAEA 2014):

- the exposure has been requested by a referring practitioner and the relevant clinical information has been supplied, or it is part of an approved health-screening program;

- the exposure has been justified by a radiological medical practitioner, or it is part of an approved health-screening program;

- a radiological medical practitioner has taken responsibility for the exposure and authorized the procedure;

- the patient or their representative has been informed as appropriate of the diagnostic benefits and radiation risks involved.

The employer has a responsibility to ensure that a register is maintained of those health professionals with responsibilities for medical exposures within a facility.

Optimization is not merely the process of minimizing dose but of creating a balance between the exposure that is given and the benefit that is received. Therefore, the lowest dose option may not always be that which is chosen. For optimization to be effective, many factors should be considered – not just the final exposure. For example:

- is diagnostic radiology equipment calibrated and maintained according to the relevant standards?

- are diagnostic radiology techniques and protocols optimized?

- are there separate procedures in place for pediatric patients?

- are records of patient dose kept and used to define diagnostic reference levels?

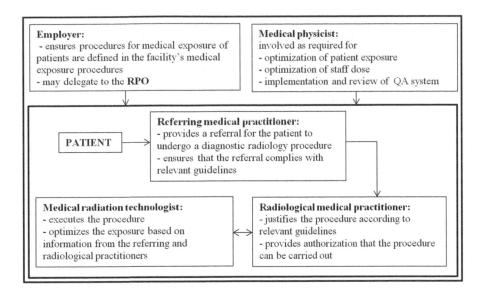

Figure 5.4 Roles and responsibilities

- is there a QA procedure in place to continually assess procedures for the medical exposure of patients?

- are unintended and accidental exposures reported so that lessons may be learnt from them?

For optimization to be successful, all the factors that may affect exposure and image quality must be monitored and reviewed on a regular basis.

5.3.1.1 Diagnostic Reference Levels

Diagnostic reference levels (DRLs) were introduced in ICRP Publication 73 (ICRP 1996). The ICRP also published Supporting Guidance 2 (ICRP 2001), to provide additional advice on the subject of DRLs. It is important to note that DRLs are not linked to any of the other limits or constraints set out in the ICRP recommendations. In Publication 73, a DRL is defined as a type of dose investigation level that indicates the need for review of doses if it is constantly exceeded. Therefore, for ease of use, DRLs must be defined in quantities that are easily measured.

A DRL may be applied as follows: A Local, National or Regional DRL is set. Dose data for a local reference group of patients who underwent the same procedure and are of similar build is then compared to the set DRL. It is important that like parameters are compared. For example, if a National DRL is set based on the mean dose for standard patients undergoing a standard procedure across several different facilities, the mean dose of standard patients for the same standard procedure should be compared to the National DRL. Individual patient doses should not be compared to a DRL.

Following this approach will enable the comparison of doses for certain procedures to local and national DRLs. It will therefore be possible to identify if doses for procedures are significantly higher, or lower, than the set DRL. Based on this knowledge, procedures may be investigated with a view to optimization.

DRLs should be reviewed periodically on a national and a local level. This will enable continuous optimization of patient exposure. Dose data is usually obtained through the radiology information system (RIS) or by manual record. Increasingly integrated dose audit software is also available for the collection and analysis of patient doses data. This software will enable the storage of patient doses directly from diagnostic radiology equipment into a database without the requirement of any data input from the medical technician. This data can be analyzed electronically to calculate and compare to local DRLs.

Application of DRLs to fluoroscopy procedures may not be straightforward because of the varying complexity between patients. However, where possible, attempts should be made to monitor and compare such procedures as part of an optimization program. DRLs are not relevant to the management of deterministic effects such as radiation erythema in interventional procedures. The potential for erythema and other deterministic effects to occur must be monitored in real-time through the dose area product (DAP). Conventionally, a DAP of 200 Gy cm^2 is considered as the threshold at which deterministic effects to the skin might be observed (ICRP 2013).

The ICRP's additional advice on DRLs (ICRP 2001) provides a useful summary of DRL units and definitions across several different agencies and countries.

5.3.1.2 Optimization of Techniques and Protocols

Dose audit and DRLs are instrumental in identifying where there is potential for optimization of techniques and protocols. The successful optimization of techniques and protocols will involve the medical radiologist practitioner, the medical technician, and the medical physicist.

5.3.1.3 Specific Issues

When considering the protection of patients and optimization of dose in diagnostic radiology, there are several important considerations. There are differences between procedures that must be taken into account, and it is also important to remember that while DRLs are defined for the "standard patient," all patients are different. Therefore, both the patient and the procedure must be considered in the optimization of a particular exam.

Computed Tomography

The increasingly widespread use of computed tomography (CT) systems in diagnostic radiology has been well documented. In their 2000 report (UNSCEAR 2000), the United Nations Scientific Committee on the Effects of Atomic Radiation (UNSCEAR) found that 34% of the collective dose due to medical exposures arose from CT examinations; in their 2008 report (UNSCEAR 2010) this had risen to 47%. This is supported by data from the UK, published by what was then the National Radiological Protection Board (NRPB) in their 2011 report (Hart and Wall 2002). The 2008 UNSCEAR report found that CT examinations accounted for 2-14% of the total diagnostic radiology procedures undertaken, depending on the health-care level of the country.

CT examinations contribute significantly to collective patient dose, and this contribution continues to increase as CT technology advances and becomes more accessible. Therefore every effort must be made to ensure exposures are optimized. This may be achieved as described in Section 5.3.2:

- using local DRLs to monitor dose levels;

- optimizing protocols and procedures for any examinations that consistently exceed national DRLs;

- ensuring that there are specific pediatric CT protocols;

- continuing to monitor and optimize where possible;

- maintaining equipment;

- training staff.

The 16th Report of the Committee on Medical Aspects of CT in the Environment (COMARE) discusses ionizing radiation dose from CT in the UK. The report includes guidance for reduction and optimization of dose resulting from the medical use of CT (COMARE 2014).

High-dose procedures

Along with CT, interventional fluoroscopic procedures such as those used in angiography and cardiology may also produce high doses. Indeed, the doses in some individual interventional procedures may be high enough to cause deterministic effects. Of particular importance when considering these procedures is the entrance surface dose (ESD), also referred to as skin dose.

The ESD is particularly difficult to calculate or measure due to the changing projection angle in fluoroscopy. Modern equipment is required to display an indication of the ESD that has been delivered. As previously mentioned, it is difficult to set DRLs for fluoroscopy procedures and, in particular, interventional procedures. For the purposes of optimization, it is recommended that a constraint be set to alert users to particularly high ESDs, based on the 200 Gy cm^2 DAP threshold for deterministic effects (Section 5.3.2.1). Any procedural ESD exceeding the constraint should then be investigated and dose calculations made to establish the possibility of skin effects occurring. The patient should be monitored to establish whether any deterministic effect occurs. IAEA Basic Safety Report No. 59 further discusses the methods that may be used to set guidance levels in interventional procedures (IAEA 2009).

Developments in the equipment used for interventional fluoroscopy have significantly improved their capabilities in relation to dose optimization. Appropriate use of this equipment by adequately trained operators is the key to ensuring dose optimization in increasingly complex radiological interventions. Techniques for optimization include (Axelsson 2007):

- use of flat panel detectors;

- use of pulsed modes rather than continuous mode;

- use of frame averaging;

- X-ray tube positioned underneath patient couch so that couch provides shielding from the primary beam. This is important for reduction of dose to patient and staff;

- use of last image hold to view the last image rather than continuous screening;

- using low-angle projections to spread the surface area of the skin on which the beam is incident;

- minimizing the use of high-dose-rate projections;

- minimized use of magnification modes – only when necessary;

- largest possible tube-to-patient distance;

- smallest possible patient-to-detector distance;

- use of beam filtration;

- optimized use of automatic brightness control;

- removal of anti-scatter grid where necessary, e.g., for pediatric patients.

Pediatric patients

Pediatric patients are more radiosensitive and therefore it is a requirement of the BSS that the justification and optimization of exposures for pediatric patients should receive greater attention. IAEA Safety Report Series No. 71 discusses how this may be achieved for different modalities (IAEA 2012). Procedures and protocols for all radiation modalities must be adjusted to ensure that radiation exposures are appropriately modified to account for patient size, which will vary considerably in the first 15 years of life.

The equipment available and techniques employed must be taken into account to achieve optimization. Lower-dose scans should be used where possible, with CT only used when essential. Lower factors, automatic exposure control, and a high-frequency generator should be used to ensure exposures are as short as the required image quality allows. The use of an anti-scatter grid for pediatric patients is not usually recommended, as smaller patients produce less scatter and therefore the increase in dose resulting from the use of such a grid is not justified. Gonad shielding should always be used where possible, field sizes should be kept as small as possible, and extra filtration may be added to reduce dose. Movement is a significant problem in pediatric radiology and it may be necessary to restrain the patient to prevent movement and potential repeated exposures.

DRLs for pediatric exposures should be set based on patient size rather than age. Therefore national DRLs may give reference doses for the average-sized 0, 1, 5, 10, and 15 year old. When analyzing local data it is important to understand how the national data has been produced and therefore how local data may be compared to it. This information should be given in the relevant national report.

Based on equipment requirements and dose audit, separate pediatric protocols must be set within a diagnostic radiology department. Equipment manufacturers will usually have a set of protocols that may be applied and optimized as dose data is collected. Procedures should be in place to ensure that additional requirements such as the use of gonad shielding, restraints, and when to ask the pregnancy question are defined. All those involved in pediatric diagnostic radiology must have relevant training.

Comforters and carers

The exposure of comforters and carers as part of medical exposures is recognized in ICRP Publication 103, with a specific dose constraint set at 5 mSv per episode. These individuals are those who care for and comfort patients and as a result are exposed to ionizing radiation. This category includes family or close friends who are exposed knowingly and willingly to ionizing radiation whilst holding a patient during a diagnostic radiology exposure. Procedures should be in place to ensure that adequate information is given to comforters and carers and suitable personal protective equipment should always be provided.

5.3.2   Training

It is the shared responsibility of the government and employer to ensure that health professionals with responsibility for medical exposures, such as the referring medical practitioner, the radiological medical practitioner, the medical radiation technologist, and the medical physicist, have completed the education and training required and are specialized in that role. The IAEA Safety Report Series No. 39, Appendix II provides an example training outline.

The BSS provides a summary of the basic qualifications required by those working in these roles. This includes a relevant degree, a course on radiation protection with content approved by the regulatory body, and practical training under the supervision of an accredited professional. Those individuals with responsibility for protection and safety in ionizing radiation safety must ensure that they are appropriately accredited and can provide evidence of accreditation. Relevant continuous professional development must be attended as recommended by the regulatory or professional body.

It is the responsibility of the employer to provide adequate professional development opportunities for all other workers involved in the use of ionizing radiation. The employer must also keep a record of the training undertaken and ensure that it is periodically renewed. Different levels of training may be required depending on the role of the employee. For example, those working within a controlled area may require more extensive training than those who do not.

There are some significant issues concerning the medical exposure of patients, which require more in depth training. These are discussed below.

5.3.2.1 Unintended and Accidental Medical Exposures

As part of the protection and safety program, procedures must be in place within a facility to minimize the possibility of any unintended or accidental exposures. These measures include robust procedures for the correct identification of patients, maintenance and calibration of radiological equipment, and staff training. However, the medical exposure of patients is vulnerable to human error, and therefore, within any department incidents may occur.

The BSS requires that investigations should be carried out as follows:

- for any exposure that is substantially greater than intended or for procedures that repeatedly exceed guidance levels;

- for any unintended dose arising from equipment failure, error, accident, or other unusual occurrence related to equipment that may lead to the patient receiving an exposure substantially greater than intended;

- in the event that deterministic effects arise as the result of a certain treatment. The definition of "substantially greater than intended" is stated by regulatory authorities in National legislation.

In the case that any of the above should occur, it is necessary that an estimate of the dose to the patient is produced. There are several resources available for this purpose: PCXMC software (STUK 2015), which may be purchased, and the ImPACT CT dose calculation spreadsheet (ImPACT 2015), which uses Public Health England (PHE, formerly HPA and NRPB) dose distribution data and is available for download. A written copy of the report should be sent to the relevant regulatory authority and to the patient's doctor.

Employees should be aware of the need for reporting unintended exposures and accidents to the relevant department for a dose assessment. Training should also ensure that employees are aware of the risks associated with ionizing radiation and certain individuals must be trained in delivering information about risk to patients. It is imperative for optimization that training and any other associated actions are carried out to address the root cause of any incident.

5.3.2.2 Pregnancy Procedures

It is essential to avoid the unintended exposure of an unborn child. Several useful guidance documents have been published regarding the effects of ionizing radiation on the fetus at different stages of development (HPS 2009, EC 1998). ICRP Publication 64, Pregnancy and Medical Radiation, provides a comprehensive discussion of the relevant considerations and recommendations (ICRP 2000). Every facility must have procedures relating to the exposure of pregnant patients and it is important that training is carried out to ensure that all workers understand and follow these.

When exposing females of childbearing age to ionizing radiation, it is essential to establish their pregnancy status prior to exposure. Signs should be displayed in waiting areas to prompt female patients to inform the medical technician or their medical practitioner if they believe they may be pregnant. Pregnancy does not necessarily indicate that an exposure should not go ahead; however, pregnant patients have the right to know the risks associated with any exposure the fetus may receive as a result of their own exposure.

The radiation risk to the fetus is dependent upon the level of development of the fetus. The fetus is most radiosensitive between 8 and 26 weeks. In general, for fetal doses below 1 mGy, the risk may be considered negligible; for doses greater than 1 mGy more careful consideration must be given to justification. There is no evidence for malformations or a decrease in IQ below a fetal dose of 100 mGy. Medical practitioners should be aware of the approximate fetal doses from conventional diagnostic radiology examinations and the implications of that dose.

Prenatal doses from most routine diagnostic radiology procedures present no increased risk to the fetus. However, some high-dose CT and interventional procedures may exceed 1 mGy fetal doses. In these cases the medical practitioner must adequately justify the procedure and ensure that the risks are discussed with the patient. The fetal dose and risk should be assessed following the exposure by a suitably qualified individual, usually a medical physicist.

It may be the case that the patient is not aware of their pregnancy or fails to disclose their pregnancy, and is exposed to ionizing radiation. In this case, an assessment of the fetal dose should be carried out by a suitably qualified individual. The relevant regulatory authority may also need to be notified depending on the National definition of doses greater than intended. Termination of pregnancy is not justified for fetal doses less than 100 mGy based on the associated radiation risk. For fetal doses between 100 and 500 mGy, decisions should be made based on the specific circumstances.

5.4 MAGNETIC RESONANCE IMAGING SAFETY ISSUES

The International Commission on Non-Ionizing Radiation Protection (IC-NIRP) publish international recommendations for non-ionizing radiation protection. With specific reference to Magnetic Resonance (MR), ICNIRP have published guidance for time-varying electromagnetic fields, static magnetic fields, and low-and high-frequency electromagnetic fields (ICNIRP 1998, IC-NIRP 2009, ICNIRP 2014).

The major protection issue for employees and patients is the static magnetic field associated with MR scanners. Correct equipment use is also an important factor in the safe application of MR. Protection of the public should be of less concern than when dealing with ionizing radiation, particularly as access is carefully restricted to MR units.

When designing an MRI facility, the hazards that should be considered include strength of fringe fields and generation of currents within the body, the projectile effect and torque the magnet may exert on ferromagnetic items including implants, the induction of currents in implanted devices, and interaction of the magnetic field with electrical equipment such as that used for monitoring.

5.4.1 Facility Design and Layout

5.4.1.1 Safety Assessment

As for ionizing radiation, a prior risk assessment must be carried out prior to the use of any MR installation. This assessment will take into account the hazards present and consider how they may be reduced. Based on the safety assessment, a controlled area should be defined. This will contain all fringe field lines above a certain strength (usually 0.5 mT) and should be access controlled at all times.

5.4.1.2 Fringe Field Lines and Boundaries

Strong static magnetic fields may cause biological effects and projectile hazards, and they may also affect the operation of surrounding equipment and implanted medical devices such as pacemakers. The majority of MR scanners operate at 1.5 tesla (T), but some units operate at 3 T or above. There are fringe fields associated with all magnets; the strength of these fields depends on the magnet strength, magnet design, and shielding.

All employees must understand fringe fields and it is good practice to have a plot of the 0.5 and 3 mT field lines displayed within the department (MHRA 2014). While fringe-field plots are provided by manufacturers, on installation it is prudent to measure these. The 0.5 mT field line should not extend beyond the controlled area.

5.4.1.3 Access

There are several categories of people who may require access to the MR controlled area: patients, volunteers, staff, and carers. All of these groups should be screened according to local procedures prior to entering the MR controlled area. Some experienced members of staff may be designated as "authorized personnel" and only they may be allowed unrestricted access to the controlled area. Members of the public should not be able to gain access to the MR controlled area.

The controlled area should have restricted access; for example, via key codes or plastic swipe cards. Methods of gaining entry should be non-magnetic and should not contain information that may be affected by the MR field. The controlled area must be demarcated with suitable warning signs displayed at all entrances. Those entering the MR controlled area must be warned of the associated hazards.

5.4.1.4 MRI Safe Equipment

A policy should be in place to ensure that only MR safe equipment is taken into the controlled area. This includes the purchasing, servicing, and routine testing of all equipment to be used in the MR environment. In order to reduce the possibility of incidents, it is recommended that all MR safe equipment be clearly labelled as such.

Equipment specified as suitable for MR use may be tested under different conditions from those in the MR controlled area, e.g., lower field strength. In these circumstances the equipment should be labelled as MR conditional, with a description of the conditions under which it was tested. These pieces of equipment should be kept under review and monitored for safety whenever significant changes are made that may affect their behavior in the MR controlled area.

Where it is impractical to label equipment, e.g., single-use disposable items, procedures should be in place to ensure that these are MR safe.

5.4.1.5 Radiofrequency Fields

In MR imaging radiofrequency (RF), coils used are used to transmit and receive RF pulses to and from the surface of the body. RF coils are essential to provide uniform excitation of protons in transmit mode and to ensure higher sensitivity of RF signals in receive mode. However, the oscillating electric and magnetic fields that make up the RF fields cause heating in body tissue. This is of particular importance if there are metal objects in the imaging field, e.g., implants, as metal absorbs RF energy very efficiently. The 2004 ICNIRP report states that no adverse health effects are expected if the increase in core body temperature is less than 1°C (ICNIRP 20014).

Some patients may be particularly susceptible to changes in temperature; for example, pregnant women, children, and those suffering from hypertension. Drugs administered to a patient may also cause an increased sensitivity to temperature increase. The ICNIRP report further states that in the case of infants and those with circulatory impairment, the temperature increase should not exceed 0.5°C.

There is a high risk of localized skin burns associated with RF coils – the temperature of the coils and associated cables will increase due to induced currents. Generally, burns can be avoided by careful screening and positioning of the patient. Care should be taken to ensure that cables are not looped or in contact with the patient. Cables should run parallel to and as close to the magnet bore as possible, although not touching the bore. The patient's skin must not be in contact with the magnet bore and insulating foam pads should be used to ensure that there are no conductive loops across the patient, e.g., between thighs or between arm and trunk.

5.4.1.6 Quench

MR magnets are maintained in a low-energy-consumption, superconducting state. This is achieved by surrounding the magnets with cryogen such as liquid helium or liquid nitrogen. When the quench valve is opened, the cryogen boils off rapidly at room temperature. The quench pipe delivers the cryogen waste gas out of the MR controlled area and to an outlet where it can be safely dispersed, usually outside.

Quench pipes require careful design and maintenance. On installation the quench pipe should be checked to ensure that the vent outlet is positioned safely and that there is no possibility of ingress into the pipe. Quench pipes should be sized according to the recommendation of the manufacturer to ensure that they are capable of dealing with the pressure that will be produced following a quench.

All employees should understand the quench process and why it occurs. Only adequately trained employees should be involved in the processes that may lead to a quench occurring, and should be aware of the procedure to follow in the event that it does or if a controlled quench is required.

5.4.2 Policies and Procedures

Policies and procedures must be clearly defined for the MR environment. These should include Local Rules for the controlled area and systems of work for exceptional occurrences, such as ferromagnetic materials entering the controlled area under supervision. Specific protocols should also be in place for scanning patients who are or may be pregnant. Pregnant employees should not remain in the scan room during scanning for the course of their pregnancy.

Suitable training procedures must also be in place, based on available recommendations and guidance. Where there is a high turnover of employees, for example, rotations of doctors, all new employees must be screened and made aware of the complexities of the environment they will be working in.

Screening of equipment, staff, and patients for ferromagnetic objects is a particularly important consideration when working with MR, and comprehensive screening policies and procedures must be in place. Entrance of non-MR safe equipment into the MR environment may not only damage the MR equipment, as it is drawn to the bore, but may also cause serious or in some cases fatal injury to any patient on the scanner. Examples of ferromagnetic items that have been inadvertently brought into an MR suite include fire extinguisher, scalpel, hemostat, syringe, scissors, stethoscope, pen, phone, laptop computer, tool box, water bottle, IV pole, wheelchair, gurney, keys, flashlight, clipboard, axe, gun, handcuffs, cleaning bucket, mop, watch, and credit cards. Patients must also be screened extensively to ensure that any ferromagnetic implants are detected. Implants that may be acted upon and moved by the MR field, with potentially fatal consequences, include: aneurysm clip, shrapnel, cochlear implant, prostheses, artificial heart valve, stent, permanent denture, defibrillator, pacemaker, medical infusion pump, drug delivery patch, tattoo, or breathing apparatus. These lists are highly detailed although incomplete; however, the outcome of ignoring any of these items may be devastating.

5.5 REFERENCES

Axelsson, B. 2007. Optimisation in fluoroscopy. Biomed Imaging Interv J 3(2): e47.

Committee on Medical Aspects of Radiation in the Environment. 2014. Patient radiation dose issues resulting from the use of CT in the UK, COMARE 16th Report.

European Commission. 1998. Radiation protection 100: Guidance for protection of unborn children and infants irradiated due to parental medical exposures, European Commission.

European Economic Community. 1993. Council Directive 93/42/EEC of June 1993 concerning medical devices.

Hart, D., Wall, B.F. 2002. Radiation Exposure of the UK Population from Medical and Dental X-ray Examinations, NRPB Report NRPB-W4.

Health Protection Agency. 2009. The Royal College of Radiologists, College of Radiographers, Protection of Pregnant Patients during Diagnostic Medical Exposures to Ionising Radiation, HPA RCE-9.

ImPACT. 2015. CTDosimetry; software http://www.impactscan.org (last accessed January 25, 2015).

International Atomic Energy Agency. 1999a. Assessment of Occupational Exposure Due to External Sources of Radiation, IAEA Safety Standards Series No. RS-G-1.3, IAEA, Vienna.

International Atomic Energy Agency. 1999b. Occupational Radiation Protection, IAEA Safety Standards Series No. RS-G-1.1, IAEA, Vienna.

International Atomic Energy Agency. 2002. Radiological Protection for Medical Exposure to Ionizing Radiation, IAEA Safety Standards Series No. RS-G-1.5, IAEA, Vienna.

International Atomic Energy Agency. 2006. Applying Radiation Safety Standards in Diagnostic Radiology and Interventional Procedures Using X-Rays, Safety Reports Series No. 39, IAEA, Vienna.

International Atomic Energy Agency. 2009. Establishing Guidance Levels in X-Ray Guided Medical Interventional Procedures: A Pilot Study, Safety Reports Series No. 59, IAEA, Vienna.

International Atomic Energy Agency. 2012. Radiation Protection in Paediatric Radiology, Safety Reports Series No. 71, IAEA, Vienna.

International Atomic Energy Agency. 2014. Radiation Protection and Safety of Radiation Sources: International Basic Safety Standards, No. GSR Part 3, IAEA, Vienna.

International Commission on Non-Ionizing Radiation Protection. 1998. ICNIRP Guidelines for limiting exposure to time-varying electric, magnetic and electromagnetic fields (up to 300 GHz). Health Physics 74 (4): 494-522.

International Commission on Non-Ionizing Radiation Protection. 2009. ICNIRP Guidelines on limits of exposure to static magnetic fields. Health Physics 96 (4): 504-514.

International Commission on Non-Ionizing Radiation Protection. 2014. ICNIRP Guidelines on limiting exposure to electric fields induced by movement of the human body in a static magnetic field and by time-varying magnetic fields below 1 Hz. Health Physics 106 (3): 418-425.

International Commission on Radiation Units and Measurements. 1993. Quantities and Units in Radiation Protection Dosimetry, Rep. No. 51, ICRU, Bethesda, MD.

International Commission on Radiological Protection. 1996. Radiological Protection and Safety in Medicine, ICRP Publication 73, Pergamon Press, Oxford and New York.

International Commission on Radiological Protection. 2000. Pregnancy and Medical Radiation, ICRP Publication 84, Elsevier, The Netherlands.

International Commission on Radiological Protection. 2001, Radiation and Your Patient: A Guide for Medical Practitioners, ICRP Supporting Guidance 2, Elsevier, The Netherlands.

International Commission on Radiological Protection. 2008. The 2007 Recommendations of the International Commission on Radiological Protection, ICRP Publication 103, Elsevier, New York.

International Commission on Radiation Units and Measurements. 2011. Fundamental Quantities and Units for Ionizing Radiation, ICRU Report 85a-Revised, Journal of the ICRU 11.

International Commission on Radiological Protection. 2012. ICRP Statement on Tissue Reactions and Early and Late Stage Effects of Radiation in Normal Tissues and Organs: Threshold Doses for Tissue Reactions in a Radiation Protection Context, ICRP Publication 118, Elsevier.

International Commission on Radiological Protection. 2013. Radiological Protection in Cardiology, ICRP Publication 120, Ann. ICRP 42(1).

International Organization for Standardization. 1975. Basic Ionizing Radiation Symbol, ISO 361, ISO, Geneva.

Medicines and Healthcare Regulatory Authority. 2014. Safety Guidelines for Magnetic Resonance Imaging Equipment in Clinical Use, MHRA.

National Council on Radiation Protection and Measurements. 1995. Use of Personal Monitors to Estimate Effective Dose Equivalent and Effective Dose to Workers for External Exposure to Low-LET Radiation, NCRP Report 122, NCRP, Bethesda, MD.

National Council on Radiation Protection and Measurements. 2004. Structural Shielding Design for Medical X-Ray Imaging Facilities, NCRP Report 147, NCRP, Bethesda, MD.

Radiological Protection Institute of Ireland. 2009. The Design of Medical Facilities where Ionising Radiation is used, RPII 09/01.

Shleien, B., Slaback, L.A., Jnr., Birky, B.K. (eds). 1998. Handbook of Health Physics and Radiological Health (Third Edition), ISBN 0-683-18334-6, Lippincott Williams & Wilkins, Udgivet.

STUK (Radiation and Nuclear Safety Authority in Finland). 2015. PCXMC - A PC based Monte Carlo program for calculating patient doses in medical x-ray examinations; software available at http://www.stuk.fi (last accessed January 25, 2015).

Sutton, D. G., Martin, C.J., Williams, J.R., Peet, D.J. 2012. Radiation Shielding for Diagnostic Radiology, 2nd Edition, British Institute of Radiology, London.

United Nations Sources and Effects of Ionizing Radiation. 2000. Report 2000, Vol. 1: Sources, Scientific Committee on the Effects of Atomic Radiation (UNSCEAR), UN, New York.

United Nations Sources and Effect of Ionizing Radiation. 2010. Report 2008, Scientific Committee on the Effects of Atomic Radiation (UNSCEAR), UNSCEAR 2008, UN, New York.

V

Radiation Protection in Nuclear Medicine

CHAPTER 6

Radiation Protection in Nuclear Medicine

Cecilia Hindorf

Radiation Physics, Skåne University Hospital, Lund, Sweden

Lena Jönsson

Medical Radiation Physics, Lund University, Lund, Sweden

CONTENTS

I N NUCLEAR MEDICINE radioactive substances are administered to patients with the purpose to diagnose or to treat a disease. After the administration of the radiopharmaceutical, the patient is imaged by using either a scintillation camera (SPECT) or a camera for positron emission tomography (PET). The cameras depict images of organ function; for example, ventilation of the lungs, perfusion of the myocardium, the filtration of the kidneys, or the glucose metabolism. Many different radionuclide therapies are given but the most widely applied therapeutic procedure is the treatment of thyrotoxicosis (hyperthyroidism) with radioactive iodine (^{131}I-NaI).

The staff of the nuclear medicine department is exposed to radiation via the handling of radiopharmaceuticals, injecting the patient with radiopharmaceuticals, spending time close to the patient after injection, and calibration and quality control of the gamma camera or PET scanner. The patient will in turn irradiate family members and himself/herself after the administration of the radiopharmaceutical, and the irradiation will persist until the radiopharmaceutical either has decayed or is excreted from the body via biological processes. The key words here are to keep the doses as low as reasonably achievable (ALARA) to staff and family members within nuclear medicine by decreasing exposure time, increasing distance from the radiation source, and use of shielding.

Nuclear medicine deals with open radioactive sources, i.e., radioactive sources in liquid or gaseous form, which means that there is one more precaution to be considered, the risks of internal and external contaminations. Dispersal of the radioactive substance can lead to an accidental intake of the radionuclide, giving an unnecessary absorbed dose to the person. It can also give rise to inaccurate measurements if detectors or other pieces of equipment are contaminated.

In recent years, nuclear medicine has moved into the PET era. More cyclotrons for the production of positron-emitting radionuclides have been installed and the development of hybrid imaging systems, combining functional and morphological imaging, have led to a significant increase in the number of PET/CT and PET/MR scanners. The use of hybrid systems combining functional imaging and CT means an increased dose to the patient; therefore, it is important to justify the use of these hybrid systems. Radiation protection for and optimization of CT examinations are not discussed here.

A current topic of interest across all areas of medicine is development of a safety culture to minimize risk of harm to patients, workers, and the public. In the discipline of nuclear medicine, safety culture can be defined as a set of core values and behaviors that result from a commitment by physicians, medical physicsts/radiation protection officers, allied health workers, and management to emphasize safety over competing goals to ensure protection of people. The safety culture acts as a guide as to how employees will behave in the workplace (Glendon et al. 2006). Healthcare team members should work together to create a safe and efficient clinical environment. Department and team leaders should demonstrate a commitment to safety in their decisions and behaviors. For example, the Department Chair should fully support prompt, full disclosure of a misadministration. Physicians, medical physicists/radiation protection officers, and technologists should identify, evaluate, and correct issues that impact the safety of patients, e.g., seating of PET patients in a public waiting room resulting in unacceptable radiation exposure to members of the public is promptly corrected. A safety-conscious work environment is maintained so that personnel feel free to raise safety concerns without fear of retaliation or intimidation, and trust and respect are highly valued within the department regardless of employee rank. Department leadership, together with the medical physicist/radiation protection officer, should promote a positive safety culture by fostering these traits as they apply to their organization.

This chapter is based mainly on international recommendations given by the International Commission on Radiological Protection (ICRP), the International Atomic Energy Agency (IAEA), the World Health Organization (WHO), and the European Commission (EC), but other sources of information may also be helpful to the reader, e.g., the Institute of Physics and Engineering in Medicine Report 109 (IPEM 2014).

6.1 FACILITY DESIGN

The design of a nuclear medicine department should be carefully planned. The flow of patients and staff, and the internal transportations of radioactive material (e.g., vials from the radiopharmacy, prepared syringes for patients, and waste) should be carefully thought out when designing the facility. All transport routes should be indicated clearly on a drawing when the facility is being planned. Rooms should be placed in an order from lower to higher activities and from lower to higher photon energies, i.e., so that PET uptake rooms are placed as far as possible from scintillation cameras, well-type scintillation detectors, and uptake probes.

The shielding in the walls, doors, floor, and ceiling must be calculated based on the radionuclides to be used, the administered activities, the number of patients, and the length of time they will be in the department (i.e., in waiting rooms, injection rooms, uptake rooms, and camera rooms). In some cases shielding in walls can be replaced by local shielding; for example, shielding in the walls of a fume hood can be replaced by use of lead bricks or L-shields. Often vials could be placed in extra thick lead or tungsten pots to enhance protection. The weight of a lead-shielded fume hood can be considerable, up to approximately 2,500 kg or more, so the strength of the floor needs to be checked before the hood is installed. In addition, fume hoods usually do not include any upward protection. The more localized the protection, the smaller the volume of lead needed, which greatly affects the weight. Lead bricks should be angled in a way so that no gaps in the protection will be available. Lead in walls, floors, and ceilings could be replaced by an amount of concrete that provides the same amount of protection if acceptable to the building architects.

The shielding for cyclotron units (IAEA 2012), PET facilities, and facilities for iodine therapy (^{131}I) requires special attention since the photon energy is high. The required thickness of lead must be calculated with an attenuation coefficient valid for broad-beam geometry; calculations assuming a narrow-beam geometry will underestimate the radiation transmission (Madsen et al. 2006). There is commercial and free software for download that can be used to calculate the shielding necessary. After construction the transmission needs to be measured to verify that the radiation protection fulfills the requirements of radiation protection for staff, patients, and the general public. The amount of shielding, the calculated transmission, and the verified transmission should be documented and archived. The legislation in some countries has constraints on dose rate, while the legislation in other countries is based on dose per year, which requires assumptions on time spent close to the source.

The cameras in a nuclear medicine facility need to be protected from ambient radiation such as activity in patients in adjacent rooms. Likewise, camera rooms need to be shielded to protect people outside the room. It is advisable to plan for a control room to protect staff for PET and hybrid cameras. A small window made of lead glass could be installed in the wall between the control and the camera room for surveillance of the patient during the scanning. However, the surveillance of the patient could also be performed via surveillance cameras.

In connection to the entrance to a radiopharmacy or a laboratory, extra space for change of shoes and protective clothing should be planned. A radiopharmacy or a laboratory also requires space for an emergency shower and an emergency eye shower.

Facilities in which work with unsealed radioactive sources takes place have areas classified as controlled or supervised, depending on the kind of work performed (IAEA 2014). Rooms where radiopharmaceuticals are prepared, stored, and injected are classified as controlled areas. In addition, according to Carlsson and Le Heron (2014), gamma camera rooms and waiting areas should be classified as controlled areas because of the potential risk of contamination.

All entrances to the facilities should be labeled adequately with warning symbols and information on the classification and type of radiation source on site. Special considerations should be given to locks as all rooms in which radioactive material is kept have to be locked at all times. Access to the area should be regulated administratively, e.g., a work permit should be required, and physically, with locks and interlocks. In a supervised area the occupational exposure conditions should be kept under review but normally no specific measures for protection are needed (IAEA 2014).

6.2 RADIATION PROTECTION OF NUCLEAR MEDICINE STAFF

6.2.1 Safe Handling of Radionuclides

All who work with ionizing radiation must have the appropriate education, and everyone, regardless of their level, should undergo the requisite training, both theoretical and practical, on handling ionizing radiation with respect to their duties. In addition, those at the facility who do not work with radiation, e.g., secretaries and cleaners, should undergo training and receive information appropriate for their position. Additional training and updated information should be supplied at appropriate intervals. The training shall include information about risks at work with ionizing radiation and operational protection to minimize the risks. Staff involved in work with ionizing radiation should be trained in handling specific task-related issues, radioactive waste, and decontamination work with respect to surfaces, equipment, and people (IAEA 1999a, IAEA 2013).

Children and fetuses are more sensitive than adults to radiation, making it particularly important to inform the employee to report pregnancy at an early stage, to minimize exposure to the unborn child. It is also important that the woman inform her employer that she intends to breastfeed her child if she works with open radioactive sources, to take into account the risk of internal contamination and possible transfer of the radionuclide to the milk (EC 2013). A female worker breastfeeding a baby should not be involved in nuclear medicine work if there is a significant risk for radionuclide intake or of contamination (IAEA 2014).

Nuclear medicine workers receive radiation exposure from external sources including patients and from any unintended internal contamination. The magnitude of the external dose is determined mainly by the time spent close to the radiation source and by the physical characteristics of the radionuclide, i.e., the activity and half-life of the radionuclide, the emitted energy, and the type of radiation. To keep the absorbed dose from external irradiation as low as reasonably possible, the fundamental measures of radiation protection, i.e., time, distance, and shielding, should be applied. The time spent close to the radiation source should be minimized and the work should be done behind radiation shielding, keeping the distance to the source by using remote tools.

Internal irradiation from an internal contamination depends on the physical and chemical properties of the radionuclide and the biokinetics and effective half-life of the radiopharmaceutical. All work procedures should be designed to prevent external and internal contamination of the body. Radiation protection in nuclear medicine work includes use of protective clothing such as a radiation protection apron, gloves, and protective glasses if there is a risk of splashing into the eyes. To reduce radiation dose, nuclear medicine workers should be well prepared and practice new and difficult procedures using nonradioactive material, to be able to handle radioactive sources in a safe, fast, and methodologically correct manner.

It is advised that all radiopharmaceuticals be handled in a fume hood, a laminar airflow cabinet, or glovebox, depending on the work. However, work with volatile radiopharmaceuticals must be done in a fume hood. The fume hood must be shielded and the work should be performed behind a barrier of lead or lead glass for photon emitting radionuclides and acrylic (e.g., Perspex®) for beta emitters and the fume hood or the walls of the room should be shielded as discussed above.

In laboratories where radioactive gases or aerosols are handled, the radioactive material should be handled in a fume hood with an appropriate ventilation system (IAEA 2005). Radioactive aerosols and gases are used for lung ventilation studies in nuclear medicine and the administration to the patient normally is performed in a separate room. When the patient inhales the radioactive gas or aerosol, some of the substance can leak from the patient's mouth and this airborne contamination can cause internal contamination of the nuclear medicine worker. An air extractor combined with a plastic cap can be placed around the patient's face to reduce the air contamination. It may be necessary to make measurements of the air concentration of the radioactive substance. It is advisable to perform the ventilation in a room separate to the gamma camera room to avoid contamination of the camera detectors via the fans. If work includes aseptic conditions, there may be additional requirements set by national authorities that must be met, for example, special clothing or laminar air-flow cabinets.

Work with vials and syringes should be performed using remote tools such as long tongs and tweezers. Appropriate syringe shields should be used during withdrawal and injection of the radiopharmaceutical to reduce the absorbed dose to the extremities, i.e., fingers and hands. Syringe shields of lead, lead glass, or tungsten should be used for photon emitting radionuclides and in the case of pure beta emitters, acrylic shielding of appropriate thickness should be used to stop the particles and prevent production of bremsstrahlung (IAEA 2013). The syringe should not be filled to more than half of its total volume to keep a large distance between the radioactive source and the fingers.

In working close to the radiation source during handling and dispensing the radionuclide but also during and after injection of the patient, the use of radiation protection aprons or mobile lead screens should be considered (IAEA 2013). Aprons used in work with ^{99m}Tc reduce external exposure by approximately half (Young 2013) and should be used, e.g., for administration of radiopharmaceuticals in case of prolonged procedures and high activity (IAEA 2005). For ^{131}I and positron emitting radionuclides, radiation protection aprons only reduce the high-photon-energy radiation by a few percent, so the apron may create a false sense of safety for therapies and PET procedures. For hybrid systems such as PET/CT and SPECT/CT, the source of the radiation dose to the staff from the CT scan is scattered radiation from the patient. All staff and those who accompany the patient should stay outside the room or behind radiation shielding during the scan. If it is necessary to stay in the room, radiation protection aprons should be worn.

To reduce the time spent close to the injected patient, interviewing should be carried out and information should be provided to the patient before the injection. After injection, the distance to the patient should be maintained in a considerate manner so as not to worry the patient.

All vials and radioactive sources should be stored in lead shields, and transport of radioactive material should be done in a shielded shipping container designed to minimize the risk of contamination or accidents. This also applies to containers for the storage of contaminated materials from patient injections and waste cans for used needles and for trash.

All work in the fume hood should be performed with a plastic-coated pad of paper, and gloves and protective clothing to minimize the risk of contamination. Areas and instruments that could be contaminated should be handled with protective gloves. It is important to wash the hands and check the hands and clothes for contamination when the work is finished. To prevent internal contamination, eating, drinking, smoking, and licking the fingers or other material in working areas with unsealed sources is forbidden.

6.2.2   Monitoring the Dose

Where individual monitoring of the staff is used the personal dosemeter should be worn on the thorax. However, there are different opinions as to whether it should be placed under or outside a radiation protection apron. The best assessment of the effective dose to the worker is obtained if one dosemeter is used under the apron and another one on an unshielded representative part of the body. In workplaces with great variation of the dose rate, direct reading dosemeters can be used for control purposes. The use of direct reading dosemeters enables a more active radiation protection than the common use of thermoluminescent dosemeters (TLDs) or film with an exchange period of one to three months (IAEA 1999b).

Adequate radiation dose measurement is necessary if there is a significant risk of high exposure of the eyes or limbs. The absorbed dose to the extremities and the eyes should be checked at regular time intervals and these doses can be measured with small TLDs placed on the fingertips and hands and in the forehead close to the eyes, to measure during certain operations or certain conditions. The TLDs can be placed on the fingertips where the highest radiation dose is assumed to be, but measurements can be made only on rare occasions since the dosemeters might interfere with the operating procedures. Ring or wrist dosemeters usually have an exchange period of one to three months and can be used to continuously monitor skin doses and provide information about changes in those doses over time. However, these measurements can underestimate the maximum skin dose by up to a factor of seven (Pant et al. 2006, Vanhavere et al. 2006, Wrzesién et al. 2008, Sans Merce et al. 2011). The ring dosemeter must always be placed on the same finger and the same place to allow comparison between measurements. Without adequate radiation protection, finger doses of nuclear medicine workers can easily exceed the annual dose limit of 500 mSv. The shielding used depends on the energy of the electrons and positrons emitted by the radionuclides. For some radionuclides, such as ^{32}P, ^{89}Sr, ^{90}Y, ^{15}O, ^{68}Ga, ^{13}N, and ^{124}I, skin doses can reach the dose limit of 500 mSv in a few minutes if the fingers are in direct contact with an unshielded syringe containing 1 GBq (Kemerink et al. 2012). The equivalent dose rate per MBq for this situation is given for some radionuclides in Table 6.1. The use of automatic dispensers and injectors will reduce the dose to the fingers. A dose reduction of 94% (Lecchi et al. 2012) has been reported for one of the automatic dispensing and injection systems available on the market that has been validated for ^{18}F-FDG. This dose reduction has been confirmed by our own measurements (unpublished data).

Table 6.1: Data for some radionuclides used in nuclear medicine

Radionuclide	Decay mode*	Half-life	Photon energy (keV)	Electron/positron particle energy (keV)**	Thickness for total particle absorption in plastic (mm)	Half value layer in lead (mm)	Equivalent dose rate from an unshielded syringe (mSv/h MBq)
^{11}C	EC, β^+	20.4 min	511	960	3.0	6	6.41
^{14}C	β^-	5730 years	-	157	0.3	-	Bremsstrahlung
^{13}N	EC, β^+	9.97 min	511	1,199	4.0	6	12.9
^{15}O	EC, β^+	2.04 min	511	1,732	6.4	6	30.5
^{18}F	EC, β^+	109.8 min	511	634	1.7	6	2.88
^{32}P	β^-	14.3 days	-	1,710	6.3	-	23.9
^{51}Cr	EC	27.7 days	5; 320	4	< 0.1	2	0.0869
^{64}Cu	EC, β^+, β^-	12.7 h	511	578; 653	1.8	6	0.579
^{67}Ga	EC	3.26 days	93; 185; 300	8; 84; 93	0.2	1	0.402
^{68}Ga	EC, β^+	67.8 min	511; 1,077	822; 1,899	7.2	6	31.4
^{89}Sr	β^-	50.7 days	909	1,492	5.3	12	16.4
^{90}Y	β^-	2.7 days	-	2,284	9.2	-	43.5
^{99m}Tc	IT	6.0 h	18; 21; 141	120; 138	0.3	< 1	0.354
^{111}In	EC	2.80 days	23; 171; 245	145; 219	0.5	< 1	1.22
^{123}I	EC	13.2 h	27; 159; 154	127; 154	0.3	1	0.605

Continued on next page

Table 6.1 — *Continued from previous page*

Radionuclide	Decay mode*	Half-life	Photon energy (keV)	Electron/positron particle energy** (keV)	Thickness for total particle absorption in plastic (mm)	Half value layer in lead (mm)	Equivalent dose rate from an unshielded syringe (mSv/h MBq)
124I	EC, β^+	4.18 days	511; 603; 1,691	23; 1,532; 2,135	8.4	8	10.7
125I	EC	59.9 days	27; 31; 36	4; 23; 31	< 0.1	< 1	0.620
131I	β^-	8.0 days	284; 365; 637	46; 248; 330; 334; 606	1.6	3	1.13
133Xe	β^-	5.2 days	31; 35; 81	26; 45; 75; 346	0.8	< 1	-
153Sm	β^-	1.95 days	41; 47; 103	21; 55; 95; 634; 703; 807	2.4	< 1	0.241
186Re	β^-, EC	3.78 days	59; 63; 137	63; 124; 134; 939; 1,077	3.4	< 1	0.380
198Au	β^-	2.7 days	412	8; 285; 329; 397; 961	3.0	4	3.63
201Tl	EC	3.04 days	71; 135; 167	16; 84; 153	0.3	< 1	0.285

*EC = electron capture, IT = internal transition

**Maximum energy is presented for β-particles. For photons or particles emitted in <1%, the energies are omitted. Data are from Delacroix et al. (2002).

In facilities where radioactive substances are handled, the dose rate should be monitored using instruments appropriate for the type of radiation used in the room, and contamination monitors and equipment for the management of radioactive waste should be available.

6.2.3   External and Internal Contaminations

Routine monitoring of the surface or the equipment can be performed with a suitable radiation monitor, followed by a wipe test if elevated readings are obtained. Regular monitoring should be made of the relevant surfaces and areas to check for possible contamination. All equipment including vial and syringe shielding should be checked for contamination before reuse. Facilities classified as a controlled area must be monitored for possible contamination on a regular basis.

Staff should also be checked for internal contamination on a regular basis. All efforts should be made to prevent internal contamination of workers. In the event of internal contamination, measurements could be carried out by direct measurements using a whole body counter or by indirect measurements of excreted activity in, e.g., urine or feces. A gamma camera without a collimator could be used as a whole body counter with appropriate calibration (ICRP 2007a). An estimation of the absorbed dose could be made using techniques for internal dosimetry (Bolch 2009) and with the aid of ICRP 53 (ICRP 1988).

In the event of a suspected spill or contamination, the area should be monitored to ensure that no dispersal of the radionuclide has occurred. If a contamination has occurred the first step is to stop the work and get assistance. Injured and contaminated persons should be treated first and then the room can be decontaminated. If no persons are contaminated, the activity should be localized and the spread of the contamination measured to isolate the area so as to prevent further spread of activity. Protective clothing, gloves, and shoe covers should be used to avoid further contamination. The contamination should be covered with absorbent paper to absorb the liquid and after absorption the contaminated area is washed by wiping toward the center of the area to avoid further spreading of the activity. Decontamination chemicals should be used in the case of a severe spill. All contaminated material should be placed in a waste container for shielded storage and decay. In the case of a major contamination, the room should be closed, warning signs posted, and the radiation protection officer or medical physicist should be contacted immediately.

In the case of a contaminated person, contaminated clothes should be removed and the contaminated part of the skin wiped with a cotton swab moistened with water and liquid soap using long forceps if high activity. The remaining activity is measured and a careful washing of the skin repeated until no more activity can be removed. If a larger skin area is contaminated, the person should shower for about 10 minutes. If activity has splashed in the eyes, the eyes should be flushed with isotonic saline or water. A small contamination can result in a high equivalent dose rate and dose rates of 0.25 mSv/h-kBq for ^{99m}Tc and 0.8 mSv/h kBq for ^{18}F have been reported (Carlsson and Le Heron, 2014).

6.2.4   Radioactive Waste

Radioactive waste must be managed safely to protect both humans and the environment from potential hazards. In nuclear medicine the radioactive waste comprises various types of waste, from high to low activity and in liquid, solid, and gaseous form. Depending on the half-life and activity of the radioactive waste, it may be stored for decay and then sent to the public general waste treatment site with or without combustion or poured into the sewer as liquid waste, or it may be sent to national facilities for radioactive waste management. The waste disposal may also be restricted in other ways than due to radioactivity, e.g., chemical, biological, or flammable form, and this should also be accounted for in the planning of the waste handling. All handling of radioactive waste should be in compliance with relevant regulations and should be planned for when designing a new facility or starting any new method or project including radionuclide use. Records of the waste must be kept to identify the origin of the waste, radionuclide, and activity (IAEA 2013).

All waste must be separated according to the type of waste and stored in separate containers labeled with radionuclide, physical form, activity, and external dose rate. The waste containing short-lived radionuclides should be stored in a suitable space or room, locked and properly marked until the activity level has decreased to permit transport to a public waste facility. The waste containers must be leak-proof and suitable for the purpose in terms of size and shielding. Biological waste should be stored in a refrigerator or freezer to decay before being sent for incineration. Liquid waste containing short-lived radionuclides that are allowed to be flushed into the sewer system should be stored for decay until the activity is within the limits authorized by the regulatory authority.

Patients in nuclear medicine diagnostics can use ordinary toilets and there is no need for collection of excreta. For therapy patients there are different solutions to this problem in different countries. Dedicated toilets can be equipped with delay tanks or an active treatment system. Some countries' regulations may also allow direct release into the sewage system.

6.2.5   Special Considerations for Handling PET Radionuclides

In general, PET workers receive higher radiation doses than those working with other nuclear medicine examinations, with respect to the whole body radiation dose and of extremity dose. For positron emitting radionuclides, radiation shielding must be adapted to the high photon energy of 511 keV. When handling PET radiopharmaceuticals, radiation shielding L-blocks of lead or tungsten and syringe shielding should be used. Transport and storage containers must be made of lead or tungsten and sufficiently thick to limit the dose rate to a reasonable level. Long tongs and forceps should be used as in conventional nuclear medicine. The vial with the PET radiopharmaceutical initially often contains a much higher activity than ^{99m}Tc-labeled radiopharmaceuticals due to the short half-life of the PET radionuclides, so it is even more important to keep the fingers and hands far from the vial.

Time spent close to the patient, e.g., when escorting the patient to and from the examination room and when positioning the patient on the scanner couch, contributes to the whole-body radiation dose of the worker. It is important to maintain the distance from the patient after injection to minimize the external irradiation. Thus, it is best to provide all information to the patient before the injection and to maintain adequate distance from the patient in the scanner room.

Automatic systems for dispensing and injecting the radiopharmaceutical reduce the hand dose significantly. However, these systems may not be validated for all radiopharmaceuticals. In the case of ^{82}Rb, the half-life (1.273 min) is so short that the eluate is automatically transferred via i.v. line from the shielded generator to the patient without the presence of a technologist (IAEA 2008).

6.3 RADIATION PROTECTION OF THE PATIENT

All nuclear medicine exposures must be justified and the nuclear medicine specialist has the ultimate responsibility for the justifications. The principle of justification of an exposure means that the benefit from an examination or treatment shall outweigh the risks from the exposure (ICRP 2007a). No dose limits exist for patients as long as the procedure is justified. All justified exposures must be optimized, i.e., the administered activity must be optimized to produce an image quality good enough to obtain the necessary diagnostic information. If the administered activity to the patient is too low or imaging time is too short, this can result in loss of diagnostic information and the examination has to be repeated, thus increasing exposure of the patient and the workers, and too-high activity results in an unnecessarily high absorbed dose. Diagnostic reference levels (DRL) for nuclear medicine examinations are given as administered activity for routine conditions and are used to facilitate the optimization and to find an appropriate level of activity (ICRP 2001).

The amount of activity needed for a certain examination depends on the type of investigation, i.e., static, dynamic, or tomographic, and the count statistics needed to yield the desired diagnostic information. For static and tomographic studies, the activity can be reduced and the time per frame can be prolonged to some extent. For dynamic studies, the time per frame is determined by organ function and the biokinetics of the radiopharmaceutical.

Although absorbed doses to children from nuclear medicine examinations in general are low, it is always important to balance the activity administered and the examination time against the image quality needed. In some situations with children, it may be more important to reduce the examination time to minimize the risk of movement during imaging or reduce the need for sedation (Treves et al. 2014).

To ensure accurate diagnostic or therapeutic results, the patients need to receive careful instructions concerning any preparations before the examination. Incorrect or missing preparations can lead to the examination being repeated, giving the patients an extra irradiation; or even worse, a misinterpreted investigation. Patient preparation can be for instance, avoidance of certain medicines or fasting and diet restrictions. The patient should be requested to tell if they are or suspect they are pregnant, or if they are breastfeeding. The identity of the patient must be checked along width the type of examination, radiopharmaceutical, and activity prior to administration of the radiopharmaceutical, to avoid misadministrations. For therapeutic procedures this is of even greater importance due to the much higher absorbed dose given to the patient.

The patients should be encouraged to increase their fluid intake the first one or two days after injection and to void frequently to increase the rate of excretion and thereby decrease the absorbed dose. Potassium iodide (KI) and potassium chlorate ($KClO_4$) can block the uptake of radioactive iodine and ^{99m}Tc-pertechnetate. These can be given even a few hours after the radionuclide has been administered and still reduce the absorbed dose to the patient's thyroid (WHO 1999). The blocking might also be done to avoid disturbance of the thyroid uptake in the image. The use of laxatives can be used to improve the image for examinations of the bowel if the radionuclide is excreted via the bowel. The increased elimination rate may also reduce the absorbed dose to the patient (ICRP 1987).

The risk of extravasal administration can be minimized by making sure the cannula is properly placed and the line is flushed with saline before and after administration. If extravasal administration is suspected, the injection site may be flushed with isotonic saline solution, the arm elevated, and local heat applied to increase the rate of resorption. Planar spot images over the injection site should be acquired to quantify the amount of activity and the volume of the extravasal fluid present at the injection site. The dose to the tissue should be estimated as one side effect from an extravasation is early erythema, which appears above 2 Gy (ICRP 1992).

6.3.1 Children and Fetuses

Children and fetuses are more sensitive to radiation than adults; therefore, the examination or treatment of children and pregnant or breastfeeding women must be well justified. In these cases, the examination must be optimized to minimize the patient dose while still providing an image quality good enough to ensure a diagnostic result. The optimal amount of administered activity in pediatric nuclear medicine cannot always be based on activities for adult patients corrected for body mass. Use of special dosing charts is recommended because of the differences in biokinetics, organ size, attenuation, and examination time between children and adults. North American and European guidelines have been developed. Gelfand et al. (2011), Treves et al. (2011), Lassmann et al. (2008), and Grant et al. (2014) compared them to show the differences and similarities.

The dose constraint for the general public is 1 mSv and this value has been considered as a reasonable basis for constraining medical exposures to protect the unborn child. Deterministic effects to the fetus are very infrequent below 100 mGy (ICRP 2007a) and the lifetime cancer risk for the fetus for an absorbed dose below 100 mGy is about the same as for irradiation in early childhood. Doses in diagnostic nuclear medicine are normally well below 100 mSv, so there should be no thought of the need for an abortion (EC 1998b).

Before a nuclear medicine examination, a woman of childbearing age should be asked whether she is or could be pregnant or if she is breastfeeding. If a pregnancy cannot be ruled out, if no other examination without the use of ionizing radiation can replace the nuclear medicine investigation, and if it cannot be postponed, the absorbed dose to the embryo/fetus has to be given special consideration. The fetus will be irradiated from photons emitted from radionuclide uptake in maternal organs and tissues and possibly from placental transfer and distribution in the fetal tissues of the radiopharmaceutical. The placental transfer depends on the chemical and biological properties of the radiopharmaceutical and some are known to cross the placenta. Increased hydration and frequent voiding is especially important for pregnant patients and can reduce the irradiation of the embryo or fetus substantially. The administered activity to the pregnant woman can be reduced and the time per frame can be increased for most examinations, except for dynamic studies. After a pregnant woman has had a nuclear medicine examination, the absorbed dose to the unborn child should be estimated unless the dose to the uterus is below 1 mSv (EC 1998b). Stabin (2014) summarizes absorbed dose calculations for several radiopharmaceuticals at four different stages of pregnancy. The dose to the fetus for most of the radiopharmaceuticals used for diagnostic imaging range from 2×10^{-3} to 5×10^{-1} mGy/MBq. The absorbed dose to the embryo or fetus was calculated with respect to activity distribution in the mother and in the fetus for those radiopharmaceuticals that are known to cross the placenta, i.e., ^{18}F-FDG, ^{67}Ga-citrate, all iodine radionuclides in the form of sodium iodide, ^{99m}Tc-DMSA, ^{99m}Tc-DTPA, ^{99m}Tc-glucoheptonate, ^{99m}Tc-HDP, ^{99m}Tc-MDP, ^{99m}Tc-MAA, ^{99m}Tc-pertechnetate, ^{99m}Tc-PYP, ^{99m}Tc-RBC, and ^{99m}Tc-sulfur colloid. The absorbed dose to the embryo or fetus is less than 20 mGy for those activities commonly used in diagnostic nuclear medicine, according to Stabin (2014), with the exception of ^{67}Ga-citrate and ^{99m}Tc-pertechnetate above 900 MBq.

A lung scan is a common examination for pregnant and nursing patients. For many patients, the perfusion study can be performed with reduced activity the first day, and, if necessary, the ventilation scan can be performed the second day. According to Stabin (2014), the absorbed dose from a perfusion/ventilation scan that uses 200 MBq ^{99m}Tc-MAA and 40 MBq ^{99m}Tc-DTPA aerosol results in an equivalent dose of about 1 mSv to the embryo or fetus.

Therapeutic activities of ^{131}I in the pregnant woman will accumulate in the thyroid of the fetus if it is more than 8 weeks old and this will likely result in ablation of the fetal thyroid due to the absorbed doses up to 1 Sv/MBq in this small organ (ICRP 2004). There is still the possibility of reducing the absorbed dose within 12 h after administration of the radiopharmaceutical by partially blocking the thyroid of the fetus using potassium iodide (ICRP 2004). According to WHO, one or two doses of stable iodine to reduce the uptake of radioiodine are not expected to give any negative consequences for fetal development (WHO 1999).

For some nuclear medicine investigations and therapies, the woman is advised to avoid pregnancy for a period followed the administration of the radiopharmaceutical (EC 1998b) so as not to exceed this constraint (Table 6.2).

Table 6.2: Period of time for avoidance of pregnancy after the nuclear medicine examination

Radiopharmaceutical	Nuclear medicine examination or treatment	Activities up to (MBq)	Time* (months)
^{32}P-phosphate	Polycythemia vera	200	3
^{59}Fe (i.v.)	Iron metabolism	0.4	6
^{75}Se-selenonorcholesterol	Adrenal imaging	8	12
^{89}Sr-chloride	Bone metastases	150	24
^{90}Y-colloid	Arthritic joints	400	0
^{90}Y-colloid	Malignancy	4,000	1
^{131}I-iodide	Thyroid metastases	>30	4
^{131}I-iodide	Hyperthyroidism	800	6 (at least)
^{131}I-iodide	Thyroid cancer	6,000	6 (at least)
^{131}I-MIBG	Tumor imaging	20	2
^{131}I-MIBG	Pheochromocytoma	7,500	3
^{153}Sm-colloid	Bone metastases	2,600	1
^{169}Eu-colloid	Arthritic joints	400	0
^{198}Au-colloid	Malignant disease	10,000	2

Data from ICRP 2004, ARSAC 2006, EU1998b.

*If the references provide different data for the specified activity, the longest time is given in the table.

6.3.2 Breastfeeding

Special care must be taken with the breastfeeding patient because some of the radiopharmaceuticals used in nuclear medicine are known to be secreted in breast milk, exposing the child to radiation. The need for the diagnostic examination has to be weighed against the irradiation of the child; if possible, the examination should be postponed.

Several factors affect the excretion of the radiopharmaceutical into breast milk, making it difficult to predict the activity of the radionuclide in the milk and the absorbed dose to the child. Typical estimates range from 0.3% to 5% of injected activity, and 10% or more has been reported to accumulate in human breast milk for ^{99m}Tc-pertechnetate, ^{131}I-NaI, and ^{67}Ga-citrate. (Stabin and Breiz 2000).

The absorbed dose calculations for the nursing infant are based on the assumptions that the radionuclide is in the same chemical form as when injected into the lactating patient, and that the biodistribution and retention is the same in the infant as in the adult patient. Data presented in the literature show large variations in absorbed dose among patients; however, the ICRP recommends interrupting breastfeeding for a period of time, depending on the radionuclide, to keep the absorbed dose to the infant below 1 mSv (Table 6.3). For many of the ^{99m}Tc-radiopharmaceuticals no interruption is needed, but to be on the safe side, ICRP (2008) advise an interruption of 4 h while discarding one meal. The relatively long interruption recommended for most ^{123}I-labelled substances is due to the risk of contamination from other iodine isotopes present in the radiopharmaceutical. In practice, an interruption of more than 3 weeks means cessation of breastfeeding. After therapy with radioiodine, breastfeeding should cease completely because the absorbed dose to the infant's thyroid would affect organ function and induce a higher risk of thyroid cancer (ICRP 2004). In addition, the patient should be restricted with respect to proximity to the infant after examination with certain radiopharmaceuticals to reduce the absorbed dose to the child (Table 6.3).

Table 6.3: Recommended breastfeeding interruption times and the need to restrict close contact with an infant after nuclear medicine examinations

Radiopharmaceutical	Restriction proximity
No interruption	
^{14}C-labeled Triolein Glycocholic acid Urea	

Continued on next page

Table 6.3 — *Continued from previous page*

Radiopharmaceutical	Restriction proximity
^{99m}Tc-labeled*	
DISIDA	
DMSA	
DTPA	
ECD	
Phosphonates (MDP)	1 h
Gluconate	
Glucoheptonate	
HM-PAO	
Sulfur colloid	
MAG3	
MIBI	4 h
PYP	
RBC (in vitro)	2 h
Technegas	
Tetrofosmin	4 h
^{11}C-labeled	
^{13}N-labeled	
^{15}O-labeled	
^{18}F-FDG	
^{51}Cr-EDTA	
^{81m}Kr gas	
^{111}In-octreotide	42 h
^{111}In-WBC	
^{133}Xe	
12-h interruption	
^{99m}Tc-labeled	
MAA	
Microspheres (HAM)	
Pertechnetate	
RBC (in vivo)	2 h
WBC	
^{123}I-iodo hippurate	
^{125}I-iodo hippurate	
^{131}I-iodo hippurate	
48-h interruption	
^{201}Tl-chloride	

Continued on next page

Table 6.3 — *Continued from previous page*

Radiopharmaceutical	Restriction proximity
>3-week interruption	
^{123}I-BMIPP	
^{123}I-HSA	
^{123}I-IPPA	
^{123}I-MIBG	
^{123}I-NaI	
^{125}I-HSA	
^{131}I-MIBG	
^{131}I-NaI**	6 h
^{22}Na	
^{67}Ga-citrate	3 days
^{75}Se-labeled agents	

Data from ICRP (2008) and ARPANSA (2006).

* Interruption not essential unless free pertechnetate in the radiopharmaceutical

**Post ablation

6.4 SPECIAL CONSIDERATIONS FOR RADIONUCLIDE THERAPY

Radionuclide therapy, also known as isotope therapy, can be used for the treatment of benign or malignant diseases, and some examples are given in Table 6.4. The treatment of the patient is delivered by alpha or beta particles emitted from the radionuclide. Some radionuclides also emit photons, which can be used for imaging. However, they will also cause radiation protection issues. The combination of high-energy photons and high-administered activities, as in the case for ^{131}I, put high demands on radiation protection. The most common radiopharmaceutical for therapy is ^{131}I-NaI, which will be focused on in this section.

Table 6.4: Common radionuclide therapy treatments: disease to be treated, the radiopharmaceutical to use, and an approximate level of activity to be administered

Disease	Radiopharmaceutical	Approximate activity (MBq)
Thyrotoxicosis (hyperthyroidism)	^{131}I-NaI	200–600
Thyroid cancer	^{131}I-NaI	1,100–7,400

Continued on next page

Table 6.4 — *Continued from previous page*

Disease	Radiopharmaceutical	Approximate activity (MBq)
Skeletal metastases	^{153}Sm-EDTMP	2,000–4,000
Skeletal metastases	^{223}Ra-Cl $_2$	3–4
Neuroendocrine tumors	^{177}Lu-peptide	7,400
Liver tumors	^{90}Y-microspheres	1,000–2,000
Polycythemia vera	^{32}P (phosphate)	200–300
Neuroblastoma	^{131}I-mIBG	3,700–7,400
B-cell lymphomas	^{90}Y-monoclonal antibody	1,850

Radionuclide therapy with ^{131}I-NaI may be performed by drinking a liquid or taking a capsule. A capsule with the correct activity is delivered to the hospital ready for use in a lead-shielded pot. Liquid ^{131}I-NaI has to be dispensed within a fume hood to reach the correct activity and volume for therapy. The dose to the fingers will be greater when liquid ^{131}I-NaI is used. In addition, transporting the liquid from the dispensing room to the therapy room has a larger risk than giving a capsule. However, it must be determined that the patient is able to swallow the capsule without chewing it. The patient may drink the liquid ^{131}I-NaI with a straw to diminish the risk of contamination. For a therapy procedure, the regulations of some countries may require that more than one person should verify the activity measurement (ICRP 2007b).

It is of the utmost importance that the patient receives written instructions on radiation protection before the therapeutic administration. The instructions should include guidance on risks of contamination, toilet visits, the importance of distance and time, the importance of avoiding close contact with small children, restrictions on visiting restaurants, the cinema, and the theater, and the period of time for keeping extra precautions.

When a child undergoes radionuclide therapy, a parent or grandparent stays in the therapy room to take care of the child during the isolation period. The beds must be placed as far apart as possible and radiation protection shields put around the child's bed. The adult should be monitored with an electronic dosemeter during the isolation period.

Radionuclide therapy can be performed on an in- or an out-patient basis. After the administration of the therapy, the patient will remain a radiation source for persons in their vicinity until the radiopharmaceutical has decayed or it has been excreted from the body, via the urine and feces.

The individuals that can come into contact with a treated patient can be divided into three groups: 1) staff at the hospital, 2) family members, and 3) the general public. Staff and family members are aware that the patient has been treated with a radioactive substance and that they therefore are willing to be exposed to the extra radiation, while the third group are not aware that they have been exposed to radiation and therefore, they should be regarded as persons from the general public (EC 1998a). Dose constraints apply to family members, which include children, adults, elderly people, and pregnant women. Different dose constraints may apply to these different groups. Family members who do not accept exposure to extra radiation are regarded as members of the general public.

The release of a patient from the hospital can be based on a maximum dose rate at 1 m distance, a maximum residual activity, or whether the estimated dose to family members will be less than a certain value (5 mSv according to the ICRP and 1100 MBq according to the IAEA (ICRP 2004)). It should be pointed out that the methods for treatment of thyroid diseases and the regulations for release of patients differ considerably between countries.

The isolation room must be monitored for contamination by a medical physicist using a radiation protection survey instrument before the room undergoes ordinary cleaning and is used by other patients. Special attention should be payed to work surfaces, bed linen, laundry, light switches, remote controls, and door knobs. Radioactive waste must be taken care of according to the hospital's standard procedures.

The results of studies on the effective dose to caregivers of patients after therapy show that the dose was less than 5 mSv, irrespective of the constraints used for release of the patient. Some studies reported that the dose to the caregivers was less for thyroid cancer patients than for patients treated for thyrotoxicosis on an outpatient basis, despite the higher activity administered to the thyroid cancer patients. This was probably the result of the high dose rate of the patient immediately after returning home when treated on an outpatient basis. (Stefanoyiannis et al. 2014).

Computer-simulated effective doses have been reported to be more than two-fold lower than values based on point sources in air, which indicates that the medical physicist provides too restrictive guidance on release of patients that have received ^{131}I therapy for thyroid cancer or thyrotoxicosis (Han et al. 2014).

The hospital should offer a travel document stating that the patient underwent radionuclide therapy if the patient has to travel within a certain time period after the therapeutic procedure. The document should include administered activity, the radiopharmaceutical used, administration date, and contact details for the hospital ward that delivered the therapy. Table 6.5 gives the time period for which such a document is needed. It is calculated using the amount of radiopharmaceutical administered, the radiopharmaceutical's decay rate, and the detection level of the detectors at airports and harbors (ICRP 2004). This is also applicable for some nuclear medicine examinations listed in Table 6.5.

Table 6.5: The time period for when a travel document is needed for radionuclides used in diagnostics and therapy in nuclear medicine

Radionuclide	Physical half-life	Administered activity (MBq)	Travel document is needed within
Examination			
^{13}N	10 min	740	Not needed
^{18}F	110 min	400	24 h
^{68}Ga	68 min	400	24 h
^{99m}Tc	6 h	700	4 days
^{111}In	67 h	300	6 weeks
^{123}I	13 h	150	1 week
^{131}I	8 days	0.4	1 month
Therapy			
^{32}P	14.3 days	350	7 months
^{90}Y	64 h	1,850	2 months
^{131}I	8 days	7,400	6 months
^{153}Sm	47 h	4,000	1 month
^{177}Lu	6.7 days	7,400	4 months
^{223}Ra	11.4 days	7	4 months

If a patient undergoing nuclear medicine therapy requires emergency treatments, for example emergency surgery or suffers a heart attack, they should be treated as any other patient, i.e., as not containing any radioactivity (EC 1998a). However, special consideration might be needed when such a patient dies shortly after a radionuclide therapy. If an autopsy has to be performed on the body and during the funeral service, an expert on radiation protection should be consulted to keep doses as low as reasonably achievable. Legislation may set out restrictions for cremation or burial (EC 1998a).

6.5 REFERENCES

Administration of Radioactive Substances Advisory Committee. 2006. Notes for Guidance on the Clinical Administration of Radiopharmaceuticals and Use of Sealed Radioactive Sources. https://www.gov.uk/government/uploads/system/uploads/attachment/_data/file/304835/ARSAC_Notes_for_Guidance.pdf (accessed January 2, 2015).

Australian Radiation Protection and Nuclear Safety Agency (ARPANSA). 2008. Radiation protection in nuclear medicine. Radiation Protection Series No. 14.2. http://www.arpansa.gov.au/pubs/rps/rps14_2.pdf (accessed December 13, 2014).

Bolch W.E., Eckerman K.F., Sgouros G., Thomas S.R. 2009. MIRD pamphlet No.21: a generalized schema for radiopharmaceutical dosimetry - standardization of nomenclature. J Nucl Med 50:477–84.

Carlsson S.T., Le Heron J.C. 2014. In Nuclear medicine physics: A handbook for teachers and students, ed. D. L. Bailey, J. L. Humm, A. Todd-Pokropek, A. van Aswegen, 73–116. Vienna: IAEA.

Delacroix D., Guerre, J.P., Leblanc, P., and Hickman C. 2002. Radionuclide and radiation protection data handbook 2002. http://www.deakin.edu.au/research/integrity/radiation/Radionuclide-Handbook.pdf (accessed December 13, 2014).

European Commission (EC). 1998a. Radiation protection 97: Radiation protection following iodine-131 therapy (exposures due to out-patients or discharged in-patients). http://ec.europa.eu/energy/nuclear/radiation_protection/doc/publication/097_en.pdf (accessed December 30, 2014).

European Commission. 1998b. Radiation Protection 100: guidance for protection of unborn children and infants irradiatioed due to parental medical exposures. http://ec.europa.eu/energy/nuclear/radiation_protection/doc/publication/100_en.pdf (accessed January 12, 2015).

European Commission. 2013. The Council Directive 2013/59/EU-RATOM of 5 December 2013 laying down basic safety standards for protection against the dangers arising from exposure to ionising radiation, and repealing Directives 89/618/Euratom, 90/641/Euratom, 96/29/Euratom, 97/43/Euratom and 2003/122/Euratom. http://eur-lex.europa.eu/LexUriServ/LexUriServ.do?uri=OJ:L:2014:013:0001:0073:EN:PDF (accessed December 13, 2014).

Gelfand M.J., Parisi M.T., and Treves S.T. 2011. Pediatric radiopharmaceutical administered doses: 2010 North American consensus guidelines. J Nucl Med 52:318–22.

Glendon A. I., Clarke S. G., McKenna E. F. 2006. Human Safety and Risk Management, CRC Press. Florida.

Grant F.D., Gelfand M.J., Drubach L.A., Treves S.T., Fahey F.H. 2014. Radiation doses for pediatric nuclear medicine studies: comparing the North American consensus guidelines and the pediatric dosage card of the European Association of Nuclear Medicine. Pediatr Radiol Nov 1 (Epub ahead of print).

Han E.Y., Lee C., Mcguire L., Bolch W.E. 2014. A practical guideline for the release of patients treated by I-131 based on Monte Carlo dose calculations for family members. J Radiol Prot 34:N7–17.

Institute of Physics and Engineering in Medicine. 2014. Radiation Protection in Nuclear Medicine. IPEM Report 109. Institute of Physics and Engineering in Medicine.

International Atomic Energy Agency. 1999a. Safety Guide. Occupational radiation protection. Vienna: International Atomic Energy Agency.

International Atomic Energy Agency. 1999b. Safety Guide. Assessment of occupational exposure due to external sources of radiation. Vienna: International Atomic Energy Agency.

International Atomic Energy Agency 2005. Safety Report Series No. 40. Applying radiation safety standards in nuclear medicine. Vienna: International Atomic Energy Agency.

International Atomic Energy Agency 2008. Safety Report Series No. 58. Radiation protection in newer medical imaging techniques: PET/CT. Vienna: International Atomic Energy Agency.

International Atomic Energy Agency 2012. Cyclotron produced radionuclides: Guidance on facility design and production of [18F]Fluorodeoxyglucose (FDG). Vienna: International Atomic Energy Agency.

International Atomic Energy Agency. 2013. Occupational protection. Training material developed in collaboration with World Health Organization (WHO), Pan American Health Organization (PAHO), International Labour Organization (ILO), International Organization for Medical Physics (IOMP). https://rpop.iaea.org/rpop/rpop/content/additionalresources/training/1_trainingmaterial/nuclearmedicine.htm (accessed December 27, 2014).

International Atomic Energy Agency 2014. Safety Standards. Radiation protection and safety of radiation sources: International basic safety standards. Vienna: International Atomic Energy Agency.

International Commission on Radiological Protection. 1987. Protection of the Patient in Nuclear Medicine (and Statement from the 1987 Como Meeting of ICRP). ICRP Publication 52. Ann. ICRP 17 (4).

International Commission on Radiological Protection. 1988. Radiation Dose to Patients from Radiopharmaceuticals. ICRP Publication 53. Ann. ICRP 18 (1–4).

International Commission on Radiological Protection. 1992. The Biological Basis for Dose Limitation in the Skin. ICRP Publication 59. Ann. ICRP 22 (2).

International Commission on Radiological Protection. 2001. Radiation and your patient - A Guide for Medical Practitioners. ICRP Supporting Guidance 2. Ann. ICRP 31 (4).

International Commission on Radiological Protection. 2004. Release of Patients after Therapy with Unsealed Radionuclides. ICRP Publication 94. Ann. ICRP 34 (2).

International Commission on Radiological Protection. 2007a. The 2007 Recommendations of the International Commission on Radiological Protection. ICRP Publication 103. Ann. ICRP 37 (2–4).

International Commission on Radiological Protection. 2007b. Radiological Protection in Medicine. ICRP Publication 105. Ann. ICRP 37 (6).

International Commission on Radiological Protection. 2008. Radiation Dose to Patients from Radiopharmaceuticals - Addendum 3 to ICRP Publication 53. ICRP Publication 106. Ann. ICRP 38 (1–2).

Kemerink G.J., Vanhavere F., Barth I., Mottaghy F.M. 2012. Extremity dose of nuclear medicine personnel: a concern. Eur J Nucl Med Mol Imaging 39:529–32.

Lassmann M., Biassoni L., Monsieurs M., Franzius C. 2008. EANM Dosimetry and Paediatrics Committees. 2008. The new EANM paediatric dosage card: Additional notes with respect to F-18. Eur J Nucl Med Mol Imaging 35:1666–8.

Lecchi M., Lucignani G., Maioli C. 2012. Validation of a new protocol for 18F-FDG infusion using an automatic combined dispenser and injector system. Eur J Nucl Med Mol Imaging 39:1720–9.

Madsen M.T., Anderson J.A., Halama J.R., et al. 2006. AAPM Task group 108: Pet and PET/CT shielding requirements. Med Phys 33:4–15.

Pant G.S., Sanjay K.S., Rath G.K. 2006. Finger doses for staff handling radiopharmaceuticals in nuclear medicine. J Nucl Med Technol 34:169–73.

Sans Merce M., Ruiz N., Barth I., et al. 2011. Extremity exposure in nuclear medicine: Preliminary results of a European study. Radiat Prot Dosimetry 144:515–20.

Stabin M.G., Breitz H.B. 2000. Breast milk excretion of radiopharmaceuticals: mechanisms, findings and radiation dosimetry. J Nucl Med 41:863–73.

Stabin M.G. 2014. Radiation dose concerns for the pregnant or lactating patient. Semin Nucl Med 44:479–88.

Stefanoyiannis A.P., Ioannidou S.P., Round W.H., et al. 2014. Radiation exposure to caregivers from patients undergoing common radionuclide therapies: a review. Radiat Prot Dosimetry (November):1–10. http://rpd.oxfordjournals.org/content/early/2014/11/26/rpd.ncu338.full .pdf+html

Treves S.T., Parisi M.T., Gelfand M.J. 2011. Pediatric radiopharmaceutical doses: New guidelines. Radiology 261:347–9.

Treves S.T., Falone A.E., Fahey F.H. 2014. Pediatric nuclear medicine and radiation dose. Semin Nucl Med 44:202–9.

Vanhavere F., Berus D., Buls N., Covens P. 2006. The use of extremity dosemeters in a hospital environment. Radiat Prot Dosimetry 118:190–5.

World Health Organization. (1999). Guidelines for iodine prophylaxis following nuclear accidents. Update 1999. Geneva: World Health Organization.

Wrzesién M., Olszewski J., Jankowski J. 2008. Hand exposure to ionising radiation of nuclear medicine workers. Radiat Prot Dosimetry 130:325–30.

Young A.M. 2013. Dose rates in nuclear medicine and the effectiveness of lead aprons: updating the department's knowledge on old and new procedures. Nucl Med Commun 34:254–64.

VI

Radiation Protection in Radiation Oncology

CHAPTER 7

Radiation Protection in Radiation Oncology

Raymond K. Wu

Department of Radiation Oncology & Cyberknife, University of Arizona Cancer Center, Phoenix, Arizona, USA

Richard J. Vetter

Mayo Clinic, Rochester, Minnesota, USA

CONTENTS

RADIATION ONCOLOGY is the sub-discipline of oncology that focuses on the use of radiation to treat cancer. This medical specialty is concerned with prescribing radiation to treat disease (radiation therapy) as opposed to the medical specialty of radiology where radiation is used to diagnose disease. Thus, the quantum energy of the radiation and the absorbed doses applied in radiation therapy are orders of magnitude greater. Radiation therapy is applied to the human body in several different ways to treat cancer. The most common method is directing a beam of radiation on the tumor, i.e., the target tissue. Beams of different dimensions, shapes, and intensities are directed on the lesion from various directions to maximize dose to the lesion and to minimize dose to normal tissues. In curative radiation therapy the objective is to eradicate the tumor while sparing critical structures and minimizing damage to other healthy tissues following strict guidelines. This beam of radiation is most often generated by a powerful X-ray machine called a linear accelerator. Other types of machines can also generate beams of radiation for therapy and are discussed below. While no longer commonly used in countries with more resources, teletherapy machines using high activity ^{60}Co sealed source is used in some practices around the world to deliver the beam of radiation. In developed countries ^{60}Co is also used in a self-shielded device called the Gamma Knife® to treat various lesions of the brain. Another form of radiation therapy that uses smaller quantities of various radionuclides to treat cancer is called brachytherapy and will be described below. In a modern radiation oncology department, there are many potential safety concerns. We will focus only on those in which radiation is the primary contributing factor.

7.1   EXTERNAL BEAM RADIATION THERAPY

Radiation therapy equipment used in radiation oncology departments includes fluoroscopic or CT simulators, teletherapy machines such as cobalt 60 units, linear accelerators, and brachytherapy units such as High Dose Rate Remote Afterloaders (HDR). Some departments may have stereotactactic radiotherapy or radio surgery units such as the Gamma Knife®, the Cyberknife®, the TomoTherapy®, or the Vero®, and, recently more hadron therapy machines.

Patients may receive radionuclide or brachytherapy treatments, which are important modalities commonly employed in a radiation oncology practice. Many radioactive isotopes used in radiation oncology treatments are readily available commercially. Table 7.1 shows a list of radioactive isotopes with data relevant to medical physicists and staff for radiation protection considerations. Radioactive isotopes used in nuclear medicine departments for diagnostic purposes are not listed. We will look at the radiation safety concerns in the following sections.

7.1.1   Simulation, Planning, and External Radiation Therapy

Before a patient can be treated with external beam radiation therapy, a treatment plan must be developed. Part of the planning includes simulating the treatment beam using fluoroscopy or CT imaging. Recent advances in high resolution imaging, computer controlled field shaping, and real-time field alignment capabilities of a modern medical linear accelerator system contribute to the improved quality of radiation oncology treatments. Such technologies have the potential to improve treatment outcome if employed properly, or if not, introduce risks that may harm the patient.

Modern radiation oncology equipment design allows the use of imaging devices to set up the field alignments as frequently as for each session. The field alignments are confirmed by the radiation therapist or the radiation oncologist before turning the beam on. Such a procedure is called Image Guided Radiotherapy (IGRT). In IGRT procedures, if the tumor volume has been outlined accurately by the treatment planning team consisting of the radiation oncologist, the medical physicist, and the dosimetrist, the treatment fields can be locked in to the tumor volume immediately before beam-on of the treatment fields. Errors are introduced if the tumor volume is outlined incorrectly at the treatment planning phase. The technology of real-time tumor volume tracking has been routinely used by the Cyberknife® using the Synchrony® and Lung Optimized Treatment (LOT) modes (a suite of software tools to optimize lung treatments by accounting for lung motion). The infrared camera and external markers are used with radiographic or fluoroscopic images to track the tumor motion in real time. The radiation fields may be moved in synchronization with the tumor by moving the multileaf collimator (MLC) in most linear accelerators including the TomoTherapy®, or moving the accelerator for the case of the Cyberknife®. Recently the Vero® made by BrainLab introduced a new design that may provide more dimensions of motion for tracking and beam alignment purposes.

It has become a routine practice that patients targeted for definitive treatments will receive CT scans with slice thickness thin enough to spare the adjacent organs at risk with respect to the tumor target. Quite frequently MRI and PET scans will be required and fused with the CT images to visualize the tumor locations. Slice thickness will have an impact on the clarity of the Digitally Reconstructed Radiographs (DRR) used for IGRT procedures mentioned above. Thinner slices will help with these needs, while the extra dose given to obtain the CT and PET scans should be an important consideration. The clinical medical physicists are instrumental in implementing radiation protection guidelines to balance the benefit of tumor targeting and the extra radiation the patient will receive.

7.1.2 Radiation Safety for External Beam Treatment Procedures

Before each external beam treatment can begin, the treatment planning process has to be carried out by the medical dosimetrist and the medical physicist. The approved plan is then exported to the treatment machine for the radiation therapists to deliver the treatment in multiple treatment sessions, which we refer to as fractions.

7.1.3 Treatment Planning

The treatment planning process is complex, and depends on clinical needs, such as how close the organ at risk is to the tumor target. The team of medical physicist and medical dosimetrist are trained to generate the radiation dose distribution that gives the optimal highest dose to the tumor target with an acceptable dose to the surrounding tissue, with particular attentions paid to minimizing the dose to critical organs. There are many ways to produce an acceptable dose distribution plan. One of the treatment planning methods to achieve this goal is introduced below as an example.

7.1.4 Intensity Modulated Radiation Therapy

Most often, radiation fields aimed towards the tumor target are non-uniform in radiation fluence, so that critical organs within the field will receive a lower dose than the target volume. When multiple fields are used, the acceptable treatment plan, which shows the summed total of the doses contributed by all the fields, will be high as prescribed for the tumor target, and as low as required to spare critical organs. Such a treatment planning method or algorithm is called Inverse Planning, and the treatment procedure based on such a planning method is called Intensity Modulated Radiation Therapy (IMRT). The non-uniform radiation fluence is generated by the MLC as described in Section 7.1. Often when IMRT is used, the radiation beam is on for a longer time because only small-size beamlets are irradiating the patient. Since photon radiation higher than 10 MV produces neutrons, with higher neutron production from higher photon energies, a majority of IMRT uses lower photon energies such as 6 MV.

7.1.5   Treatment Delivery

When conventional external beam radiation treatments are delivered, careful alignments are made manually to assure the patient is positioned exactly as simulated and as planned in the treatment planning computer. Imaging means are used to confirm the alignment and may be used as the baseline for comparison to images obtained in subsequent treatment sessions. The process to achieve the correct alignment is not straightforward, but is expected routinely for every treatment fraction. A team of well-trained radiation oncologists, medical physicists, and radiation therapists (technologists) is necessary to produce this result for every fraction. Likewise for brachytherapy, a team of highly trained staff works together to deliver each treatment. In the United States, for high-dose-rate treatments by an HDR machine or by a Gamma Knife® unit, an authorized medical physicist (AMP) must be present in addition to the radiation oncologist who is qualified as an authorized user (AU). For the case of HDR treatments, the AMP must be physically present continuously during the treatment, and the AU must be present when the treatment is initiated. The AU or a physician under the supervision of the AU, who has been trained in the operation and emergency response of the HDR unit, must be physically present continuously during the treatment. The meaning of physical presence is "within hearing distance of normal voice" as defined in the NRC regulation. For the case of Gamma Knife® treatments, the AMP and the AU, not even another physician, must be physically present continuously.

7.1.6   Radiation Safety Considerations for Radiation Therapy Treatments

To deliver radiotherapy treatments safely is highly dependent on whether the staff is well trained, and whether appropriate quality assurance equipment is available. A comprehensive radiation safety program for a radiation oncology department must consider the safety of staff as well as that of the patient. Although the dose prescribed for the tumor target is high, it does not mean the dose to tissues outside the tumor target should not be kept low if it is reasonably achievable. When the target volume prescribed to receive a certain dose fails to receive the dose, it is as much a failure in radiation safety as for treatment outcome. When a patient is receiving neutron dose to the total body due to high-energy photon radiation incorrectly chosen for IMRT planning, that is also a failure in radiation safety. In the United States the radiation protection regulations in most states stipulate that when 20% overdose or underdose, delivered to the patient, it is considered a misadministration. For treatments given in 1 to 3 fractions, a difference of 10% is a misadministration. Within 24 hours after discovery of a misadministration, the responsible person of the department is required to take actions including verbally notifying the regulatory agency, the referring physician, and if appropriate, the patient. The responsible person is required to provide a written report detailing the misadministration to the regulatory agency within two to three weeks. Similar requirements are applicable to misadministration of treatments using radionuclides.

7.1.7 Image-Guided Radiotherapy

Nowadays imaging technologies are built into treatment delivery machines to allow better alignment of the radiation beam with the tumor volume, with ease. The use of portal imaging with the MV beam, orthogonal kV images, or kV/MV imaging, is expected to help with beam alignments. Cone Beam CT is another IGRT modality using the relatively large field size kV beam to generate a CT dataset. The automatic process will use the CBCT images and the DRR generated by the treatment planning computer, to recommend appropriate shift and rotation of the treatment couch to improve the field alignment. All such imaging technologies will be beneficial but will give more radiation dose to the patient. The doses for such CBCT procedures have been reported to range from 2 mGy for a low-dose head view to 19.4 mGy for a high quality head image (Alaei 2010).

7.2 TREATMENT VAULT DESIGN FOR CONVENTIONAL LINEAR ACCELERATORS

One of the common design features in most radiation oncology departments is the thick radiation barriers employed to provide shielding for exposure (air kerma) levels of up to 12 Gy per minute from X-ray beams generated by linear accelerators. This radiation dose rate is equivalent to 720 Sv/hr and has to be reduced to the order of μSv/hr to be considered safe for the staff and the public.

7.2.1 General Considerations

A busy radiation oncology department with the treatment vaults properly constructed will have a work environment free of known radiation exposure risks to staff and the public. The methods employed to provide proper designs and barrier thicknesses are well established and may be found in various textbooks and reports. NCRP Report No. 151, Structural Shielding Design and Evaluation for Megavoltage X- and Gamma-Ray Radiotherapy Facilities (NCRP 2005), is a frequently referenced publication on this topic. The report gives a coherent set of recommendations based on the relevant NCRP publications, including Report No.147 (NCRP 2004) for lower-energy, Report No. 102 (NCRP 1989) for higher-energy facilities, Report No. 79 (NCRP 1984) on neutron contamination, and other long-standing recommendations of the Council. Report No. 151 paves the way for smooth transition from the relatively higher permissible dose of decades earlier to the stricter design dose limits of today. It is expected that the barrier thicknesses calculated using the Report 151 approach and the latest design dose limits will be just as conservative as the traditional calculation approach of the NCRP Report 49. The report also gives the recommendation on meeting the in-any-hour requirement originated from Cobalt Teletherapy days, but is still relevant for linear accelerator vaults.

The report introduces a term Rw, which is the time-averaged dose-equivalent rate (TADR) averaged over a week.

W is the workload per week (Gy week^{-1})

U is the Use factor

The weekly TADR R_W is convenient for the purpose of evaluating whether a location where the occupancy factor T is known meets the design dose limit P. If $R_W \times$ T is less than P, the barrier is adequate for radiation shielding.

7.2.2 Dose Limit in-Any-Hour R_h

The concept of in-any-hour R_h is relevant for linear-acelerator-generated radiation to assure adequate shielding if the workload or the occupancy factor is exceptionally low. The requirement to meet the 20 μSv in-any-hour limit for public areas is important for Cobalt-60 teletherapy machines (USNRC 2015). (It is not the same as 20 μSv per hour, which is the instantaneous dose rate). It is impractical and unnecessary to impose a limit of 20 μSv per hour if the radiation source is the linear accelerator, because the dose output per minute can be very high, but the beam is on for a very short fraction of an hour. In NCRP 151it is recommended that R_h not exceed 20 μSv for linear accelerator vault designs.

Another important publication on radiation facilities shielding is the IRR1999 Approved Code of Practice and Guidance published by the British Health & Safety Commission (BHSC 2015)). The method to calculate the barrier thickness is the same but the design dose limits are quite different compared with the NCRP approach. IAEA Safety Reports Series No. 47 highlights the differences between these approaches and worked examples (IAEA 2006).

7.2.3 Difference in Design Dose Limits

The IAEA Report No 47 summarized the recommended/legal effective dose limits and design effective dose limits for the IAEA (Basic Safety Standard), the United States (NCRP), and the United Kingdom (IRR1999) in Table 2 of the report (IAEA 2006).

The UK limits are the IDR (Instantaneous Dose Rate) and TADR (Time-Averaged Dose Rate) values listed in the table. The TADR2000 is the time-averaged dose rate estimated over 2000 hours, which takes into account the workload, use, and occupancy factors. According to the IRR1999, if the IDR is less than 7.5 μSv per hour, and the TADR is less than 0.5 μSv per hour or the TADR2000 is less than 0.15 μSv per hour, the area does not need to be supervised. 0.15 μSv per hour is equivalent to 0.3 mSv per year. This is significantly stricter than the U.S. limit for public areas, which only needs to meet the limit of 1 mSv per year with workload, use, and occupancy factors taken into consideration.

For both U.S. NCRP and U.K. IRR there are additional requirements to avoid under protection due to exceptionally low workload and use factor specifications. The IRR requirement for a public unsupervised area is that the IDR shall be below 7.5 μSv per hour. This is also much stricter than the 20 μSv in-any-hour requirement as defined in NCRP 151.

Due to many factors, such as expectations of freedom from risks not directly related to quantifiable parameters such as workload and annual dose levels, each jurisdiction has its reasons for the determination of how conservatively the radiation barriers should be designed. The reader is advised to adhere to the local regulations when performing tasks of radiation barrier designs. Radiation monitoring badges may be helpful in confirming whether the design is effective.

7.2.4   Shielding Evaluation Survey

At installation time, a shielding evaluation survey should be performed as soon as the treatment machine can produce radiation, and before the engineer continues the installation work. After the installation, and before the department is open for occupancy, a comprehensive shielding evaluation survey should be performed. If the photon modality is higher than 10 MV, the survey measurements should be performed for X-ray and neutron using both types of survey meters that had been calibrated. A fast response survey meter may be used to spot weaknesses in shielding followed by another survey meter to obtain more accurate readings. Floor plans for floors both above and below, as well as for the floor of the vault, are helpful. Special attention should be paid to door frames, particularly at heights where the ventilation ducts are passing through. It is most important to seek out the primary beam to make sure there is no misplaced or misaligned barrier, and to determine the oblique beam direction towards adjacent buildings close by. For the machine leakage, it is convenient to use radiochromic film taped around the head of the machine, and to give a large dose to confirm that there is no missing shielding block (Figure 7.1).

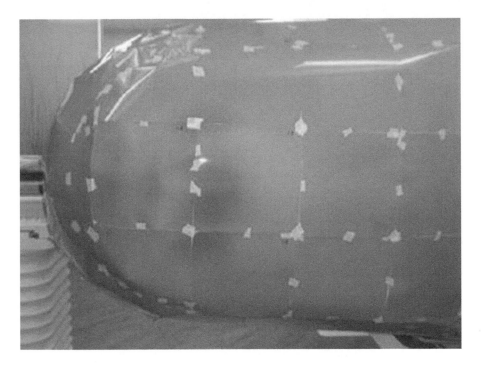

Figure 7.1 Radiochomic film taped to head of linear accelerator to monitor for leakage radiation

7.3   RADIATION SAFETY FOR BRACHYTHERAPY PROCEDURES

7.3.1   Brachytherapy

In addition to external beam radiation therapy for treatment of cancer, patients may receive treatment with radioactive sources inserted directly into or in contact with tumors. This technique is called brachytherapy ("brachy" meaning "short" in Greek). Many radioactive isotopes used in brachytherapy and radionuclide (radiopharmaceutical) therapy are readily available commercially. Table 7.1 shows a list of radioactive isotopes and their properties, commonly used for radiation oncology treatments.

One of the important steps before administering radiation treatments is to verify the identity of the patient, and the treatment site for the tumor. In some radiation oncology treatment centers, such a step is referred to as the time-out procedure. The same procedure should be implemented for external beam treatments and surgical procedures. Another radiation safety procedure common to all sealed sources used in brachytherapy applications is the periodic inventory and wipe test. The NRC requires each licensee who uses a sealed source to perform wipe tests at intervals not to exceed 6 months to confirm that the source is not leaking. To be of value, the wipe test and equipment must be capable of detecting the presence of very low activity, e.g., 185 Bq (0.005 microcuries) of radioactive material on the test sample. The test analysis must be performed by a person approved by the NRC or an Agreement State (see "Regulatory Structures and Issues in North America" for definition of Agreement State) to do the analysis. The source inventory must be performed semi-annually unless the sources are in storage.

7.3.2 Brachytherapy Using Permanent Implants

Three radionuclides shown in Table 7.1 are commonly used as permanent implants by virtue of their low dose to individuals other than the patient: ^{103}Pd, ^{125}I, and ^{131}Cs. The radiation dose to radiation oncology and other medical personnel during implantation is very low, with most of the dose coming from use of a fluoroscope to image the target during implantation (Schwartz et al. 2003). Therefore, the primary precautions during implantation are: wearing a lead apron to protect against the fluoroscopy X-rays, maintaining an inventory of the radioactive seeds to ensure none are lost, and conducting a radiation survey of the operating room after implantation, to confirm no seeds have been left behind.

Table 7.1: Radionuclides used for brachytherapy and radionuclide therapy

Symbol	Primary Emission	Energy (or max energy) keV	Half-life	10 half-lives
^{60}Co	Gamma	1170–1330	5.26 yrs	53 yrs
^{89}Sr	Beta	Ave 1.463 MeV	50.5 days	1.4 yrs
^{90}Sr	beta/gamma	546 (up to 2.27 MeV)	28.5 yrs	285 yrs
^{103}Pd	Gamma	21	17 days	170 days
^{125}I	Gamma	27–36	60.2 days	20 months
^{131}I	Gamma	364, 637	8.04 days	2.7 months
^{131}Cs	beta/gamma EC	29	9.7 days	3.25 months
^{137}Cs	Gamma	510β, 1180β, 662γ	30 yrs	300 yrs
^{192}Ir	gamma/beta	380γ	73.83 days	2 yrs
^{223}Ra	Alpha	5.78 MeV	11.43 days	114 days

Perhaps the most common permanent implant is ^{125}I for treatment of localized prostate cancer. Worldwide, 1.1 million new cases of prostate cancer are diagnosed each year, and more than 300,000 of those cases are fatal (WHO 2014). Since its introduction in the mid-1980s, prostate brachytherapy has become a well-established treatment option for patients with early, localized disease. In the United States alone, more than 50,000 prostate cancer patients a year are treated using this method (PBAG 2015). In Europe several thousand cases are treated annually, and this number is increasing rapidly (ICRP 2005). Recently prostate seed implant procedures have not been increasing as rapidly as in the past due to the availability of IMRT and IGRT.

No adverse effects to medical staff or the patient's family associated with permanent brachytherapy have been reported (ICRP 2005), which shows that this procedure is safe. However, several issues have been addressed by ICRP (2005) in response to issues raised by members of the medical community and patients' families. These include dose to people close to the patient, management of excreted or expelled sources, and cremation of the patient's body.

7.3.3 Dose to People Approaching the Implanted Patient

Available data derived from direct measurements and from calculations indicate that the annual dose to the family or household members is well below the ICRP-recommended limit of 1 mSv in almost all cases (see Chapter 3 for discussion of ICRP limits); thus, it will not reach the level of 5 mSv set for comforters and carers of such patients. However, patients and caregivers should be provided with written information designed to keep their doses as low as reasonably achievable (ALARA). ICRP (2005) provides an example of the minimum recommendations to be provided to prostate cancer patients who have been treated with permanently implanted seeds. In summary, these instructions state:

- Since radiation decreases significantly with distance from the patient's body, no significant dose levels can be detected beyond 1 meter. In addition, ^{125}I seeds lose about 50% of their activity every 2 months and ^{103}Pd lose 50% every 17 days. Thus, the patient cannot harm anyone by briefly hugging, kissing, shaking hands, or being in the same room with them. For peace of mind, the patient should not hold young children on their lap or sit close to them for long periods for the first half-life of the radionuclide.

- Since the radioactive material is contained in a sealed capsule, the patient cannot contaminate anyone or anything. Thus, linen, tableware, dishes, and toilet facilities may be used by others without taking special precautions.

- Rarely a seed may be passed in the patient's urine. Thus, it is recommended that the patient strain their urine for 3 days after the procedure.

If a seed is found in the sieve, it must not be touched with fingers or hands; use a spoon or tweezers to pick it up and place it in a small covered container, and return it to the radiation oncology physician. If a seed is passed after 3 days, it can be flushed away in the toilet without any significant risk.

- Very rarely a seed may be passed during ejaculation. Thus, resumption of sexual activity is not recommended for 1 week after the implantation, and a condom should be used for at least the first five ejaculations. If a seed is found in the ejaculate, it should be treated the same as a seed passed in the urine as discussed above.

7.3.4   Cremation

Cremation is relatively uncommon in many countries but frequent in others, e.g., India and China, and is the rule in Japan. Cremation of a patient previously implanted with a permanent radioactive seed raises concerns about the radioactivity in the patient's ashes and airborne release of radioactivity during cremation. Cremation can be allowed without special precautions if 12 months have elapsed since implantation of ^{125}I, and 3 months for ^{103}Pd. Prior to these time periods it is recommended that the implanted prostate gland be removed from the patient and that the organ (and seeds) be stored according to national, state, or local regulations. Cremated remains should be placed in a sealed metal container for a minimum of 1 year after cremation and should not be scattered in the environment until a minimum of 10 half-lives have elapsed since implantation (ICRP 2005).

7.3.5   Subsequent Pelvic or Abdominal Surgery

In rare cases (less than a few percent), surgery may be required to address complications. This must be done by a surgeon who is fully aware of the brachytherapy implantation technique. Seeds should be identified, placed in a container, and returned to the radiation oncology practice (ICRP 2005).

7.3.6   Fathering of Children

Fathering a child following a permanent radioactive seed implants should be avoided. However, a few cases of fatherhood have been reported after permanent implantation, so patients must be aware of this possibility and take all necessary precautions if indicated (ICRP 2005).

7.3.7   Triggering of Radiation Detection Monitors

During the first few months after an implant, some sensitive radiation monitors can be triggered by a patient who has undergone a permanent implant. These monitors are typically located at the entry or exit of nuclear plants and nuclear research centers, in some waste areas, and in some scrap metal factories. Also, some airports, border crossing, bridges, and tunnels have been equipped with such high-sensitivity detectors. Since a patient could trigger such a security monitor, he should be given a personal wallet card or letter from the hospital explaining the implant, to avoid any problem with security (ICRP 2005).

7.3.8   Brachytherapy Using High-Dose-Rate (HDR) Afterloaders

High-dose-rate remote afterloading procedures (HDR) are recommended for staff safety because the radiation oncologist and staff can take time to position the applicators without the presence of radionuclides. The high activity radiation source or sources are usually inserted into the applicators remotely and controlled by a computer. The HDR treatment room should be properly shielded with concrete wall or lead bricks, and the door should be lead lined unless there is a maze design. To avoid the construction cost, HDR treatments may be delivered in a linear accelerator vault. A qualified medical physicist should be engaged to specify the barrier thicknesses with special attention given to floors above and below the room, particularly at oblique angles. The treatment room should be equipped with audio and visual patient monitoring equipment. Emergency procedures should be established in case the high-activity radioactive source does not return to the safe. After the HDR treatment, a radiation survey of the patient should be performed to confirm no radioactive element is left in the patient.

7.3.9   Low-Dose-Rate and Manual Loading Procedures

While HDR has replaced low-dose-rate afterloaders and manual loading procedures in many radiation oncology treatment centers, these techniques are still in use. For example, ^{192}Ir strands for sarcoma and other conditions, ^{125}I or ^{103}Pd implants for prostate, and ^{131}Cs for brain or other sites are not uncommon, and some treatment centers still maintain an inventory of ^{137}Cs needles and tubes for manual afterloading procedures. Due to the need to deliver the prescribed dose with relatively low activity sources, the patient will need to stay in a properly shielded hospital room for several days if ^{192}Ir, ^{60}Co, and ^{137}Cs sources are used. If a shielded room is not available, a corner hospital room may be used only if the adjacent rooms including those on floors above and below are surveyed and monitored to confirm the radiation levels are meeting the guidelines for uncontrolled areas. For low energy sources such as ^{125}I, ^{103}Pd, and ^{131}Cs, the radiation safety concerns are easier to take care of. Since they are usually permanent implants, the home-based safety precautions should be explained to the patients and family members and should take into account recovery or disposal considerations if the implanted seeds may exit the patient's body. In most cases, the distance and time factors are adequate to maintain the safety level of exposures to other people around the patient.

7.3.10   Other Brachytherapy Procedures

Beta sources such as the BetaCath® for cardiovascular conditions and minia-ture X-ray sources such as the Xoft® for similar and other conditions are examples of other radiotherapy modalities in use. Due to their low penetrat-ing powers and easily transportable designs, proper precautions should be implemented to guard against improper use by un-trained personnel.

7.4   RADIATION SAFETY FOR RADIONUCLIDE THERAPY PROCEDURES

7.4.1   General Considerations

In some hospitals the therapeutic application of nuclear medicine is located in the Radiation Oncology Department. Therefore, a brief discussion of this application will be provided here. Whether the administration of therapeu-tic radionuclides occurs in Radiation Oncology or Nuclear Medicine, qualified medical physicists should be involved in the design of procedures for handling the radionuclide treatments appropriately. These procedures should include every attempt to avoid radioactive spills and unintentional exposure to pa-tients and staff and to other people. Only designated rooms should be used for the administration of radionuclides or the hospitalization of patients. A tutorial on the administration of radioiodine therapy and release of patients has been prepared by Karesh (2014).

The United Nations Scientific Committee on the Effects of Atomic Radi-ation (UNSCEAR) has estimated that nearly 400,000 therapeutic adminis-trations of radiopharmaceuticals are carried out annually (UNSCEAR 2000). Patients may be released after calculation shows that the exposure to the fam-ily members and the general public is below the limit specified by the national or state authorities.

7.4.2   Thyroid Ablation

Some radionuclide procedures are performed in Radiation Oncology Depart-ments rather than Nuclear Medicine services due to the high activities used. Some radiation oncologists prefer to do the procedure for reasons such as more comprehensive care and easier follow-up. As an example, ^{131}I thyroid ablation procedures are performed by radiation oncologists in some hospitals. The patient is given the isotope in a capsule or as a liquid after the time-out procedure that also includes a blood test to rule out pregnancy, if applicable. The patient stays in a room prepared with absorbent paper material on the floor, and linen on the chair and bedding (Figure 7.2). This step will reduce the work required to remove any contamination from the floor and furniture. The acceptable wipe count level may be different in different jurisdictions, but all removable contamination, as demonstrated by a wipe test, must be removed before the room can be released for general use. For patients with young children at home or patients who need special assistance, the medical physicist or the radiation oncologist may choose to keep the patient in the hospital for a few days until the exposure is below the limit set by authorities. The patient and family members should be given instructions to minimize radiation exposure to other people, such as using the time factor, avoiding contact, and staying at a distance. Movable radiation shields (Figure 7.3) should be used in the patient's room to provide additional protection of staff and other patients.

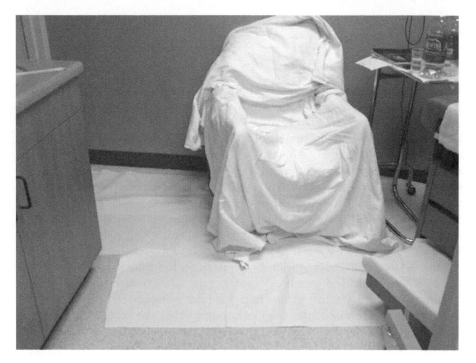

Figure 7.2 Floor of radioiodine hospital room covered with absorbent paper, and chair covered with sheet, to facilitate decontamination

The release calculation for thyroid ablation in the United States is described in NRC Regulaory Guide 9-39 (USNRC 1997). Taking into considerations whether the patient can or cannot sleep alone, live alone, maintain a safe distance from others, have sole use of a bathroom, and other conditions for certain numbers of days after discharge, the medical physicist can calculate the estimated dose received by an individual most likely to be exposed based on the estimated thyroid uptake fraction. If the estimated dose to the highest-exposed individual is less than 5 mSv, the patient may be discharged. A good radiation safety practice hinges on good instructions given to the patient, and relies on the cooperation of the patient. If the patient and staff place high-activity wastes, such as face towels, napkins that have been used to wipe the mouth on the first day, and pillow cases, etc., in separate containers at home or in the hospital and dispose them appropriately, the risk of ^{131}I contamination getting into the thyroid of staff or family members will be greatly minimized.

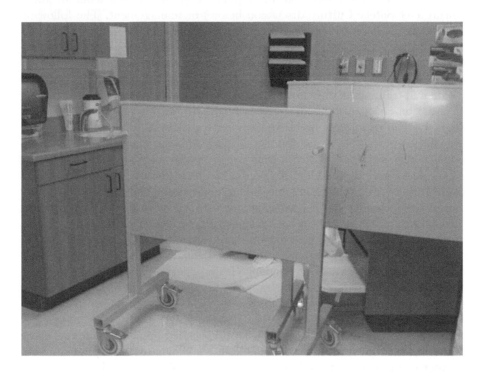

Figure 7.3 Bedside shields used to protect nurses and other patient-care staff from radioiodine gamma ray exposure during patient-care activities

7.5 PREVENTION OF UNINTENDED EXPOSURES

During radiation therapy, patients are exposed to an intense beam of radiation (external beam). With brachytherapy, radioactive sources are placed in direct contact with tissue, to deliver very high doses of radiation. Underexposure can have serious consequences related to treatment outcome, and overexposure can have serious consequences related to tissue damage. From the point of view of radiation safety, the second situation is of primary interest and will be discussed here.

ICRP Publication 86 and Publication 112 provide valuable information to help prevent unintended exposures in radiation therapy (ICRP 2000, ICRP 2009). These documents describe case histories of some severe unintended exposures, discuss the causes for these events, summarize the consequences, and provide recommendations on measures to prevent such events. In many cases a single cause cannot be identified. Often a combination of factors contributed to the event, e.g., a combination of inadequate staff training, lack of independent checks, lack of quality control procedures, or absence of overall supervision. These combinations often point to an overall deficiency in safety culture (see Safety Culture discussion below) or management. The following factors are common to many unintended exposures.

Inadequate training has contributed to many incidents. One major factor related to unintended exposures is erroneous calibration of radiotherapy beams and brachytherapy sources, the root cause often being the lack of education and training, mainly in the field of medical physics. Insufficient training of radiation oncologists has resulted in misinterpretation and incorrect treatment. Lack of training of brachytherapy nurses has resulted in accidents with brachytherapy sources. Lack of understanding of beam parameters by a maintenance engineer leads to devastating consequences for many patients (ICRP 2000). The International Organization of Medical Physics recommends that a medical physicist's minimum educational qualification should be a university degree or equivalent (level corresponding to a Master's degree) majoring in medical physics or an appropriate science subject (IOMP 2010).

Other issues resulting in accidental exposures include deficiencies in procedures and protocols, equipment faults, inadequate communication of essential information, lack of independent checks, inattention to detail and unawareness, unsecured long-term storage or abandonment of radiation therapy sources, field or beam calibration problems, computer problems, imaging errors, and treatment set-up errors (ICRP 2000, ICRP 2009).

Prevention of errors associated with unintended exposures such as those above is often obvious after the event, but implementation of procedures to prevent recurrence requires determined efforts by physicians, physicists, management, and others. Prevention can be enhanced by development, monitoring, and enforcement of radiation safety regulations, and implementation of best practices, such as quality ensurance equipment checks and other checks to assure that the target and surrounding tissues will receive the correct radiation dose. A high priority should be given to ensuring that professionals involved in calibration and treatment have received appropriate education and training as well as continuing education on new treatment techniques. All radiation oncology departments should adopt a written, comprehensive, and functional quality assurance program. Finally, all radiation oncology departments should make every effort to instill and maintain a radiation safety culture that clearly places the needs of the patient ahead of the needs of the department.

7.6 SAFETY CULTURE IN RADIATION ONCOLOGY

Safety culture in the workplace reflects the attitudes, beliefs, perceptions, and values shared in relation to safety. It is expressed through employee behavior and work activities, i.e., safety culture acts as a guide as to how employees will behave in the workplace (Glendon et al. 2006). While some leaders within radiation oncology departments may depend heavily on a culture of compliance to protect patients, workers, and the public, Classic et al. (2014) point out that compliance contributes to a radiation safety strategy, but it is not a vision and will not create or maintain a safety culture. In their report, Safety is No Accident, the American Society of Radiation Oncology (ASTRO 2012) emphasized that safe delivery of radiation therapy requires the concerted and coordinated efforts of many individuals. They further point out that safety and efficiency are linked in that inefficiencies lead to staff frustration, rushing, and sometimes cutting corners, which can lead to errors. Consequently, team members should work together to create a safe and efficient clinical environment. In their policy statement on safety culture, the NRC defines safety culture as the core values and behaviors resulting from a collective commitment by leaders and individuals to emphasize safety over competing goals, to ensure protection of people and the environment (NRC 2011). The NRC policy statement identifies nine traits of a safety culture. Commensurate with a modern radiation oncology practice, these traits influence the safety of patients, employees, and the public:

- Leadership safety values and actions: leaders demonstrate a commitment to safety in their decisions and behaviors. For example, the Radiation Oncology Department Chair will fully support prompt, full disclosure of a misadministration.

- Problem identification and resolution: issues that could impact safety are promptly identified, fully evaluated, and promptly addressed and corrected. For example, a treatment plan that could result in unacceptable toxicity is promptly corrected to serve the best interests of the patient.

- Personal accountability: individuals take personal responsibility for safety of themselves, co-workers, patients, and the public. For example, a technologist will immediately report a misadministration.

- Work processes: planning and controlling work activities is implemented so that safety is maintained. For example, a new plan to accommodate treating more patients per day will not sacrifice patient safety to improve throughput.

- Continuous learning: opportunities to learn about ways to ensure safety are sought out and implemented. For example, routine seminars will include lectures on how to assure or improve safety.

- Environment for raising concerns: a safety-conscious work environment is maintained so that personnel feel free to raise safety concerns without fear of retaliation, intimidation, harassment, or discrimination. For example, work relationships should not intimidate a technologist to not report a potential error.

- Effective safety communications: oral and written communications maintain a focus on safety. For example, routine newsletters should include columns to encourage safe practices.

- Respectable work environment: trust and respect are highly valued within the department regardless of employee rank. For example, technologists should be encourages to approach a radiation oncologist about a potential safety issue.

- Questioning attitude: all workers avoid complacency, continually challenge existing conditions and activities, and strive to identify discrepancies that might result in error or inappropriate action.

Radiation oncology personnel, whether performing or overseeing use of radiation sources, should take the necessary steps to promote a positive safety culture by fostering these traits as they apply to their organizational environments. The maturity of safety cultures among radiation oncology departments across the world is likely to reflect considerable diversity, with some having spent significant time and resources in the development of a positive safety culture.

7.7 TERRORIST THREATS AND THE IMPACT ON RADIATION ONCOLOGY

Many radiotherapy treatment units use high-activity radionuclide sources. They may be potential targets for misguided individuals to acquire illegally, with the purpose of producing massive harm to a community. It is no longer a good assumption that these individuals will use common sense and avoid doing harm to themselves while acquiring the high-activity radiation sources. For example, heavy shielding and remote handling tools, weighing hundreds to thousands of pounds, would be needed to move ^{60}Co sources from a teletherapy machine, but it is not safe to assume that terrorists could not obtain and use such equipment to remove and steal the ^{60}Co sources. It is mandatory that owners of equipment with high-activity sources establish proper policies and procedures to safeguard the radiation sources. Radiation oncology facilities that possess radioactive sources should have detailed procedures for responding to unlawful or terrorist events, making timely notifications to appropriate authorities, and providing accurate information about the radiation source involved. These facilities should review and exercise their programs periodically and coordinate their planned actions with law enforcement and regulatory authorities.

The IAEA categorized radionuclide-specific activity levels for the purposes of emergency planning and response (IAEA 2003). The purpose of the categorization system was to provide a basis to be used as an input to many activities relating to the safety and security of radioactive sources, and to develop security emergency response plans accordingly. For example, users of a

Gamma Knife® unit in the United States must pass a criminal history check by the local police department and provide finger-prints for a national criminal history check by the Federal Bureau of Investigation. The Gamma Knife® unit must also be monitored continuously against unauthorized access. When the spent sources are returned to the manufacturer, secured transportation arrangements must be met to make sure only authorized couriers are allowed to pick up the shipment. Readers are advised to monitor the IAEA website (www.IAEA.org) for updates on recommendations and for training materials on how to avoid or respond to a terrorist radiological event.

7.8   PERSONAL MONITORING

A radiation oncology department usually employs many supportive staffs in additional to the radiation oncologists and medical physicists. All radiation workers including the radiation therapists, medical dosimetrists, radiation oncology nurses, radiation oncologists, and medical physicists should be badged and monitored. Other staff who may receive more than a certain threshold such as 10% of the non-occupationally exposed limits should also be badged. Front desk staff such as receptionists, and clerical and administrative staff are not radiation workers. Dietitians, financial consultants, and medical oncology staff who may receive occasional exposures from patients with radioactive implants may need monitoring as determined by the radiation protection officer. Badge readings in modern radiation oncology departments typically are very low, indicating that the level of radiation exposure in a typical modern radiation oncology department is not a major concern in radiation safety. This can be credited to adequate design and shielding of radiation treatment rooms and to staff following standard operating procedures.

7.9   CONCLUSION

Cancer patients receive radiation treatments by a variety of methods including those that generate intense beams of radiation focused on the target tissue and radionuclides applied directly to or near the target. The magnitude of the radiation doses used to treat these patients can, if misapplied, result in significant risk to the patient. Likewise, radiation oncology personnel are at risk of receiving high doses of radiation from these treatments and must pay attention to detail to protect themselves as well as their patients. Some radiation safety can be engineered into the radiation oncology facilities, e.g., construction of shielded walls to reduce the intensity of treatment beams to levels ALARA outside the treatment rooms. Radiation oncology personnel must be well trained to reduce the likelihood of treating the patient to an absorbed dose that is either too low to be effective or too high, resulting in damage to the patient. All radiation oncology departments should develop and maintain a culture of safety to minimize the risk of injury to patients, staff, and the public.

7.10   REFERENCES

Alaei, P. 2010. Review of the Doses from Cone Beam CT and Their Inclusion in the Treatment Planning. American Association of Medical Dosimetrists 35th Annual Meeting - June 16, 2010. http://www.medicaldosimetry.org/pub/39774274-2354-d714-51f0-8be87ec1b43b. Accessed January 31, 2015.

American Society for Radiation Oncology (ASTRO). 2012. Safety is No Accident. https://www.astro.org/ uploaded-Files/Main_Site/Clinical_Practice/Patient_Safety/Blue_Book/Safetyisno Accident.pdf. Accessed January 31, 2015.

British Health and Safety Commission. 2015. IRR1999 Approved Code of Practice and Guidance. Health and Safety Executive, Ionizing Radiations Regulations. London. http://www.legislation.gov.uk/ uksi/1999/3232/contents/made. Accessed January 31, 2015.

Classic, K.C. et al., 2014. Safety and Radiation Protection Culture. In: Radiological Safety and Quality, Lau, L. and Ng K.-H., eds. Springer, New York.

Glendon, A. I., Clarke, S. G., McKenna, E. F. 2006. Human Safety and Risk Management, CRC Press. Florida.

International Atomic Energy Agency. 2003. Categorization of radioactive sources. http://www-pub.iaea.org/MTCD/publications/ pdf/te_1344_web.pdf. (Accessed January 29, 2015).

International Atomic Energy Agency. 2006. Radiation Protection in the Design of Radiotherapy Facilities. Safety Reports Series No. 47. http://www-pub.iaea.org/MTCD/publications/PDF/Pub1223_web.pdf. Accessed January 31, 2015.

International Commission on Radiological Protection. 2000. Prevention of Accidental Exposures to Patients Undergoing Radiation Therapy. Annals of the ICRP, 30: (3) 2000. Elsevier, Oxford.

International Commission on Radiological Protection. 2005. Radiation Safety Aspects of Brachytherapy for Prostate Cancer Using Permanently Implanted Sources. Annals of the ICRP, 35: (3) 2005. Elsevier, Oxford.

International Commission on Radiological Protection. 2009. Preventing Accidental Exposures from New External Beam Radiation Therapy Technologies. Annals of the ICRP, 39: (4) 2009. Elsevier, Oxford.

International Organization of Medical Physics. 2010. Basic Requirements for Education and Training of Medical Physicists. IOMP Policy Statement No. 2. http://www.iomp.org/sites/default/files/ iomp_policy_statement_no_2_0.pdf. Accessed January 31, 2015.

Karesh, S.M. 2014. Radionuclide Therapy with I-131 Sodium Iodide: Patient and Dose Administrator Precautions and Administration Methods. http://www.nucmedtutorials.com/dwradnuctherapy12/i131.html. Accessed January 31, 2015.

National Council on Radiation Protection and Measurements. 1984. Neutron Contamination from Medical Electron Accelerators. NRCP Report No. 79. NCRP Bethesda, Maryland.

National Council on Radiation Protection and Measurements. 1989. Medical X-Ray, Electron Beam and Gamma-Ray Protection for Energies up to 50 MeV (Equipment Design, Performance and Use). NRCP Report No. 102. NCRP Bethesda, Maryland.

National Council on Radiation Protection and Measurements. 2004. Structural Shielding Design for Medical X-Ray Imaging Facilities. NRCP Report No. 147. NCRP Bethesda, Maryland.

National Council on Radiation Protection and Measurements. 2005. Structural Shielding Design and Evaluation for Megavoltage X- and Gamma-Ray Radiotherapy Facilities. NRCP Report No. 151. NCRP Bethesda, Maryland.

Prostate Brachytherapy Advisory Group (PBAG). 2015. www.prostate brachytherapyinfo.net. Accessed January 29, 2015.

Schwartz, D.J., Davis, B.J., Vetter, R.J., et al. 2003. Radiation exposure to operating room personnel during transperineal interstitial permanent prostate brachytherapy. Brachytherapy, 2:98–102.

United Nations Scientific Committee on the Effects of Atomic Radiation. 2000. Sources and Effects of Ionizing Radiation. 2000 Report to the General Assembly with Annexes, United Nations, Vienna.

U.S. Nuclear Regulatory Commission. 1997. Release of Patients Administered Radioactive Materials. Regulatory Guide 8.39. http://www.orau.org/ptp/PTP — Library/library/NRC/Reguide/08-039.pdf. Accessed January 31, 2015.

U.S. Nuclear Regulatory Commission. 2011. Safety culture. http://www.nrc.gov/about-nrc/safety-culture.html. Accessed January 29, 2015.

U.S. Nuclear Regulatory Commission. 2015. Standards of Protection against Radiation, Title 10 Code of Federal Regulations, Part 20. http://www.nrc.gov/reading-rm/doc-collections/cfr/part020/. Accessed January 31, 2015.

World Health Organization. 2014. World Cancer Report. Chapter 1.1. ISBN 9283204298. WHO, Geneva.

VII

Regulatory Philosophy and Control

CHAPTER 8

Regulatory Structures and Issues in Africa

Taofeeq A. Ige

Federation of African Medical Physics Organization (FAMPO)

CONTENTS

R EGULATIONS AND REGULATORY CONTROLS particularly with respect to the use of radiation in all areas of human endeavor in Africa have and will continue to be a topical issue now and in the foreseeable future. Regulatory controls may be defined as the limitations imposed on the activities of users of radiation or firms in compliance with the requirements of a regulatory agency (Black 2015). The current chapter shall be restricted to the regulatory activities with respect to nuclear safety and radiation protection in the region.

Africa is the world's second largest and most populous (1 billion population: 15% of the world's human race) continent, with 54 fully recognized sovereign states, out of which 49 are United Nations (UN) and 34 are AFRA (African Regional Co-operative Agreement for Research, Development and Training Related to Nuclear Science and Technology) member states.

The first semblance of regulatory activities in the continent probably came about in the early 1960s with various precursors in the different countries of Algeria, Egypt, Ghana, Nigeria, and South Africa. For instance, the setting up of the Federal Radiation Protection Service (FRPS) in Nigeria in 1964 was due to the nuclear test conducted by the French authorities in the Sahara Desert and the attendant migration of the effluents and fall-outs southwards. Also, the implementation of the nuclear program in the then-apartheid enclave may have necessitated the setting up of the Atomic Energy Commission of the Republic of South Africa.

In most of the African states, radiation safety predominates over nuclear safety, since most of the applications are in the medical and industrial (including mining) sectors. There are only eight Member States with nuclear research reactors (Algeria, Democratic Republic of Congo, Egypt, Ghana, Libya, Morocco, Nigeria, and South Africa), whereas all the countries have one or several diagnostic radiology and to a lesser extent, nuclear medicine and radiotherapy facilities. There are also a number of countries with fledgling mining activities, including Gabon, Malawi, Namibia, Niger, and United Republic of Tanzania, among others.

The International Atomic Energy Agency (IAEA) has also made it a mandatory requirement for member states that access technical assistance in the various uses of nuclear technology to put in place a regulatory body: the IAEA General Conference Board Resolution (IAEA 2001). Member States therefore need an effective infrastructure for radiation, transport, and waste safety (hereafter known in short form as "radiation safety"). Article III.A of the IAEA statute mandates the agency to establish standards for safety and to assist Member States with their application of those standards.

Many Member States therefore utilize IAEA safety standards as the basis for a legal framework, and many also benefit from IAEA assistance to establish or strengthen their national radiation safety infrastructure. In fact, the approval of projects and the procurement of items such as radiation sources and related equipment are contingent upon Member States applying IAEA safety standards and having an adequate infrastructure for radiation safety in place (IAEA 2001).

The importance of having an effective national radiation safety infrastructure has been noted by the Standing Advisory Group on Technical Assistance and Cooperation (SAGTAC), who recommended that strengthening radiation safety should be given a high priority by the IAEA. The Standing Advisory Group on Technical Assistance and Cooperation (SAGTAC) was established in 1996 to advise the Director General on the Agency's technical cooperation (TC) strategy and policies (IAEA 1996).

8.1 REGULATORY STRUCTURE

Regulation of nuclear and radiation safety can be carried out by one or more bodies. The typical regulatory structure in the region has a Chief Executive Officer or Director-General with a management team that is responsible to a board. The membership of the board is appointed by the President or the Prime Minister. This goes to underscore the importance and the relevance with which regulatory bodies are accorded.

Over the past 10 years, several countries and particularly IAEA member states (MS) in the region have promulgated laws on nuclear safety and radiation protection but its implementation in most of the member states in the entire region has been fraught with challenges. It is pertinent to note that the African Nuclear Weapon Free Zone (NWFZ) otherwise known as the "Pelindaba" treaty, provides for the regulation of nuclear safety and radiation protection in Africa. The treaty has been in force since 2010.

As mentioned earlier, the emphasis in the region has been geared more towards radiation safety, especially in terms of the medical, occupational and public exposure arising from the preponderance of thousands of X-ray facilities and some radiotherapy and nuclear medicine infrastructure. There is also a wide use of industrial radiography and nuclear gauges, especially in the petroleum industries. Nearly all the North African member states except Morocco (Algeria, Egypt, Libya, and Tunisia) have very elaborate oil production capabilities and facilities extending to the West African coast (Gulf of Guinea) and onto the Eastern and Southern parts of the continent as in the countries of Angola, Uganda, and South-Africa.

8.2 REGIONAL BODIES

8.2.1 African Commission on Nuclear Energy (AFCONE)

The treaty of Pelindaba established the African Commission on Nuclear Energy (AFCONE) for the purpose of ensuring states parties' compliance with their undertakings. The treaty mandates AFCONE, inter alia, to collate states parties annual reports, review the application of peaceful nuclear activities and safeguards by the IAEA, bring into effect the complaints procedure, encourage regional and sub-regional cooperation, as well as promote international cooperation with extra-zonal states for the peaceful applications of nuclear science and technology. The AFCONE Secretariat is based in Pretoria, South Africa.

AFCONE plays a key role in advancing the peaceful application of nuclear science and technology in Africa and in bringing much-needed support to states parties to fully benefit from nuclear science and technology applications in the areas of health, agriculture, and energy. AFCONE is also actively engaged in global and regional efforts towards disarmament and non-proliferation.

AFCONE consists of twelve states parties that serve for a three-year term.

States parties that are members of AFCONE are elected by the Conference of States Parties with due regard to equitable regional representation and national development in nuclear science and technology. The 3rd Conference of States Parties (CSP) held in Addis Ababa, from May 29 to 30, 2014, elected the following countries to the membership of AFCONE for a three-year term: Algeria, Cameroon, Ethiopia, Kenya, Libya, Mali, Mauritius, Senegal, South Africa, Togo, Tunisia, and Zimbabwe.

Members of AFCONE are represented through Commissioners, who are high caliber professionals with experience in the area of nuclear science and technology, diplomacy, and security. AFCONE Commissioners meet in annual Ordinary Sessions to discuss all aspects relating to the implementation of the AFCONE program of work (AFCONE 2014).

AFCONE may therefore be considered to be a key player in the regulation of nuclear safety in the region arising from the anticipated peaceful uses of nuclear energy and the non-proliferation efforts that it is expected to promote.

AFCONE does not currently have a web address but some details about the organization may be found at - http://Peaceau.org/en/page/78.

8.2.2   Forum of Nuclear Regulatory Bodies of Africa (FNRBA)

The preparatory meeting for the establishment of the Forum of Nuclear Regulatory Bodies of Africa (FNRBA) was held on October 2, 2008 at the Vienna International Center (VIC). The meeting was organized on the margins of the IAEA 52nd General Conference (GC 52) as a follow up to the request of the Provisional Steering Committee and communication sent by the Division for Africa to the heads of regulatory authorities and national liaison officers.The meeting was chaired by the Chairperson of the then Provisional Steering Committee, Prof. S.B. Elegba (Nigeria). In attendance were heads of national regulatory authorities for nuclear and radiation safety in Africa and other representatives from African countries.

The forum was thus created through the coordinating efforts of the Nigerian Nuclear Regulatory Authority (NNRA) and the then-management of the agency which had been empowered by the act establishing the authority to liase with and foster cooperation with international and other organizations or bodies concerned with the regulation of nuclear safety, security, and safeguards. The forum is to further develop and strengthen national regulatory infrastructures and continuously improve regulatory performance through self assessment and the promotion of regional cooperation among African nuclear regulatory bodies.

The benefits envisaged from the establishment of FNRBA include the advancement of regional cooperation and the associated advantages of:

- Promotion of a common understanding of radiation and nuclear safety regulatory issues.

- Facilitation of information exchange on the African continent.

- Development and strengthening of radiation and nuclear safety infrastructure across the African region.

- Addressing the present and future challenges with respect to radiation and nuclear safety.

- Creating a uniform frontier of coordinating support and partnership initiatives in Africa.

The FNRBA has embarked on Seven Thematic Areas of Operation in Nuclear Safety and Security and Radiological Protection. In this regard, the FNRBA established the following Thematic Working Groups (TWGs): Upgrading Safety in Radiotherapy, Upgrading Safety in Uranium Mining and Milling, Regulatory Framework for Licensing of Nuclear Power Plants, Upgrading Safety in Nuclear Research Reactors, Upgrading Legislative and Regulatory Infrastructure, Education and Training, and Knowledge Management, and Upgrading Safety of Radioactive Waste Management Infrastructure.

The Thematic Working Group on Regulatory Infrastructure for Nuclear Power Plants (TWG-NPP) has been established in accordance with Article 8 of the Charter of the FNRBA. Membership of the WGNPP is voluntary and based on the needs of the FNRBA members. The following ten countries indicated that they will be actively participating in the WGNPP: Democratic Republic of Congo (DRC), Egypt, Libya, Morocco, Namibia, Nigeria, Senegal, Tanzania, Tunisia, and South Africa. Working Group members will commit to active participation in the work and commit to continuing exchange of information. South Africa will coordinate the activities of the WGNPP.

FNRBA has cooperative agreements with the Republic of Korea and with the U.S. Nuclear Regulatory Commission in areas of radiation and nuclear safety (FNRBA 2010).

The web address of FNRBA is http://gnssn.iaea.org/fnrba/EN.

8.2.3 Federation of African Medical Physics Organizations (FAMPO)

FAMPO is the newest regional chapter of the International Organization of Medical Physics (IOMP) and currently boasts more than twenty member organizations, both individual and national (NMOs) among which are Algeria, Burkina Faso, Cameroon, Cote d'Ivoire, Egypt, Ethiopia, Ghana, Kenya, Libya, Madagascar, Mauritania, Mauritius, Morocco, Namibia, Niger, Nigeria, Senegal, South Africa, Sudan, Tunisia, Uganda, United Republic of Tanzania, Zambia, and Zimbabwe.

The second article in the FAMPO Constitution with respect to name, scope, and extent of activities states that the federation extends its activities throughout Africa and local islands in the region. The membership adopted shall be from among IOMP member organizations and medical physicists in Africa.

Taking cognizance of the dearth of medical physics professionals that is even far more acute in the region, individual membership in the countries lacking critical mass to form a body or organization shall be recognized. There are currently less than five hundred clinically qualified medical physicists in the region with a one-billion population base. And only three countries (Egypt, Morocco, and South Africa) account for more than 50% of this number.

In a recent baseline data effort by the executive committee of the organization, it emerged that more than 75% of the group are in government employment, although more private enterprises are gradually coming forth. The radiotherapy medical physics sub-discipline has the highest numbers in the group; however, diagnostic radiology and nuclear medicine sub-specialization are under gradual implementation. Also some medical physicists work in more than one of the three sub-disciplines (DR, NM, and RT) and they do move between them as well. The survey also revealed the fact that a preponderance of medical physicists are in academia and research in some countries, while some are engaged in commerce, services, and consulting (trade sector), radiation protection (health physics), regulatory bodies, health ministries and department, of health in some of the countries. In Africa, currently about 30% of the medical physicists group are females.

The web address of the organization is www.federation-fampo.org.

8.3 REGULATORY CHALLENGES AND ISSUES

The regulatory challenges in the region shall be discussed in the context of Nigeria, which is used as a case study because the situation in the different member states appears to be similar and follows a familiar pattern.

8.3.1 Medical Exposure Issues

Radiation safety, particularly with respect to patients, has become a public health concern in the region as in other parts of the world especially with certain diagnostic radiology, radiotherapy, and nuclear medicine procedures. In recent years, there has been a significant increase in population exposure (dose) from non-natural ionizing radiation sources, most of which arises from the use of computed tomography (CT), fluoroscopy, and interventional procedures (Hammou 2014).

Most countries have basic diagnostic radiology equipment, but more often this is old and close to being obsolete. Hence they are mostly not calibrated and poorly functioning. Spare parts are usually scanty and often unaffordable. Quality assurance (QA) programs on the equipment are many times non-existent.

Most equipment is procured with minimal or no training component at all, hence trouble-shooting and repair by in-house personnel or technicians/engineers are largely absent. In effect, there is usually no maintenance of these facilities. The ambient temperature that guarantees optimal functionality of these machines barely exists because the air-conditioning is also not regularly maintained and even the electricity supply in some of the countries is poorly regulated, leading to voltage surges and consequent damage to these few facilities.There are often no guidelines and written instructions on procedures, either because they are not available or due lack of use.

The totality of all the above-mentioned challenges in terms of equipment, lack of trained manpower and human resources, as well as few documented medical/radiological procedures as to the quality of patient care are dire, and make the service delivery sub-optimal, leading to compromised and non-efficacious outcomes with respect to the population's living standards in the region.

On a good note, it is heartwarming to report that standardization in radiotherapy dosimetry has been achieved in Africa where most hospitals use a common dosimetry code of practice (IAEA-TRS-398). There is also a remarkable improvement in the results emanating from the IAEA/WHO TLD Intercomparison dose audit of radiotherapy beams in the region. This is comparable with what is obtained from some of the other advanced economies and this feat was achieved with continuous assistance and guidance of the IAEA.

8.3.2   Occupational Exposure and Safety

The new IAEA Safety Standards Series GSR Part 3 states inter alia (under occupational exposure) that the government or regulatory body shall establish and enforce requirements to ensure that protection and safety is optimized, and the regulatory body shall enforce compliance with dose limits for occupational exposure. Also, the regulatory body shall establish and enforce requirements for the monitoring and recording of occupational exposures in planned exposure situations (Requirements 19 and 20) (IAEA 2014).

The reliability of personnel dose results and records from the licensed service providers is one of the major issues and challenges currently being experienced in the region, especially from the private establishments that are contracted to carry this out in some member states. This normally comes about because of the lack of strong scientific skills and backgrounds of these entities.

Dose-records intercomparison nationwide from similar and varied institutions and establishments that would have possibly corroborated the results that are churned out by these service providers are often lacking. There is also the non-standardization of facilities that are employed by the various service providers in the country, both in terms of technology and regulatory oversight.

The rather poor radiation safety awareness on the part of both medical and non-medical staff appears to exacerbate this occupational safety challenge in our region partly because the existing radiation safety infrastructure system in many of our countries does not comply with the basic safety standards.

Also, the absence of academic, postgraduate, and continuing education and training in radiation protection among medical and paramedical staff in most member states goes a very long way in negatively impacting the immediate medical and health communities, as well as society at large.

8.3.3 Public or Population Exposure

Requirement 29 of the International Basic Safety Standards enunciated the responsibilities of the government and the regulatory body specific to public exposure, and further reiterates that the government or the regulatory body shall establish the responsibilities of relevant parties that are specific to public exposure, shall establish and enforce requirements for optimization, and shall establish, and the regulatory body shall enforce compliance with, dose limits for public exposure.

The implementations of this requirement in the region, just like the previous exposure issues, have largely been ignored. For example, air and foodstuff monitoring, which greatly impacts population exposure, among other factors, are largely non-existent in most of the member states.

8.4 PANACEA TO THE CHALLENGES

The recognition of medical physics as a profession in Africa is still a thorny issue, and many countries including Cameroon, Egypt, Morocco, Senegal, Sudan, and Tunisia, among the other countries, continue to struggle with this serious obstacle. A major factor that inhibits recognition is that academic and clinical training criteria for the profession are not clearly defined in many countries (Meghzifene, 2012).

About nine countries in the region — Algeria, Egypt, Ghana, Libya, Morocco, Nigeria, South Africa, Sudan, and Tunisia currently run post-graduate academic programs in medical physics. This very important milestone (academic: M.Sc. and Ph.D.s in medical physics) appears to have been sustainable over the years; however, it will require further consolidation with the gradual and methodical introduction of structured clinical training to the overall national curriculum, with a view to increasing the availability of more competent human resources that will, for example, mitigate the severe or fatal misadministration of radiation doses to patients in radiation oncology (Meghzifene, 2012).

In addressing the concerns raised with respect to the above-mentioned issues emanating from the International Basic Safety Standards in the region, among other solutions to the challenges, especially with respect to medical exposure, there is the need to produce competent manpower in the form of clinically qualified medical physicists (CQMP).

For this reason, Nigeria approached the IAEA under the Technical Assistance initiative of the agency, and this resulted in the birth of a National Project with the title Developing the National Capacity to Train Medical Physicists to Support Radiotherapy Facilities in Tertiary Hospitals in Cancer

Management (NIR/6/023) which was started in September 2012 with seven trainees (residents).

The program is currently implementing the IAEA Training Course Series No. 37 (2009) (IAEA 2009), which provides guidelines for the clinical training of medical physicists specializing in radiation oncology, similar to the earlier pilot-testing of this training program in the Philippines and Malaysia also in collaboration with this agency. Member States are encouraged to adapt the IAEA guidelines to national conditions and needs when establishing programs. Moreover, national centers, even with limited radiation medicine facilities, are further encouraged to initiate programs using the resources that are available. This may be limited to partial fulfillment of the program only, which is supplemented by regional cooperative efforts, in order to develop the comprehensive set of competencies.

The CQMP requires the academic and clinical training components. Nigeria is one of the eight countries in the region that have had a post-graduate medical physics program for quite some number of years now, and this fits perfectly well with the recent recommendations from the published IAEA Human Health Series 25 (Roles and Responsibilities, and Education and Training Requirements for Clinically Qualified Medical Physicists, Human Health Series No. 25, IAEA, Vienna (2013)), which, inter alia, states that successful completion of a postgraduate academic program in medical physics leads to partial fulfillment of the requirements to be recognized as a clinically qualified medical physicist (CQMP) (IAEA 2013a).

The academic program needs to be complemented by a structured clinical training program in order to develop the skills and competencies necessary to practice in the clinical environment. The aim of a supervised hospital-based clinical training program is to provide a resident (also known as a registrar, intern, or trainee) with the opportunity to develop the skills and competencies required to practice independently. Figure 8.1 schematically shows the recommended education requirements for recognition as a CQMP (IAEA 2013a).

In Africa, a 1-year clinical training program in radiotherapy medical physics, and 6 months each in diagnostic radiology and nuclear medicine medical physics, is currently recommended, noting that the relevant radiation protection aspects are to be included within each of the three components.

For the foreseeable future, in order to expedite affordable capacity building for clinical medical physics in the region and to promote the establishment of national programs, clinical training in only one discipline will most likely be considered acceptable (1 year clinical training in radiotherapy medical physics, 6 months in nuclear medicine medical physics, or 6 months in radiology medical physics). However, 1 year clinical training, providing competence in both nuclear medicine and radiology, is highly desirable to strengthen the role of medical physics in imaging in the region.

All programs should be accredited, and successful completion of such a program should result in appropriate recognition by the national responsible authority. Medical physicists and other health professionals such as medical

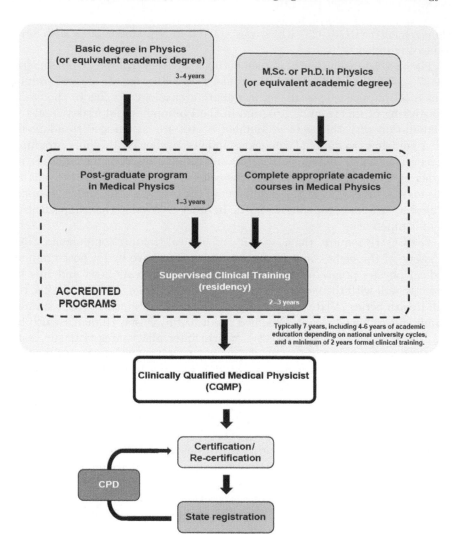

Figure 8.1 Minimum requirements for the academic education and clinical training of a Clinically Qualified Medical Physicist (IAEA 2013a)

practitioners, for instance, undergo clinical training programs that are conducted in hospitals and regulated by Ministry of Health authorities. Degree qualifications awarded as a result of academic programs, on the other hand, generally fall under a Higher Education authority.

In most countries where formal programs exist, residents are required to undergo clinical training in a full-time capacity, and a qualification is then awarded on completion of a clinical training program.

Experiential training requires that residents are privy to patients and their records, and therefore awareness and adherence to national health professional ethical standards in clinical practice becomes necessary. Residents should be required to compile a logbook or portfolio that reflects the competencies attained during their clinical training. In principle the portfolio itself is evidence of having undergone a clinical training program.

Ideally, continuous evaluation mechanisms should be developed to monitor the resident's progress during the program. A formal, independent assessment of the resident should take place at least at the end of the training program, to confirm successful completion of the clinical training program. This should result in recognition and national registration as a CQMP (IAEA 2013b).

The report of the task force meeting convened under the Regional Project - RAF/6/044 (Strengthening Medical Physics in Support of Cancer Management - Phase II), which produced "A Regional Clinical Training Program for Radiotherapy Medical Physics" as part of Recommendations for Medical Physics Education in AFRA Member States, further highlighted in clear terms what resources in terms of supervision, facilities, and references as well as infrastructure and equipment will be required to fully and successfully execute this laudable program in the region (IAEA 2013b). It states that the IAEA Basic Safety Standards (IAEA 2014) require that medical physicists with responsibilities for medical exposures are specialized "in the appropriate area" and as such, "meet the respective requirements for education, training and competence in radiation protection." In addition, "for therapeutic uses of radiation, the requirements of these Standards for calibration, dosimetry and quality assurance, including the acceptance and commissioning of medical radiological equipment... are fulfilled by or under the supervision of a medical physicist." In order to fulfill these requirements, an intensive program of structured learning in a conducive environment is necessary to produce a CQMP. Although a busy clinical environment can often be perceived by residents to offer fewer individualized opportunities for professional development, the importance of being exposed to a range of cases, techniques, and technologies that are appropriate to regional needs must be emphasized. As a result, this document does not recommend clinical training in advanced or emerging technologies, but concentrates on providing residents with a thorough grasp of safe and effective evidence-based practice (IAEA 2013b).

With regards to supervision of residents, a practising CQMP with at least 5 years' experience in hospital-based independent practice should supervise the program. The maximum ratio of residents (interns or trainees) to CQMP staff should be 2:1, taking into account the workload of the facility and additional numbers of practising CQMPs. As to facilities and references, there should be Internet access for the residents, and the training program should be conducted within a licensed facility that complies with all the requirements for a comprehensive program of quality assurance for medical exposures in all aspects of radiotherapy (IAEA 2007).

In addition to numerous IAEA publications in the field, the availability of some non-IAEA references was considered essential, many of which are available for free download.

And finally, with respect to infrastructure and equipment, the following is the minimum list of equipment for sites in the region to offer the complete comprehensive clinical training program in radiotherapy medical physics:

- Positioning/immobilization systems (breast, head, and neck).

- Mould room and workshop equipment.

- Conventional/fluoroscopic radiotherapy simulation.

- CT-based 3D treatment planning, including access to a CT scanner.

- ^{60}Co teletherapy.

- Linear accelerator (LINAC) with photon and electron beams.

- Kilovoltage therapy.

- Brachytherapy low dose rate (LDR) and/or high dose rate (HDR).

- Access to systems for absolute and relative dosimetry of all treatment equipment (IAEA 2013b).

8.5   SUMMARY

In summary, the above copiously referenced document is a companion guideline for harmonized clinical training of medical physicists in Africa. Clinical training is required in order to complement the education of medical physicists who have completed a postgraduate academic program. The regional postgraduate medical physics syllabus for academic programs was published earlier in 2013 (IAEA 2013c). The syllabus is geared towards encouraging member states with some minimal human resources to initiate their own national programs, and where there is proliferation of different syllabi, towards harmonization of programs within the different member states and hopefully in the entire region.

Owing to the technological expansion and multidisciplinary nature of the field of radiation medicine, there is a need for diagnostic radiology and nuclear medicine medical physics competencies in radiotherapy medical physics. As a result, clinical training in radiotherapy medical physics will necessarily require some competencies in radiology and nuclear medicine medical physics and vice versa.

Additional regional African task force meetings are being planned to develop a companion logbook for the syllabus, and guidelines for clinical training in radiology and nuclear medicine medical physics. The document is endorsed by the Federation of African Medical Physics Organizations (FAMPO).

8.6 CONCLUSIONS AND RECOMMENDATIONS

The implementation of the NIR/6/023 program is still a work-in-progress and if successfully anchored, may serve as a catalytic model for other member states in the region. The lessons learned will definitely be shared so as to minimize "re-inventing the wheels" by other bodies who may want to go the same route to locally enhance human resource development towards the mitigation of undesirable exposure consequences arising from inadequate personnel knowledge, skills, and competences. This will equally assist in minimizing the brain-drain syndrome in the region.

There should be a closer collaboration between and among the regional bodies such as FAMPO and FNRBA with regards to tackling some of the emergent concerns with regards to the population doses and overall exposure and health issues of citizens in the African region. FAMPO should provide the necessary professional back-up support with respect to the regulatory controls that is expected to be facilitated by the FNRBA.

The regional bodies should ensure that procurement and maintenance of equipment follows international best practices and that personnel training and re-training are strongly emphasized as the technology and techniques evolve.

8.7 ACKNOWLEDGMENTS

The International Atomic Energy Agency, for its various developmental roles in the region and especially for its support of the NIR/6/023 project (among other several national and regional projects) in terms of expert visits, residents fellowship, scientific visit, for resource persons, and procurement of some equipment. Also, the Executive Committee of the regional body — FAMPO, the committee members saddled with the execution of the NIR6023 national project; and last but not least, Professor S.B. Elegba, for his input to this short treatise.

8.8 REFERENCES

African Commission on Nuclear Energy. 2014. http://Peaceau.org/en/page/78-nuclear-energy-afcone. (Accessed January 26, 2015).

Black's Law Dictionary. 2015. Free Online Legal Dictionary, 2nd Edition. http://thelawdictionary.org/. (Accessed January 26, 2015).

Forum of Nuclear Regulatory Bodies in Africa. 2010. Charter of the Forum of Nuclear Regulatory Bodies in Africa. FNRBA.

Hammou, A. 2014. Leadership and innovations to improve quality imaging and radiation safety in Africa - A Plenary Lecture Session at the Fourth Regional African congress of the International Radiation Protection Association (AFRIRPA04), Rabat, Morocco: September 13–17, 2014.

International Atomic Energy Agency. 1996. SAGTAC VI/2 Report. IAEA. Vienna.

International Atomic Energy Agency. 2001. Revised Guiding Principles and General Operating Rules to Govern the Provision of Technical Assistance by the Agency INFCIRC/267; GOV/2001/48, IAEA. Vienna.

International Atomic Energy Agency. 2007. Comprehensive Audits of Radiotherapy Practices: A Tool for Quality Improvement. Quality Assurance Team for Radiation Oncology (QUATRO), IAEA, Vienna.

International Atomic Energy Agency. 2009. Clinical Training of Medical Physicists Specializing in Radiation Oncology, Training Course Series No. 37, IAEA, Vienna.

International Atomic Energy Agency. 2013a. Roles and Responsibilities, and Education and Training Requirements for Clinically Qualified Medical Physicists, Human Health Series No. 25, IAEA, Vienna.

International Atomic Energy Agency. 2013b. A Regional Clinical Training Program for Radiotherapy Medical Physics - Report of a Task Force Meeting, IAEA, Vienna.

International Atomic Energy Agency. 2013c. A Regional Post Graduate Medical Physics Syllabus for Academic Programs - Report of a Task Force Meeting, IAEA, Vienna.

International Atomic Energy Agency. 2014. Radiation Protection and Safety of Radiation Sources: International Basic Safety Standards - IAEA Safety Standards Series No. GSR Part 3 (8). Vienna.

Meghzifene, A. Call for recognition of the medical physics profession: (http://www.thelancet.com/journals/lancet/article/PIISO140-6736(11)60311-5) - 2012.

CHAPTER 9

Regulatory Structures and Issues in Asia and Oceania

Kwan-Hoong Ng

Department of Biomedical Imaging and Medical Physics Unit, University of Malaya, Kuala Lumpur, Malaysia

Aik-Hao Ng

Ministry of Health, Putrajaya, Malaysia

Marina binti Mishar

Department of Technical Cooperation, International Atomic Energy Agency, Vienna, Austria

CONTENTS

A SIA HAS AN INCREDIBLY DIVERSE cultural, educational, social, and economic background. Some four billion (60%) of the world population reside in Asia and it boasts hundreds of languages and dialects. It is also a land of extremes and contrasts. It has the highest mountains and most of the longest rivers, highest plateaus, and largest deserts and plains of all the continents. Asia is also home to some of the world's oldest cultures. It has some of the poorest as well as some of the richest nations in the world (Ng 2008; Sun, Ng 2012).

Similarly, the regulatory structures and practices of radiation protection in Asia are also very diverse and non-homogeneous. The region has also experienced widespread introduction of technologically advanced facilities, both within health care and in other fields (Sun, Ng 2012). Amidst this backdrop, two organizations were formed: the South East Asian Federation of Organizations for Medical Physics (SEAFOMP) (Ng, Wong 2008) and the Asia-Oceania Federation of Organizations for Medical Physics (AFOMP) (AFOMP 2015).

9.1 INTRODUCTION

The aim of this chapter is to provide some background on how the international recommendations on ionizing radiation protection are being developed nationally, and to highlight their regulatory structures and implementation in Asia and Oceania (categorized in sub-regions such as South East Asia, South Asia, East and Central Asia, and Oceania; see Figures 9.1 and 9.2). The extent of implementation and enforcement vary substantially within Asia, as they are influenced by national priorities, economic status, availability of resources, cultural diversity, and educational level.

As stated in the IAEA Fundamental Safety Principles, it is the role of a government to establish and sustain an effective legal and governmental framework for radiological safety. An appropriate and effective legal and regulatory framework for the protection and safety in all exposure situations shall be established and maintained. This framework shall encompass both the assignment and the discharge of governmental responsibilities, and the regulatory control of facilities and activities that give rise to radiation risks (IAEA 2006).

Each country should form an independent regulatory body to regulate radiation-related activities. Specific responsibilities and functions of the regulatory body shall be mandated in order to establish or adopt regulations and guidelines for protection and safety and establish a system to ensure their implementation. In general, their roles should include but should not be limited to (IAEA 2014a):

- Establishing requirements for the application of the principles of radiation protection as specified in the IAEA Fundamental Safety Principles for all exposure situations, and establishing or adopting regulations and guides for protection and safety;

- Establishing a regulatory system for protection and safety;

- Adopting a graded approach to the implementation of the system of protection and safety, such that the application of regulatory requirements is commensurate with the radiation risks associated with the exposure situation;

- Ensuring the application of the requirements for education, training, qualification, and competence in protection and safety of all persons engaged in activities relevant to protection and safety;

- Ensuring that mechanisms are in place for the timely dissemination of information to relevant parties, such as suppliers and users of sources, on lessons learned for protection and safety from regulatory experience and operating experience, and from incidents and accidents and the related findings. The mechanisms established shall, as appropriate, be used to provide relevant information to other relevant organizations at the national and international level;

- In conjunction with other competent authorities, adopting specific requirements for acceptance and for performance, by regulation or by the application of published standards, for any manufactured or constructed source, device, equipment, or facility that, when in use, has implications for protection and safety;

- Establishing mechanisms for communication and discussion that involve professional and constructive interactions with relevant parties for all protection and safety-related issues;

- Establishing, implementing, assessing, and striving to continually improve a management system that is aligned with the goals of the regulatory body and that contributes to the achievement of those goals.

The regulatory systems in some of the Asian-Oceania countries have been strengthened with the aid of international organizations. For example, the IAEA has launched the Radiation Protection Advisory Teams (RAPAT) to assist member states in assessing the existing state of their radiation protection activities and in determining their immediate and future needs. Developing

Figure 9.1 A map of Asia: Eastern Asia, Central Asia, Western Asia, South Eastern Asia, and Southern Asia (IOMP 2015)

countries like China, Malaysia, the Philippines, and the Republic of Korea participated in the program between 1984 and 1987 (Rosen 1987).

The initiatives to strengthen radiation protection in the region are further emphasized by the Asian and Oceanic Association for Radiation Protection (AOARP), a regional chapter of the International Radiation Protection Association (IRPA) that consists of national associate societies in Asia and Oceania. One of the key strategies is to host the Asian and Oceanic Congress on Radiation Protection (AOCRP). The main aim of the regional congress is to provide a means whereby the members may discuss their professional experiences, exchange ideas, and acquaint themselves with the scientific and technical problems of their international colleagues (AOARP 2014).

A survey on the current status (as of January 2015) of the region's regulatory infrastructure, and implementation has been conducted, and brief accounts are given in this chapter. Links to government webpages are provided in Tables 9.1, 9.2, 9.3, and 9.4. This is not intended to give a comprehensive picture but just a snapshot of the current scenario in the sub-regions.

9.2 NATIONAL REGULATORY INFRASTRUCTURE IN THE REGION

Most countries have specific ionizing radiation legislation, but there are large variations in the degree of implementation and enforcement. Traditionally, legislation relating to radiation safety in healthcare facilities has been enforced

Figure 9.2 A map of Oceania including Pacific islands (IOMP 2015)

by the government agencies responsible for health; for example, Malaysia, the Philippines, Singapore, and Thailand.

Transformation of the legislative management and its regulatory structure in some of the countries has resulted in the concentration of the regulatory roles in a single authority body dealing with atomic energy. In countries such as Indonesia, Thailand, and Vietnam, the national atomic energy agency regulates all activities dealing with ionizing radiation, including its application in the field of medicine. In contrast, the Ministry of Health in countries such as Malaysia, China, and Japan remain in control of the use of radiation medicine and of monitoring radiation exposure levels. All legislation follows the ICRP 26 (ICRP 1977), but most countries are in various stages of revision to meet the standards as set out in the International Commission on Radiological

Protection (ICRP) 60 (ICRP 1991) and the IAEA Basic Safety Standards (BSS) (IAEA 2014b).

Many countries have implemented the Regulatory Authority Information System (RAIS), a software application developed by the IAEA, in managing their regulatory activities in accordance with the IAEA Safety Standards and Guidance, including the Code of Conduct on the Safety and Security of Radioactive Sources and Supplementary Guidance (IAEA 2014c).

9.2.1 Australia

Australia is a federation and for the purposes of radiation safety, the regulations are determined by nine independent regulatory authorities, the Commonwealth and eight States and Territories. Recommendations of the ICRP are taken up into the individual regulatory frameworks and a system of national uniformity is sought and negotiated at the Radiation Health Committee meetings, which have representation from each regulatory authority and are administered by the Australian Radiation Protection and Nuclear Safety Agency (ARPANSA). Radiation safety regulatory requirements and implementation are the responsibility of the separate authorities. The National Directory for Radiation Protection provides nationally uniform requirements for the protection of people and the environment against the exposure or potential exposure to ionizing and non-ionizing radiation and for the safety of radiation sources, including provision for the national adoption of codes and standards. Nationally agreed-upon requirements are published in the ARPANSA Radiation Protection Series of documents.

9.2.2 Bangladesh

The Bangladesh Atomic Energy Regulatory Authority (BAERA) was established on February 12, 2013 under the Ministry of Science and Technology. The Nuclear Safety and Radiation Control Act-1993 was revoked, after the enactment of the new Act No. 19 of 2012, entitled Bangladesh Atomic Energy Regulatory (BAER) Act - 2012. The BAERA is responsible for administering the BAER - Act 2012 and Nuclear Safety and Radiation Control Rules (NSRCR) 1997. The act confers all necessary powers to the BAERA to regulate the uses of atomic energy, radiation practices, nuclear installations, and the management of radioactive waste. The new BAER Act-2012 covers all the regulatory functions to enable the regulatory authority to effectively operate a regulatory program.

At present the main radiation safety regulations are the Nuclear Safety and Radiation Control Rules-1997 (NSRC Rules-1997) (SRO No. 205/1997). NSRC Rules-1997 are based on the International Basic Safety Standards for Protection against Ionizing Radiation and for the Safety of Radiation Sources, known as Safety Series No. 115 (BSS-115) and cover most of the principal elements necessary for an effective nuclear safety and radiation protection regime. These rules apply to all practices, sources and nuclear materials.

9.2.3 Cambodia

Cambodia has been working with the IAEA under project KAM9001 from 2012 to 2015 to establish the National Radiation Safety Infrastructure. This project includes drafting a comprehensive nuclear law and nuclear center for nuclear/radiation management. The Ministry of Mines and Energy (MME) has been drafting the nuclear law. The first draft was sent to the IAEA for review in January and March 2014. Currently, the draft law is being revised internally within the MME. After this review, the draft law will be sent to the Council of Ministers for inter-ministerial review and approval. Then the draft law will be sent to the National Assembly and Senate for review and approval. After the promulgation of the nuclear law, a regulatory body will be established to manage safety, security, and safeguards activities of nuclear and radioactive materials in Cambodia.

The current issue in Cambodia is that there is no regulatory body. The licensing of radioactive materials and equipment process is done on a case-by-case basis. Activities that are related to medical need receive approval or a license from the Ministry of Health. Activities that are related to industry need to have approval or a license from the Ministry of Industry and Handicraft. Activities that are related to mines and energy need to have approval or a license from the Ministry of Mines and Energy. Another issue is that there is no national inventory of radioactive materials in Cambodia.

9.2.4 Hong Kong

The Radiation Ordinance (Cap 303, Laws of Hong Kong) was enacted in 1957 to control the import, export, possession, and use of radioactive substances and irradiating apparatus. Two subsidiary regulations responsible for the control of radioactive substances and irradiating apparatus were subsequently enacted in 1965 to enable a complete system of radiological protection. The ordinance is enforced by the Radiation Board, a statutory body established under the ordinance.

The system of radiological protection adopted in the ordinance follows that of the ICRP. In particular, the principle of justification of radiological practice is applied by means of a licensing system. The licence applicant is required to provide justifications for the introduction of a practice if it has not been justified by precedents or if significant variations from precedents exist. The principle of optimization of radiological protection is exercised by means of licence assessments and conditions of licence. The dose limits stipulated in the ordinance, for workers and the public, basically follow those recently recommended by the ICRP. The ordinance has no jurisdiction over medical irradiation. Radiological procedure optimization and patient exposure protection rest with the professionals, whose conduct is regulated under their respective professional registration and disciplinary regulations.

9.2.5   India

The Atomic Energy Regulatory Board (AERB) of India, constituted in 1983, has a mandate to regulate the entire gamut of nuclear and radiation facilities and equipment. AERB has in place a firm safety and regulatory frame work based on the prime national legislation, i.e., the Atomic Energy Act of 1962, which ensures that the use of radioisotopes and radiation do not cause unacceptable impact on working personnel, the public, and the environment. It works on the principle of "regulation by graded approach" based on the hazard potential of the radioactive source and spanning the entire life cycle of radiation sources in the country, from generation, procurement, transport of radioactive/nuclear material in the public domain, to their use and ultimate disposal/ decommissioning. The statutory requirements as per the Atomic Energy Act of 1962 and the Atomic Energy (Radiation Protection) Rules of 2004, issued under the Act, are expounded in the AERB safety Codes, Standards, and Directives issued from time to time. Medical X-ray installations (CT/cath-labs/general X-ray machines) in diagnostic radiology are also under the regulatory purview of AERB.

The chairman of AERB serves as the competent authority, under the Radiation Protection Rules of 2004, for radiation protection in the country. Requirement of Licence for handling of radiation sources is one of the statutory requirements in the country. It is the responsibility of the licensee to ensure compliance with the regulatory requirements.

9.2.6   Indonesia

Radiation protection activities in Indonesia from a technical point of view are regulated by the Government Regulation (GR) No. 33 of 2007 on Ionizing Radiation Safety and Security of Radioactive Sources. However, licensing of the utilization of ionizing radiation and nuclear material is governed by GR No. 29 of 2008.

GR No. 33 is further implemented by the Chairman of Nuclear Energy Regulatory Agency (BAPETEN) Regulation (CBR) No. 4 of 2013 on Radiation Protection and Safety in the Utilization of Nuclear Energy. Since the GR is based on the BSS, the CBR as the implementing regulation should obviously not deviate from the BSS. However, the CBR adopts some major points of GSR Part 3, the revised publication of the BSS. This results in an inhomogeneous CBR since some parts refer to the BSS and some others refer to the GSR Part 3.

9.2.7 Japan

Radiation protection standards of Japan comply with those of the ICRP and have been incorporated into legislation. The Radiation Council ensures consistency between relevant pieces of radiation hazard legislation. The fundamental legislation is "The Atomic Energy Basic Act" of 1955. The three basic policies of promoting the R&D and utilization of atomic energy are "democratic," "independent," and "open to the public." Based on the concept of the policy, related acts, orders, ordinances, etc., have been established and applied in all fields on radiation uses.

The current standards incorporate basic ICRP 1990 Recommendations (Publication 60) and additional provisions:

- The standard for exposure in Radiation Controlled Areas has been set at 1.3 mSv per three months, based on the special limit for the public (5 mSv per year).

- The limit for female radiation workers has been set at 5 mSv per three months, this shorter period reflecting the need for stronger protection of a fetus before any pregnancy has been recognized.

- The dose limit relating to emergency work remains at 100 mSv, taking into account the IAEA's BSS.

After the Fukushima accident, the regulatory bodies on radiation were reorganized, and the Nuclear Regulation Agency (NRA) was established in 2012 as an extra-ministerial bureau of the Ministry of the Environment. Currently, the Radiation Council belongs to the NRA. Since 2008, the Radiation Council has conducted deliberations aimed at incorporating the ICRP's 2007 Recommendations into domestic legislation. The council has also been discussing a review of the dose limit relating to emergency work.

9.2.8 Malaysia

Malaysia governed the use of radioactive materials through the Radioactive Substances Act of 1968, which was then superseded by the Atomic Energy Licensing Act of 1984 (Act 304) after nearly two decades of implementation. Act 304 is one of the core components of the legislative framework for regulating the use of atomic energy in the country in a safe and secure manner. The legislative hierarchy of Act 304 starts with the act, then regulations, circulars/codes and standards, plus guidelines/local rules. There are several subsidiary regulations in place to control and monitor activities, which include: Radiation Protection (Licensing) Regulations 1986; Radiation Protection (Transport) Regulations 1989; Atomic Energy Licensing (Basic Safety Radiation Protection) Regulations 2010; and Atomic Energy Licensing (Radioactive Waste Management) Regulations 2011.

Under the Act, the Atomic Energy Licensing Board (AELB) is established and acts as the regulatory authority for non-medical purposes, while the use of atomic energy for medical purposes is under the auspices of the Director-General of Health. The governmental agencies that assist the appropriate authorities in implementing the Act are the Atomic Energy Licensing Board (known as Department of AELB), Ministry of Science, Technology and Innovation, and the Radiation Health and Safety Section (RHSS), Ministry of Health. The RHSS has prepared circulars, guidelines, standards, and related legislative documents on issues such as quality assurance, national diagnostic reference dose, and licensing requirements. The authorities are also taking further steps to strengthen the legal infrastructure in safety, security, and safeguards, which is in line with the IAEA recommendations.

9.2.9 Nepal

Nepal has a long history of medical radiology, dating back to 1923. Radiological personnel who work under the Ministry of Health and Population (MOHP) in diagnostic healthcare settings do not use personnel monitoring devices except in the radiotherapy section. However, some privately-owned healthcare facilities are facing calibration difficulties.

Nepal has been a member of the IAEA since 2008. IAEA has been extending technical cooperation to Nepal for peaceful applications of nuclear science and technology to contribute to the achievement of national development objectives as well as socio-economic uplift of the nation. A draft Nuclear Law was prepared and approved by the Council of Ministers on September 16, 2011, and is still under the review of the Nepal Law Commission. Nepal still does not have any radiation protection infrastructure to control the use of ionizing radiation in various fields, so there is a great need for rules, regulations, and a Radiation Protection Act governing radiation for medical use.

9.2.10 New Zealand

The Radiation Protection Act of 1965 and Radiation Protection Regulations of 1982 govern radiation safety in New Zealand. The act sets up the regulatory structure via licensing individual users of irradiating apparatus and radioactive materials, and the regulations set out general requirements for safety. The act allows conditions to be placed on licenses, and this mechanism is used to make compliance with the relevant Code of Safe Practice mandatory. Specific radiation protection requirements are laid down in the Codes of Safe Practice. The Office of Radiation Safety, Ministry of Health, is New Zealand's regulatory authority, issuing licenses, writing codes of safe practice, and overseeing compliance-monitoring audits. Until 2011, the National Radiation Laboratory (NRL) was the regulatory authority and also provided scientific services relating to ionizing radiation. Those scientific services are now provided by the Institute of Environmental Science and Research (ESR), which is a Crown research institute focused on environmental sciences.

9.2.11 Pakistan

Pakistan's nuclear regulatory infrastructure has been in place since 1965, when the first research reactor PARR-I was commissioned. The nuclear regulatory regime further improved when the first nuclear power plant was commissioned in 1971 at Karachi. A nuclear safety and licensing division was established in 1956 as the Pakistan Atomic Energy Commission (PAEC), which functioned as the de facto regulatory body till it was upgraded to "Directorate of Nuclear Safety and Radiation Protection" (DNSRP) after the promulgation of Pakistan Nuclear Safety and Radiation Protection Ordinance 1984.

Pakistan signed the International Convention on Nuclear Safety in 1994, as a result of which the Government of Pakistan had to establish an independent nuclear regulatory body entrusted with the implementation of the legislative and regulatory framework governing nuclear power and radiation use in the country; further, to separate the regulatory functions from the promotional aspects of the nuclear program. As a transitory measure, the Pakistan Nuclear Regulatory Board (PNRB), was established within PAEC to oversee regulatory affairs. Complete separation of promotion and regulatory functions and responsibilities was achieved in 2001, when the President of Pakistan promulgated the Pakistan Nuclear Regulatory Authority Ordinance No. III of 2001. Consequently, the Pakistan Nuclear Regulatory Authority (PNRA) was created, dissolving the PNRB and DNSRP. The mission of PNRA is to ensure safe operation of nuclear and radiation facilities (including medical applications) and to protect radiation workers, the general public, and the environment from the harmful effects of radiation by formulating and implementing effective regulations and regulatory guidelines based on IAEA recommendations.

9.2.12 People's Republic of China

The Law of the People's Republic of China on Radioactive Pollution Prevention and Control (2003) is the principal law for regulating radiation sources. Subsidiary regulations are set to further elaborate the regulatory requirements, such as the Radioisotope and Radiation Emitting Devices Safety and Protective Regulations (2005) and Basic Safety Standards on Ionizing Radiation Protection and Radiation Sources (2002). Specific requirements are enforced for the use of radiation in medicine, known as the No. 46 Rule on the Administration of Radiodiagnosis and Radiotherapy (2006) and States Councils Order No. 5 (1989).

The Ministry of Environmental Protection (MEP) (National Nuclear Safety Administration, NNSA) is the Chinese regulatory body for nuclear safety, radiation safety, and radiological environment management. In addition to the MEP (NNSA), the Ministry of Health (MOH) and the Health Departments of the provincial governments (DoH) have legislative responsibilities for regulating the use of medical radiation and the protection of occupationally exposed workers. Both agencies share the responsibilities of controlling justification and optimization of medical exposure; occupational exposure; health surveillance of radiation workers; and quality assurance in medical facilities and practices.

9.2.13 Philippines

There are two national radiation regulatory agencies in the Philippines. One is the Philippine Nuclear Research Institute (PNRI) of the Department of Science and Technology. PNRI used to be the Philippine Nuclear Energy Commission, which was created in 1957 to promote and regulate the peaceful uses of nuclear energy. The second one is the Center for Device Regulation, Radiation Health, and Research (CDRRHR) of the Food and Drug Administration (FDA) of the Department of Health. CDRRHR used to be called the Bureau of Health Devices and Technology (BHDT), previously named the Radiation Health Office, which was created in 1974 to regulate the use of electrical/electronic devices emitting ionizing or non-ionizing radiation. In the FDA Act of 2009, the BHDT became the CDRRHR of the FDA with the additional function of regulating medical devices and health-related devices. Both PNRI and CDRRHR have adopted into their regulations the Basic Safety Standards issued by the IAEA. Both agencies are actively involved in IAEA activities pertaining to ionizing radiation protection. The CDRRHR is also involved in World Health Organization activities in the area of non-ionizing radiation protection.

9.2.14 Republic of Korea

Radiation protection and safety in the Republic of Korea is governed by the Nuclear Safety Act and its Decree, Regulations, and Notices. The Nuclear Safety and Security Commission (NSSC) is the regulating authority. The Korea Institute of Nuclear Safety (KINS) is the dedicated technical expert organization, and carries out the safety reviews and inspections, and the development of regulatory technical standards and guidelines for the licensing of nuclear and radiation facilities to ensure radiation safety.

9.2.15 Singapore

In Singapore, the Radiation Protection Act was first enacted in 1973, and the Radiation Protection Regulations followed in 1974. This act was repealed in 1991 and re-enacted to include the control of non-ionizing radiation. The Radiation Protection (Non-Ionizing Radiation) Regulations came into force in 1992. The Radiation Protection Regulations of 1974 were amended in 2000 to take into consideration the 1990 Recommendations of the ICRP and the BSS requirements. The regulations for the transport of radioactive materials were also amended in 2000, to be in line with the IAEA Safety Standards Series

No. ST-1, 1996 Edition-Regulations for the Safe Transport of Radioactive Materials.

In July 2007, the Radiation Protection Act was repealed and re-enacted with amendments to transfer the roles and functions of the Center for Radiation Protection (CRP), as well as the administration of the Radiation Protection Act, from the Health Sciences Authority (HSA) to the National Environment Agency (NEA). With the transfer of CRP from HSA, a new department, the Center for Radiation Protection and Nuclear Science (CRPNS) was formed on July 1, 2007 under the NEA, for the administration of the Radiation Protection Act. It was renamed Radiation Protection and Nuclear Science Department (RPNSD) on August 1, 2013 to better reflect its regulatory role as a department within NEA. Presently, RPNSD is the national authority for radiation protection in Singapore.

9.2.16   Sri Lanka

The basic law controlling of use, transport, and disposal of radioactive material and the use of irradiation apparatus in Sri Lanka is the Atomic Energy Authority Act No. 19 of 1969. The Atomic Energy Authority (AEA) established under this act functions as the regulatory authority for matters connected with ionizing radiation. As the act was promulgated in 1969, the current act has no provisions for matters connected with nuclear security and safeguards; the current act has provisions for making an order for importation, exportation, production, acquisition, treatment, storage, transport, and disposal of radioactive material; making regulations for the sale and supply of radioactive material taken internally; and for the use of irradiating apparatus and radioactive material. The act also has provisions for making regulations for transport and disposal of radioactive materials.

The AEA has made an order for regulating importation of food items above the levels given in the order. The regulations on Ionizing Radiation Protection were promulgated in 2000 repealing regulations made in 1975. The new regulations are consistent with the BSS-115. The AEA issues authorizations for the use and possession of radioactive material and irradiating apparatus, and for transport and disposal of radioactive material. The AEA Radiation Protection Officers are empowered to do inspections by prior arrangements or as surprise visits. As the act is too old and does not have provisions for current international requirements, the AEA drafted a new act entitled the "Sri Lanka Atomic Energy Act," which establishes two entities: one for regulatory matters and the other for promotional matters.

9.2.17   Republic of China (Taiwan)

In 1968 the Atomic Energy Council (AEC) of Executive Yuan in Taiwan was established and the Atomic Energy Law was enacted in order to regulate nuclear power plants, the use of ionizing radiation, and radiation work practices.

In 2002, the AEC enacted the Ionizing Radiation Protection Act (IRPA), which was based on the ICRP-60 Report issued in 1990 and the BSS-115 issued in 1996. For medical surveillance, the IRPA has incorporated the Medical Exposure Quality Assurance Regulations to enhance the patient's safety.

Major features of the radiation protection regulation and control administered by the AEC include:

- A computerized trade facilitation and registration control platform established to monitor the import/export of radiation sources and the status of each licensee, to strengthen security and control.

- In cooperation with the U.S. Megaport Initiatives, inbound and outbound containers will undergo radiation detection to stop the illicit trafficking of radioactive materials. In addition, Category-I and -II high-risk radiation sources are inspected annually to assure their security.

- A national radiation detection program, which requires all steel makers with smelting furnaces to install portal-type radiation detectors to effectively prevent mistakenly smelting radiation sources in the imported metal scraps.

9.2.18 Thailand

The utilization of either radioactive materials or radiation generating devices in Thailand in the past several decades has been continuously expanding, especially in medical and industrial applications. Each year, more than 2,000 licenses have been issued to licensees for possession and use of radioactive materials or radiation generating devices, as well as for importing or exporting radioactive materials.

Overseeing these activities, the Office of Atoms for Peace (OAP), Ministry of Science and Technology, is responsible for the authorization and regulation of nuclear and radiation source utilizations in Thailand, as stated in the Atomic Energy for Peace Act B.E. 2504 and Ministerial Regulation B.E. 2550 on Requirements and Methods for Licensing of Nuclear and Radiation Sources, as well as other relevant laws and regulations. International standards on safety, security, and safeguards of nuclear and radiation sources have been adopted and integrated into these regulating instruments. OAP also has extensively cooperation with the Thailand Institute of Nuclear Technology (Public Organization) and the IAEA, in order to strengthen and enhance the effectiveness of Thailand's nuclear and radiation safety.

9.2.19 Vietnam

All activities using radioactive materials/sources and radiation equipment were controlled by the government under the Ordinance on Radiation Safety

& Control (ORSC) from 1996 to 2009, and under the Atomic Energy Law from 2009. For implementation of the Atomic Energy Law, a system of regulations and technical standards on radiation protection has been developed; for example: ordinances on exemption level, licensing, inspection and enforcement, occupational and public exposure control, safe transport of radioactive materials.

Principles and requirements for radiation protection prescribed in the law and regulations were adopted from the IAEA Safety Standards Series (Fundamental Safety Principles, BSS-115 and Safety Guides in radiation protection). Under the law, the Vietnam Agency for Radiation and Nuclear Safety (VARANS) on behalf of the Ministry of Science and Technology (MOST) is responsible for state management of radiation protection through licensing and inspection mechanisms.

Table 9.1: Summary of regulatory infrastructure and status in South East Asia

Country	Name of regulatory body	Name(s) of legislation	Legislation enacted (Year)	Website
Indonesia	Nuclear Energy Regulatory Agency	Act No. 10 Year 1997 on Nuclear Energy	1997	http://www.bapeten.go.id/
Malaysia	Atomic Energy Licensing Board Ministry of Health	Atomic Energy Licensing Act	1984	http://www.aelb.gov.my/ https://radia.moh.gov.my/
Myanmar	Department of Atomic Energy	Atomic Energy Law	1998	http://www.most.gov.mm/
Philippines	Philippine Nuclear Research Institute Department of Health	Science Act of 1958 Atomic Energy Regulatory and Liability Act of 1968	1958 1968	http://www.pnri.dost.gov.ph/ http://www.doh.gov.ph/
Singapore	Radiation Protection and Nuclear Science Department	Radiation Protection Act 2007	2007	http://nea.gov.sg/
Thailand	Thai Atomic Energy Commission for Peace	Atomic Energy for Peace Act 1961	1961	http://www.oaep.go.th/
Vietnam	Vietnam Agency for Radiation and Nuclear Safety	The Atomic Energy Law	2009	http://www.varans.vn/

Note: There is no information for the following countries: Brunei, Cambodia, and Lao People's Democratic Republic.

Table 9.2: Summary of regulatory infrastructure and status in South Asia

Country	Name of regulatory body	Name(s) of legislation	Legislation enacted (Year)	Website
Afghanistan	Afghan Atomic Energy High Commission	NA	NA	http://aaehc.afghanistan.af/
Bangladesh	Bangladesh Atomic Energy Regulatory Authority	Bangladesh Atomic Energy Regulatory (BAER) Act	2012	http://www.baera.gov.bd/
India	Atomic Energy Regulatory Board	Atomic Energy Act of 1962	1962	http://www.aerb.gov.in/
Pakistan	Pakistan Nuclear Regulatory Authority	Pakistan Nuclear Regulatory Authority Ordinance 2001	2001	http://www.pnra.org/
Sri Lanka	Atomic Energy Authority	Atomic Energy Authority Act No. 19 of 1969	1969	http://www.aea.gov.lk/

Note: There is no information for the following countries: Bhutan, Maldives, and Nepal.

Table 9.3: Summary of regulatory infrastructure and status in East and Central Asia

Country	Name of regulatory body	Name(s) of legislation	Legislation enacted (Year)	Website
People's Republic of China	National Nuclear Safety Administration State Environment Protection Administration Ministry of Health	Law of the People's Republic of China on Prevention and Control of Radioactive Pollution	2003	http://nnsa.energy.gov/ http://www.chinacp.org.cn http://www.nhfpc.gov.cn/
The Government of the Hong Kong Special Administrative Region	Radiation Board of Hong Kong	Radiation Ordinance	1957	http://www.rbhk.org.hk/
Japan	Atomic Energy Commission Nuclear Regulation Authority Minister of Economy, Trade and Industry Minister of Education, Culture, Sports, Science and Technology Minister of Health, Labor and Welfare	Atomic Energy Basic Law	1955	http://www.aec.go.jp/ http://www.nsr.go.jp/ http://www.meti.go.jp http://www.mext.go.jp/ http://www.mhlw.go.jp/

Continued on next page

Table 9.3 – *Continued from previous page*

Country	Name of regulatory body	Name(s) of legislation	Legislation enacted (Year)	Website
Republic of Korea	Minister of Education, Science and Technology Nuclear Safety and Security Commission	Nuclear Safety Act Act on Physical Protection and Radiological Emergency	2011 2008	http://www.mest.go.kr/ http://www.nssc.go.kr/
Mongolia	Nuclear Energy Agency	Radiation Protection and Safety Law Nuclear Energy Law	2001 2009	http://nea.gov.mn/
Republic of China (Taiwan)	Atomic Energy Council	Atomic Energy Law Ionizing Radiation Protection Act	1968 2002	http://www.aec.gov.tw/
Kazakhstan	Kazakhstan Atomic Energy Committee	Law on Atomic Energy Use Law on Radiation Safety of the Population Law on Licensing	1997 1998 2007	http://www.mint.gov.kz/

Continued on next page

Table 9.3 – *Continued from previous page*

Country	Name of regulatory body	Name(s) of legislation	Legislation enacted (Year)	Website
Kyrgyzstan	NA	Law No. 58 on radiation safety of the population	1999	-
		Law No. 224 on Technical Regulations concerning Radiation Safety	2011	
Tajikistan	Nuclear and Radiation Safety Agency	Law No. 42 on radiation safety	2003	http://www.nrsa.tj/
		Law No. 69 on Atomic Energy Use	2004	
Turkmenistan	NA	Law of Turkmenistan on Radiation Safety	2009	-
Uzbekistan	State Inspectorate on Safety in Industry and Mining Ministry of Health	Law No. 120-II on radiation safety	2000	http://www.sgktn.gov.uz/ http://www.minzdrav.uz/

Note: There is no information for the following country: Democratic People's Republic of Korea.

Table 9.4: Summary of regulatory infrastructure and status in the Oceania (Pacific)

Country	Name of regulatory body	Name(s) of legislation	Legislation enacted (Year)	Website
American Samoa	Food and Drug Administration	Federal Food, Drug, and Cosmetic Act	1938	http://www.fda.gov/
Australia	Australian Radiation Protection and Nuclear Safety Agency	Atomic Energy Act	1953	http://www.arpansa.gov.au/
		Australian Nuclear Science and Technology Organisation Act	1987	
		Australian Radiation Protection and Nuclear Safety Act	1998	
Fiji Islands	Radiation Health Board	Radiation Health Decree 2009	2009	-
Guam	Food and Drug Administration	Federal Food, Drug, and Cosmetic Act	1938	http://www.fda.gov/
New Zealand	Ministry of Health's Office of Radiation Safety	Radiation Protection Act	1965	http://www.health.govt.nz/

Note: There is no information for the following countries: Cook Islands, Christmas Island, Coco (Keeling) Islands, French Polynesia, Kiribati, Niue, Marshall Islands, Federated States of Micronesia, Northern Mariana Island, Nauru, New Caledonia, Norfolk Island, Palau, Papua New Guinea, Samoa, Solomon Islands, Tonga, Tuvalu, Vanuatu, Wallis and Futuna Islands.

9.3 CONCLUSION

There are many radiation protection-related challenges in the Asian Oceania region. The diversity in geographical distribution and the various stages of social-economic development have contributed to the current unbalanced scenario. Just as in other parts of the world, there is significant expansion of radiation uses for both diagnosis and therapy. This has resulted in unnecessary patient exposures/overexposures to radiation in diagnosis and therapy, one implication being the increase in occupational dose (Sun, Ng 2012).

A strong national regulatory framework and effective implementation are essential to ensure patient safety in healthcare environments. In order to fully implement the new BSS in the Asian Oceania region, capacity building and clinical training of competent medical physics and radiation protection experts have to be given greater emphasis and further reinforced. To address this complex issue, several steps could be taken, such as strengthening regulatory networks, promoting research partnerships, sharing available resources, and collaborating in regional cooperation. Meanwhile, there is still a lot of work to be done towards establishing a robust and effective radiation protection structure and an adequately trained and resourceful radiation protection workforce in the Asian Oceania region.

9.4 ACKNOWLEDGMENTS

We thank the following persons for providing us with the relevant information: Anthony Wallace (Australia), Meherun Nahar (Bangladesh), Chhom Sakborey (Cambodia), Yongkong Zhao (China), C. B. Poon (Hong Kong), Avinash U. Sonawane (India), Eka Djatnika Nugraha (Indonesia), Keiichi Akahane (Japan), Kun Woo Choo (Korea), Raju Srivastava (Nepal), Shafqat Faaruq (Pakistan), Agnette Peralta (Philippines), Wee Teck Hoo (Singapore), H. L. Anil Ranjith (Sri Lanka), R. T. Lee (Taiwan), Rungdham Takam (Thailand), and Le Quang Hiep (Vietnam).

9.5 REFERENCES

Asian and Oceanic Association for Radiation Protection. 2014. http://www.jhps.or.jp/AOARP/Home.html (accessed January 15, 2015)

Asian-Oceania Federation of Organizations for Medical Physics. 2015. Asian-Oceania Federation of Organizations for Medical Physics. http://www.afomp.org/ (accessed January 15, 2015)

International Atomic Energy Agency. 2006. Fundamental safety principles: safety fundamentals (PDF Document). http://www-pub.iaea.org/MTCD/publications/PDF/Pub1273_web.pdf (accessed January 15, 2015)

International Atomic Energy Agency. 2014a. National competent authorities responsible for approvals and authorizations in respect of the transport of radioactive material (PDF Document). http://www-ns.iaea.org/downloads/rw/transport-safety/transport-safety-nca-list.pdf (accessed January 15, 2015)

International Atomic Energy Agency. 2014b. Radiation protection and safety of radiation sources: international basic safety standards: general safety requirements. Interim edition (PDF Document). http://www-pub.iaea.org/MTCD/publications/PDF/p1531interim_web.pdf (accessed January 22, 2015)

International Atomic Energy Agency. 2014c. Regulatory Authority Information System - RAIS. http://www-ns.iaea.org/tech-areas/regulatory-infrastructure/rais.asp?s=3 (accessed 15 Jan 2015)

International Commission on Radiological Protection. 1977. Recommendations of the International Commission on Radiological Protection. ICRP Publication 26. Ann. ICRP 1: (3).

International Commission on Radiological Protection. 1991. The 1990 Recommendations of the International Commission on Radiological Protection. ICRP Publication 60. Ann. ICRP 21: (1-3).

International Organization of Medical Physics website. http://www.iomp.org (accessed January 15, 2015)

Ng, K. H. 2008. Medical Physics in Asia - Where are we going? Australas Phys & Eng Sci in Med 31(4): xi-xii.

Ng, K. H., Wong, J. H. D. 2008. The South East Asian Federation of Organizations for Medical Physics (SEAFOMP) - Its history and role in the ASEAN countries. Biomed Imaging Interv J 4(2):e21. http://www.biij.org/2008/2/e21/ (accessed January 15, 2015)

Rosen, M. 1987. Special Reports: Adequate radiation protection: a lingering problem. International Atomic Energy Agency Bulletin, Vol. 4.

Sun., Z, Ng, K. H. 2012. Use of radiation in medicine in the Asia-Pacific region. Singapore Med J 53(12): 784-788

General references for the tables

Afghan Atomic Energy High Commission. Afghan Atomic Energy High Commission, Islamic Republic of Afghanistan. http://aaehc.afghanistan.af/ (accessed 15 Jan 2015)

Atomic Energy Authority of Sri Lanka. Atomic Energy Authority of Sri Lanka. http://www.aea.gov.lk/ (accessed January 15, 2015)

Atomic Energy Authority. Radioactive Waste Management in Sri Lanka (PowerPoint Slides). http://gnssn.iaea.org/RTWS/general/Shared Documents/Waste Management/March 2014 RAS9069 WS on policy and strategy development/national presentations/Sri Lanka.pdf (accessed January 15, 2015)

Atomic Energy Council. Atomic Energy Council (AEC). http://www.aec.gov.tw/english/index.html (accessed January 15, 2015)

Atomic Energy Licensing Board. Atomic Energy Licensing Board (AELB). http://portal.aelb.gov.my/sites/aelb/en (accessed January 15, 2015)

Atomic Energy Regulatory Board. Government of India Atomic Energy Regulatory Board. http://www.aerb.gov.in/ (accessed January 15, 2015)

Badan Tenaga Nuklir Indonesia. Country Report - National Nuclear Energy Agency BATAN Indonesia (PowerPoint Slides). https://ansn.iaea.org/Common/Topics/OpenTopic.aspx?ID=13492 (accessed January 15, 2015)

Bangladesh Atomic Energy Commission. Country Presentation - Bangladesh (PowerPoint Slides). https://ansn.iaea.org/Common/Topics/ OpenTopic.aspx?ID=13497 (accessed 15 Jan 2015)

Bangladesh Atomic Energy Commission. http://www.baec.org.bd/ (accessed January 15, 2015)

Department of Atomic Energy Ministry of Science and Technology Myanmar. Regional Workshop on Development of National Policy and Strategy for Radioactive Waste Management (Myanmar) (PowerPoint Slides). http://gnssn.iaea.org/RTWS/general/Shared Documents/Waste Management/March 2014 RAS9069 WS on policy and strategy development/national presentations/myanmar.pdf (accessed January 15, 2015)

Department of Health. Republic of the Philippines Department of Health. http://www.doh.gov.ph/ (accessed January 15, 2015)

Food and Agriculture Organization of the United Nations Legal Office. FAOLEX-legislative database of FAO legal office: Mongolia. http://faolex.fao.org/cgi-bin/faolex.exe?rec_id=046829&database=FAOLEX&search_type=link&table=result&lang=eng&format_name=@ERALL (accessed January 15, 2015)

International Atomic Energy Agency. 2010. Governmental, legal and regulatory framework for safe (PDF Document). http://www-pub.iaea.org/MTCD/publications/PDF/Pub1465_web.pdf (accessed January 15, 2015)

International Atomic Energy Agency. 2011. Radiation protection and safety of radiation sources: international basic safety standards: general safety requirements. - Interim edition (PDF Document). http://www-pub.iaea.org/MTCD/publications/PDF/p1531interim_web.pdf (accessed January 15, 2015)

International Atomic Energy Agency. Control of Radiation Sources. http://www-ns.iaea.org/tech-areas/radiation-safety/source.asp?s=3&l=22 (accessed January 15, 2015)

International Atomic Energy Agency. Country Nuclear Power Profiles 2014 Edition. (accessed January 15, 2015)

International Atomic Energy Agency. Directory of National Regulatory Bodies for the Control of Radiation Sources (PDF Document). http://www-ns.iaea.org/downloads/rw/code-conduct/reg-auth-directory.pdf (accessed January 15, 2015)

International Atomic Energy Agency. Radiation protection and safety of radiation sources: international basic safety standards (PDF Document). http://www-pub.iaea.org/MTCD/publications/PDF/Pub1578_web-57265295.pdf (accessed January 15, 2015)

International Commission on Radiological Protection, 1977. Recommendations of the ICRP. ICRP Publication 26. Ann. ICRP 1 (3). (superseded by ICRP 103)

International Commission on Radiological Protection, 1991. 1990 Recommendations of the International Commission on Radiological Protection. ICRP Publication 60. Ann. ICRP 21 (1-3).

International Commission on Radiological Protection, 2007. The 2007 Recommendations of the International Commission on Radiological Protection. ICRP Publication 103. Ann. ICRP 37 (2-4).

Korea Institute of Nuclear Safety. 2012. Nuclear Regulatory Organization Changes in Korea (PowerPoint Slides). http://www.oecd-nea.org/nsd/fukushima/documents/workshop_Korea_JapanMeeting NEA2012117KOREA.pdf (accessed January 15, 2015)

Ministry of Environment Protection of the People's Republic of China. Integrated regulatory review service (IRRS) Report to The Government of the People's Republic of China (PDF Document). http://www.mep.gov.cn/ztbd/rdzl/zqyj/qtxgbg/201206/P020120625357408488821.pdf (accessed January 15, 2015)

Ministry of Health. Radiation Health and Safety Section. https://radia.moh.gov.my/project/new/radia/index.php (accessed January 15, 2015)

Ministry of Science and Technology. The Republic of the Union of Myanmar Ministry of Science and Technology. http://www.most.gov.mm/most2eng/ (accessed January 15, 2015)

Ministry of Science Technology and Environment, Government of Nepal. Status of radioactive waste management in Nepal (PowerPoint Slides). https://ansn.iaea.org/Common/Topics/OpenTopic.aspx?ID=13529 (accessed January 15, 2015)

National Environment Agency. Overview of Radiation Protection. http://app2.nea.gov.sg/anti-pollution-radiation-protection/radiation-protection/overview-of-radiation-protection. (accessed January 15, 2015)

Nuclear Energy Regulatory Agency. Nuclear Energy Regulatory Agency (BAPETEN). http://www.bapeten.go.id/ (accessed January 15, 2015)

Office of Atoms for Peace. Office of Atoms for Peace Ministry of Science and Technology. http://www.oaep.go.th/index_en.php (accessed January 15, 2015)

Pakistan Nuclear Regulatory Authority. Pakistan Nuclear Regulatory Authority. http://www.pnra.org/ (accessed January 15, 2015)

Philippine Nuclear Research Institute. Philippines' Nuclear and Radiation Safety Infrastructure and Activities. Before, During And After The Fukushima Daiichi Accident (PowerPoint Slides). https://ansn.iaea.org/Common/Topics/OpenTopic.aspx?ID=13495 (accessed January 15, 2015)

Philippine Nuclear Research Institute. Philippine Nuclear Research Institute. http://www.pnri.dost.gov.ph/ (accessed January 15, 2015)

U.S. Food and Drug Administration. What does FDA do? http://www.fda.gov/aboutfda/transparency/basics/ucm194877.htm (accessed January 15, 2015)

Vietnam Agency for Radiation and Nuclear Safety. Nuclear Power and Energy Strategy in Vietnam before and after Fukushima's accident (PowerPoint Slides). https://ansn.iaea.org/Common/Topics/OpenTopic.aspx?ID=13498 (accessed January 15, 2015)

Vietnam Agency for Radiation and Nuclear Safety. Vietnam Agency for Radiation and Nuclear Safety. http://www.varans.vn/ (accessed January 15, 2015)

World Nuclear Association. Uranium in Mongolia. http://www.world-nuclear.org/info/Country-Profiles/Countries-G-N/Mongolia/ (accessed January 15, 2015)

CHAPTER 10

Regulatory Structures and Issues in the European Union

Carmel J. Caruana

Past Chair, Education and Training Committee, European Federation of Organizations for Medical Physics (EFOMP)
EFOMP Lead for Education and Training and Role Development, "European Guidelines on the Medical Physics Expert" Project
Head, Medical Physics Department, University of Malta

CONTENTS

R ADIATION PROTECTION has been one of the issues on the minds of European legislators since the inception of the European Union (EU); indeed, the foundations of the extensive radiation protection legislation and documentation in our possession today can be found in the Euratom treaty of 1957 (EAEC 2012). From that time several directives (and revisions of said directives) concerning radiation protection have been adopted culminating in the revised European Basic Safety Standards (BSS) directive adopted in 2013 (EURATOM 2013). This chapter first describes briefly the more important milestones leading to the revised BSS and its general regulatory structure. This is followed by a description of the role of the Medical Physics Expert (MPE) in the directive, a discussion of the issues of concern to the Medical Physics (MP) profession originating from the revised European BSS, and finally how the more immediate of these concerns have been addressed in the European Commission (EC) funded project, "European Guidelines on the Medical Physics Expert" Project (EU 2014b).

10.1 HISTORY OF EU DIRECTIVES INVOLVING RADIATION PROTECTION IN MEDICINE

The following have been the principal milestones with respect to EU radiation protection legislation:

(a) 1957 The Euratom Treaty

Article 2: "....the Community shall...establish uniform standards to protect the health of workers and the general public" (EAEC 2012).

(b) 1959 First Euratom BSS Directive: covers protection of workers and the general public. There has been a regular revision of the directive, the most recent being Directive 96/29/Euratom (EUROATOM, EUROATOM 1996).

(c) 1984 First Euratom "Medical Exposure" Directive: covers medical exposures, the most recent being 97/43/Euratom (EUROATOM, EUROATOM 1997).

(d) 2013 Revised European BSS (EURATOM 2013): consolidates existing European radiation protection legislation (in particular, 96/29/Euratom and 97/43/Euratom) into a single directive that caters to the protection of patients (including carers, comforters, volunteers in medical or biomedical research and non-medical imaging exposure), workers, and the public. It includes updates necessitated by new scientific findings, ICRP Publication 103 (ICRP 2007), and the International BSS (IAEA 1996).

This chapter will henceforth focus solely on the revised BSS.

10.2 GENERAL REGULATORY STRUCTURE OF THE REVISED EUROPEAN BSS

The general structure of the revised European BSS is as follows:

Preamble

Chapter I: Subject matter and scope
Chapter II: Definitions
Chapter III: System of radiation protection
Chapter IV: Requirements for radiation protection education, training, and information
Chapter V: Justification and regulatory control of practices
Chapter VI: Occupational exposures
Chapter VII: Medical exposures
Chapter VIII: Public exposures
Chapter IX: General responsibilities of member states and competent authorities and other requirements for regulatory control
Chapter X: Final provisions
19 Appendices

Chapter I describes the subject matter ("occupational, medical and public exposures"), scope, and exclusions from the scope. Chapter II consists of a comprehensive list of definitions. The ones most relevant to this chapter are:

(a) "Medical exposure means exposure incurred by patients or asymptomatic individuals as part of their own medical or dental diagnosis or treatment, and intended to benefit their health, as well as exposure incurred by carers and comforters and by volunteers in medical or biomedical research."

(b) "Medical physics expert means an individual or, if provided for in national legislation, a group of individuals, having the knowledge, training and experience to act or give advice on matters relating to radiation physics applied to medical exposure, whose competence in this respect is recognised by the competent authority."

(c) "Non-medical imaging exposure means any deliberate exposure of humans for imaging purposes where the primary intention of the exposure is not to bring a health benefit to the individual being exposed."

(d) "Competent authority means an authority or system of authorities designated by Member States as having legal authority for the purposes of this Directive."

(e) "Diagnostic reference levels means dose levels in medical radiodiagnostic or interventional radiology practices, or, in the case of radiopharmaceuticals, levels of activity, for typical examinations for groups of standard-sized patients or standard phantoms for broadly defined types of equipment."

(f) "Radiation protection expert means an individual or, if provided for in the national legislation, a group of individuals having the knowledge, training and experience needed to give radiation protection advice in order to ensure the effective protection of individuals, and whose competence in this respect is recognised by the competent authority."

Chapter III describes the fundamental principles governing the system of radiation protection adopted: "Member States shall establish legal requirements and an appropriate regime of regulatory control which, for all exposure situations, reflect a system of radiation protection based on the principles of justification, optimisation and dose limitation." Tools for optimisation include dose constraints and reference levels (Articles 6 and 7, respectively). Chapter IV emphasizes the requirements for education and training and continuous professional development: "Member states shall ensure that arrangements are made for the establishment of education, training and retraining to allow recognition of radiation protection experts and medical physics experts..." (Article 14). "Member States shall ensure that practitioners and the individuals involved in the practical aspects of medical radiological procedures have adequate education, information and theoretical and practical training for the purpose of medical radiological practices, as well as relevant competence in radiation protection. For this purpose Member States shall ensure that appropriate curricula are established and shall recognise the corresponding diplomas, certificates or formal qualifications... Member States shall ensure that continuing education and training after qualification is provided and, in the special case of the clinical use of new techniques, training is provided on these techniques and the relevant radiation protection requirements" (Article 18). "Individuals involved in the practical aspects of medical radiological procedures" includes medical physicists and radiographers (p. 3, note 29 of the directive). It is important to note that radiation protection for medical and dental students is also being encouraged: "Member States shall encourage the introduction of a course on radiation protection in the basic curriculum of medical and dental schools" (Article 18).

Chapter V links for the first time medical and occupational/public exposure: "Practices involving medical exposure shall be justified both as a class or type of practice, taking into account medical and, where relevant, associated occupational and public exposures..." This linkage is important particularly owing to the increase in the rate of cataracts in the case of interventional procedures (ICRP 2012). Article 22, entitled "Practices involving the deliberate exposure of humans for non-medical imaging purposes," is relevant to the MP/MPE as such procedures that use medical radiological equipment would require the "appropriate involvement of the medical physics expert." Chapter VI stipulates the regulatory requirements for the protection of workers, apprentices, and students. This chapter is the relevant chapter for the self-protection of the MP/MPE as worker. Chapter VII stipulates the regulatory requirements for medical exposure. It is structured as follows: Article 55 Justification, Article 56 Optimisation, Article 57 Responsibilities, Article 58 Procedures, Article 59 Training and Recognition, Article 60 Equipment, Article 61 Special Practices, Article 62 Special Protection During Pregnancy and Breastfeeding, Article 63 Accidental and Unintended Exposures, Article 64 Estimates of Population Doses. It includes the general role of the MPE (and by association the medical physicist) in medical exposure under Articles 57(1)(b) and 58(d), and recognition of the MPE (Article 59). Chapter VIII stipulates the regulatory requirements for protection of members of the public, while Chapter IX stipulates the institutional regulatory infrastructure including information on equipment (Article 78), details on the recognition of the

MPE (Article 79), role of the RPE (Article 82) and MPE (Article 83), control of radioactive sources, significant events, emergency exposure situations, and system of enforcement. It includes a requirement for the MPE (medical exposures) and RPE (occupational and public exposures) to liaise that each other in situations which would need the expertise of both, e.g., interventional radiology, and nuclear medicine. Chapter X stipulates the requirement for transposition into state law, repeal of previous directives (February 6, 2018) and entry into force of the directive (January 2014).

10.3   THE ROLE OF THE MEDICAL PHYSICS EXPERT IN THE REVISED EUROPEAN BSS

The more pertinent articles from the revised BSS relating to the role of the Medical Physics Expert are:

Article 22: Practices involving the deliberate exposure of humans for non-medical imaging purposes

4(c) for procedures using medical radiological equipment

(i) relevant requirements identified for medical exposure as set out in Chapter VII are applied, including those for equipment, optimization, responsibilities, training, and special protection during pregnancy and the appropriate involvement of the medical physics expert.

Article 57: Responsibilities

1. Member States shall ensure that:

(b) the practitioner, the medical physics expert, and those entitled to carry out practical aspects of medical radiological procedures are involved, as specified by Member States, in the optimization process.

Article 58: Procedures

(d) In medical radiological practices, a medical physics expert is appropriately involved, the level of involvement being commensurate with the radiological risk posed by the practice. In particular:

(i) in radiotherapeutic practices other than standardized therapeutic nuclear medicine practices, a medical physics expert shall be closely involved;

(ii) in standardized therapeutical nuclear medicine practices as well as in radiodiagnostic and Interventional radiology practices, involving high doses as referred to in point (c) Article 61(1), a medical physics expert shall be involved;

(iii) for other medical radiological practices, not covered by (i) and (ii), a medical physics expert shall be involved, as appropriate, for consultation and advice on matters relating to radiation protection concerning medical exposure.

Article 83: Medical physics expert

1. Member States shall require the medical physics expert to act or give specialist advice, as appropriate, on matters relating to radiation physics for implementing the requirements set out in Chapter VII and in point (c) of Article 22(4) of this Directive.

2. Member States shall ensure that depending on the medical radiological practice, the medical physics expert takes responsibility for dosimetry, including physical measurements for evaluation of the dose delivered to the patient and other individuals subject to medical exposure, give advice on medical radiological equipment, and contribute in particular to the following:

(a) optimization of the radiation protection of patients and other individuals subject to medical exposure, including the application and use of diagnostic reference levels;

(b) the definition and performance of quality assurance of the medical radiological equipment;

(c) acceptance testing of medical radiological equipment;

(d) the preparation of technical specifications for medical radiological equipment and installation design;

(e) the surveillance of the medical radiological installations;

(f) the analysis of events involving, or potentially involving, accidental or unintended medical exposures;

(g) the selection of equipment required to perform radiation protection measurements;

(h) the training of practitioners and other staff in relevant aspects of radiation protection.

3. The medical physics expert shall, where appropriate, liaise with the radiation protection expert.

10.4 ISSUES OF CONCERN TO MEDICAL PHYSICISTS ARISING FROM THE REVISED EUROPEAN BSS

The provisions on the Medical Physics Expert in the revised BSS did go a long way into improving the situation for the Medical Physics profession. However, some areas of concern remain and are discussed in this section of the article. The most urgent of these have been addressed in the EU sponsored "European Guidelines on the MPE" project, which is discussed further on in the chapter.

(a) Insufficient guidance regarding the role of the MPE. Although the definition of the role has improved, it is still not sufficiently detailed for day-to-day practice. A more detailed list of key activities and competences is required. This has been addressed in full in the "European Guidelines on the MPE" project.

(b) Insufficient guidance regarding the qualification framework and educational/training curricula for the MPE. Although the revised BSS does stipulate the requirement for education and training for the MPE: "Member States shall ensure that arrangements are made for the establishment of education, training and retraining to allow the recognition of ... medical physics experts..." (Article 14), no specific qualifications, education, and training are mandated. This would hinder the harmonization of competence levels and hence make cross-border mobility difficult. This has also been addressed in full in the "European Guidelines on the MPE" project.

(c) The MPE as "individual" or "group of individuals". The definition of the MPE in the revised BSS refers to the MPE as "...an individual or, if provided for in national legislation, a group of individuals, having the knowledge, training and experience to act or give advice ...." (Article 4). The BSS does not offer specification regarding the "group of individuals." This has also been addressed in full in the "European Guidelines on the MPE" project.

(d) The relationship between the MPE and Radiation Protection Expert (RPE). Articles 82 and 83 do require the MPE (responsible for medical exposures) and RPE (responsible for occupational and public exposures) to liaise with each other in situations that would need the expertise of both. However, the nature of this liaison needs to be elaborated in cases when the effectiveness and safety of radiological procedures with respect to patient service may require higher occupational or public doses, e.g., interventional radiology and nuclear medicine. It is pertinent to note that there is nothing in the directive which precludes that an individual be recognized as both MPE and RPE provided he/she is qualified as both, indeed, this has traditionally been the case and would probably remain the prevalent situation in the case of small healthcare organizations.

(e) The role of the MPE in non-ionizing radiation as an alternative to ionizing radiation in medical imaging. Although Article 55 regarding justification does state that "Medical exposure shall show a sufficient net benefit, weighing the total potential diagnostic or therapeutic benefits it produces, including the direct benefits to health of an individual and the benefits to society, against the individual detriment that the exposure might cause, taking into account the efficacy, benefits and risks of available alternative techniques having the same objective but involving no or less exposure to ionising radiation," there is no further guidance in the revised BSS. This renders the role of the MPE in giving advice regarding the relative effectiveness and safety of the various non-ionizing imaging modalities ambiguous. This has been addressed in the "European Guidelines on the MPE" project.

(f) Insufficient guidance regarding the process and conditions for recognition of the MPE. Although Article 79 does state that "Member States shall ensure that arrangements are in place for the recognition of ... medical physics experts ...shall ensure that the necessary arrangements are in place to ensure the continuity of expertise of these ... experts...shall specify the recognition requirements and communicate them to the Commission...the Commission shall make the information received .... available to the Member States," there is insufficient guidance regarding the process and conditions for recognition of the MPE. Again this would hinder the harmonization of competency levels and make cross-border mobility difficult. This has also been addressed in full in the "European Guidelines on the MPE" project.

(g) No guidance whatsoever regarding the important issue of MPE staffing levels. This has also been addressed in full in the "European Guidelines on the MPE" project.

10.5 THE "EUROPEAN GUIDELINES ON THE MEDICAL PHYSICS EXPERT" PROJECT

The European Commission plays an active role beyond the formulation and overseeing of the adoption of directives. This includes support to Member States through fostering cooperation and providing practical guidelines. In 2009 the European Commission issued a tender with the intent of establishing European guidelines on the MPE (TREN/H4/167-2009 — "European Guidelines on Medical Physics Expert"). The objectives of this tender were to provide for improved implementation of the provisions relating to the MPE within Council Directive 97/43/EURATOM and the then-proposed successor, the revised European BSS. The project has been concluded and the resulting guidelines published as European Union (2014) Radiation Protection 174 European Guidelines on Medical Physics Expert (EU 2014a).

The proposed guidelines go a long way in addressing the more pressing of the issues discussed in the previous section, and include clear statements on the role of the MPE in Europe (based on the revised European BSS and comprising for the first time an agreed mission statement and defined key activities), recommendations on education and training requirements, recognition by the competent authorities, a direction for the clarification of the relationship of the MPE to the profession of Medical Physics, and recommendations for staffing levels.

As part of the project, the European Federation of Organizations for Medical Physics (EFOMP) and the other members of the consortium took the opportunity to produce detailed learning outcomes in terms of knowledge–skill–competence (KSC) inventories (circa 900 learning outcomes in total) for the specialties of Medical Physics relevant to these directives, namely: Diagnostic and Interventional Radiology, Nuclear Medicine, and Radiation Oncology. These inventories are structured within a novel curriculum framework that is founded on the aforementioned mission statement and key activities, thus directly linking curriculum content to professional role. The document hence serves both curriculum planning and promotion of professional function. In addition, the curriculum framework is designed to promote the unity of the profession and the complementarity of its scientific and healthcare professional aspects.

The rest of the chapter summarizes the recommendations of the guidelines regarding the issues as to the role of the MPE and the qualification and curriculum frameworks from the guidelines. Readers interested in other issues are referred to the guidelines document. A second European document of interest is "European Union (2014) Radiation Protection 175: Guidelines on Radiation Protection Education and Training of Medical Professionals in the European Union," which also includes a chapter on the education and training of the MP and MPE (EU 2014b).

10.5.1   Role of the MPE

To make the role more understandable to decision-makers and management of healthcare institutions, and to provide direction for role holders, a mission statement was formulated by the consortium based on the articles of the revised BSS relevant to the MPE. The mission statement is the following:

"Medical Physics Experts will contribute to maintaining and improving the quality, safety and cost-effectiveness of healthcare services through patient-oriented activities requiring expert action, involvement or advice regarding the specification, selection, acceptance testing, commissioning, quality assurance/control and optimised clinical use of medical radiological devices and regarding patient risks from associated ionising radiations including radiation protection, installation design and surveillance, and the prevention of unintended or accidental exposures; all activities will be based on current best evidence or own scientific research when the available evidence is not sufficient. The scope includes risks to volunteers in biomedical research, carers and comforters."

The guidelines also identify and define the key activities of MPEs. These are shown and defined in Table 10.1.

Table 10.1: Definition and elaboration of the key activities of the MPE

Key Activity	Main Actions
Scientific problem solving service.	Comprehensive problem solving service involving recognition of less than optimal performance or optimised use of medical radiological devices, identification and elimination of possible causes or misuse, and confirmation that proposed solutions have restored device performance and use to acceptable status. All activities are to be based on current best scientific evidence or own research when the available evidence is not sufficient.

Continued on next page

Table 10.1 — *Continued from previous page*

Key Activity	Main Actions
Dosimetry measurements.	Measurement and calculations of doses received by patients, volunteers in biomedical research, carers, comforters and persons subjected to non-medical imaging procedures using medical radiological equipment for the purpose of supporting justification and optimisation processes; selection, calibration and maintenance of dosimetry related instrumentation; independent checking of dose related quantities provided by dose reporting devices (including software devices); measurement of dose related quantities required as inputs to dose reporting or estimating devices (including software). Measurements to be based on current recommended techniques and protocols.
Patient safety/risk management (including volunteers in biomedical research, carers, comforters and persons subjected to non-medical imaging exposures).	Surveillance of medical radiological devices and evaluation of clinical protocols to ensure the on-going radiation protection of patients, volunteers in biomedical research, carers, comforters and persons subjected to non-medical imaging exposures from the deleterious effects of ionising radiations in accordance with the latest published evidence or own research when the available evidence is not sufficient. Includes optimisation, the development of risk assessment protocols, including the analysis of events involving, or potentially involving, accidental or unintended medical exposures and dose audit.
Occupational and public safety/risk management when there is an impact on medical exposure or own safety.	Surveillance of medical radiological devices and evaluation of clinical protocols with respect to the radiation protection of workers and public when impacting the exposure of patients, volunteers in biomedical research, carers, comforters and persons subjected to non-medical imaging exposures or responsibility with respect to own safety. Correlation of occupational and medical exposures — balancing occupational risk and patient needs. To this effect, the MPE shall, where appropriate, liaise with the Radiation Protection Expert.

Continued on next page

Table 10.1 — *Continued from previous page*

Key Activity	Main Actions
Clinical medical device management.	Provide technical advice and participate in the specification, selection, acceptance testing, commissioning, installation design and decommissioning of medical radiological devices in accordance with the latest published European or International recommendations. The specification, management and supervision of associated quality assurance/control programmes. Design of all testing protocols is to be based on current European or international recommended techniques and protocols.
Clinical involvement.	Carrying out, participating in and supervising everyday patient radiation protection and quality control procedures to ensure on-going effective and optimised use of medical radiological devices and including patient specific optimisation, prevention of unintended or accidental exposures and patient follow-up. Optimization of protocols before first use with patients via the use of anthropomorphic phantoms and simulation using specialized dosimetry software.
Development of service quality and cost-effectiveness.	Support the introduction of new medical radiological devices into clinical service, lead the introduction of new medical physics services and participate in the introduction/development of clinical protocols/techniques whilst giving due attention to economic issues.
Expert consultancy.	Provision of expert advice to outside clients (e.g., smaller clinics with no in-house medical physics expertise).
Education of healthcare professionals (including medical physics trainees)	Contributing to quality healthcare professional education through knowledge transfer activities concerning the technical-scientific knowledge, skills and competences supporting the clinically-effective, safe, evidence-based and economical use of medical radiological devices. Participation in the education of medical physics students and organisation of medical physics residency programmes.
Health technology assessment (HTA)	Taking responsibility for the physics component of health technology assessments related to medical radiological devices and/or the medical uses of radioactive substances/sources.

Continued on next page

Table 10.1 — *Continued from previous page*

Key Activity	Main Actions
Innovation	Developing new or modifying existing devices (including software) and improved use of protocols for the solution of hitherto unresolved clinical problems.

10.5.2   Qualification Framework for the MPE

The proposed qualification framework is shown in Figure 10.1. All qualification frameworks in Europe should be referred to the European Qualifications Framework (EQF) for lifelong learning (Council of EU, 2008). In the EQF, qualifications are set at 8 levels with level 8 being the highest expert level. Learning outcomes are expressed as inventories of knowledge, skills, and competences (KSC). Owing to the rapid expansion of medical device technology and research results, it is becoming increasingly difficult for an MPE to be competent in more than one specialty of medical physics covered by the revised European BSS (i.e., Diagnostic and Interventional Radiology, Nuclear Medicine, and Radiation Oncology); therefore, the MPE should be independently recognized in each specialty of medical physics.

The KSC for the recognition of MPE status by the competent authorities are to be gained initially through learning in an institution of higher education and structured clinical training in a residency within an accredited healthcare institution, and subsequently developed further through structured advanced experience and continuous professional development (CPD). The MPE is defined as a Medical Physicist who has achieved the highest EQF level possible (EQF level 8) in his specialty of medical physics. In view of the near impossibility of achieving EQF level 8 in all three specialties of medical physics, the "group of individuals" in the definition of the MPE is a group made up of MPEs in the three specialties of medical physics and possibly the subspecialties of each (e.g., brachytherapy, radionuclide therapy, external beam therapy, and proton therapy in the case of radiation oncology).

Explanatory notes to the qualification framework diagram plus associated rationales are shown in Table 10.2.

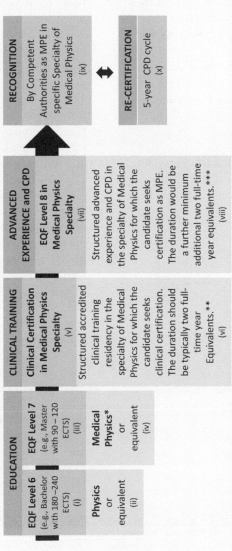

Figure 10.1 The Qualification Framework for the MPE in Europe

Table 10.2: Notes to the Qualification Framework diagram

ID	Note	Rationale
(i)	The fundamental educational level for medical physics professionals is a level 6 in physics and associated mathematics.	Medical physics professionals need to have good foundations in physics and mathematics as Medical Physics is a physical, numeric and exact science.
(ii)	"Equivalent" here meaning EQF level 6 with a high level of physics and mathematics content.	This will make it possible for graduates from other level 6 programmes which include a high level of physics and mathematics (e.g., engineering, biophysics) to enter the field.
(iii)	The educational entry level for the medical physics professional has been set at EQF level 7.	At entry level the medical physics professional needs to have highly specialised knowledge, critical awareness of knowledge issues in the field, specialised problem-solving skills, ability to manage work contexts that are complex and ability to review the performance of teams (EU 2008). Medical physics professionals require highly specialised knowledge in radiation protection and the medical devices covered by the revised BSS and specialised problem-solving and troubleshooting skills. The medical physics professional is involved in clinical contexts that may be very complex and reviews the performance of radiation protection and quality control teams in own specialty of medical physics.

Continued on next page

Table 10.2 — *Continued from previous page*

ID	Note	Rationale
(iv)	"Equivalent" here meaning EQF level 7 with a high level of physics and mathematics content plus the educational component of the core KSC of medical physics and the educational component of the KSC specific to the specialty of medical physics for which the candidate would be seeking clinical certification (as specified in this document). This additional education can be concurrent with the training.	This will make it possible for candidates with Masters in physics, biophysics, engineering, etc., to enter the field; however, such candidates need to undertake an additional educational programme which includes the educational component of the core KSC of medical physics and the educational component of the KSC specific to the specialty of medical physics for which the candidate would be seeking clinical certification.
(v)	The medical physics professional at entry level is a professional with clinical certification in medical physics, i.e., having a level of education in medical physics at a level intermediate between EQF levels 7 and 8, having typically 2 years full-time equivalent accredited clinical training and recognized as competent to act independently through enrollment in a national register for Medical Physics professionals.	The education and training to clinical certification in medical physics is a necessary foundation for further development to MPE EQF level 8.
(vi)	Structured accredited residency based training for clinically based development of the core KSC of medical physics and the KSC specific to the specialty of medical physics for which the candidate would be seeking clinical certification. The duration of this structured training is typically two full-time year equivalents.	The IAEA recommends that clinical certification would need a training period of two full-time year equivalents for any one specialty of medical physics (IAEA 2009, 2010, 2011)

Continued on next page

Table 10.2 — *Continued from previous page*

ID	Note	Rationale
(vii)	The MPE in a given specialty of medical physics is a professional with clinical certification in a specialty of medical physics who has achieved the highest level of expertise in that particular specialty. The medical physics professional through structured advanced experience, ongoing extensive CPD and commitment places the KSC at the highest possible level i.e., EQF level 8.	The qualification level for the MPE has been set at EQF level 8 because the MPE requires knowledge at the most advanced frontier of a field of work and at the interface between fields, the most advanced and specialised skills and techniques, including synthesis and evaluation, required to solve critical problems in research/innovation and to extend/redefine existing professional practice, demonstrate substantial authority, innovation, autonomy, professional integrity and sustained commitment to the development of new ideas or processes at the forefront of work contexts including research (EU 2008). To carry out activities requiring expert action, involvement or advice with authority and autonomy and which are based on current best evidence (or own scientific research when the available evidence is not sufficient), the MPE requires frontier knowledge in own specialty of medical physics and at the interface between physics and medicine. The MPE requires specialised skills and techniques in radiation protection and comprehensive experience regarding the effective and safe use of the medical devices in own specialty, and the synthesis and evaluation skills required to solve critical problems in service development, research, innovation and the extension and redefinition of existing professional practice.

Continued on next page

Table 10.2 — *Continued from previous page*

ID	Note	Rationale
(viii)	This will mean that to reach MPE status (level 8) in the specialty area requires a minimum total of 4 years equivalent clinical training (2 years equivalent of foundation training in the specialty area to clinical certification and a further 2 years equivalent of advanced, structured experience and CPD in the specialty).	It should be emphasised that the further 2 years to reach MPE status must consist of advanced, structured experience and CPD and not simply CPD designed to maintain competence. The two years minimum of advanced experience must be measured from the time when the advanced experience commences. The advanced experience and CPD might not follow immediately the 2 years of basic training if the candidate is not deemed to be sufficiently prepared. It is to be understood that senior MPEs practicing in large medical centres with a full range of devices would need more years of advanced experience than the 2 years minimum. On the other hand small facilities can be serviced by novice MPEs working under the guidance of a senior MPE.
(ix)	A person who is currently recognised as an MPE and is in possession of the core KSC of medical physics and the KSC specific to the specialty for which recognition is sought should be deemed to satisfy the requirements for recognition as an MPE if they are currently on active duty as an MPE and are deemed to have reached level 8.	This is a grandparenting clause.

Continued on next page

Table 10.2 — *Continued from previous page*

ID	Note	Rationale
(x)	This is the requirement for an MPE to maintain recognition.	A 5 year cycle for re-certification (i.e., recognition by the Competent Authorities as having maintained a level 8 in the particular specialty of Medical Physics) is recommended.

10.6   CURRICULUM FRAMEWORK FOR MPE PROGRAMS IN EUROPE

The curriculum framework consists of a structured inventory of KSC underpinning the mission and key activities of the MPE. The proposed curriculum framework is intended to be comprehensive yet concise. It is designed to make the commonalities between the various specialties of medical physics apparent and emphasize common terminology — hence facilitating collaboration between MPEs from the different specialties (e.g., in hybrid imaging, radiotherapy planning). The KSC are classified in two categories: generic skills and subject-specific KSC as required by the EC Tuning Project (Caruana 2011). Generic skills consist of transferable skills that are expected of all professionals at a particular level of the EQF. In this case the relevant levels are level 7 and level 8. Subject-specific KSC are specific to a profession and are further divided into sub-categories as determined by the particular profession. The following classification is based on proposals by EFOMP and Caruana (Caruana 2011, Christofides et al. 2009):

(a) Medical physics core KSC: these KSC are expected of all MPEs irrespective of their specialty:

i. KSC for the MPE as physical scientist: these are fundamental physics KSC expected of all physical scientists;

ii. KSC for the MPE as healthcare professional: these are KSC expected of all healthcare professionals;

iii. KSC for the MPE as expert on the clinical use of medical radiological devices and protection from associated ionizing radiations (and other physical agents as appropriate): these represent medical device and safety KSC common to all specialties of medical physics.

(b) Medical physics specialties KSC: these KSC are highly specific to each specialty of medical physics (i.e., diagnostic and interventional radiology or nuclear medicine or radiation oncology/radiotherapy) and therefore cannot be included in the core.

It is important to note that an MPE from one specialty of medical physics who is required to assume specific responsibilities from another specialty may be certified to carry out those specific responsibilities following the acquisition of the corresponding KSC. Such cases may arise, for example, in a small nuclear medicine facility that requires its nuclear medicine MPE to take responsibility for the management of quality control testing of the CT component of a PET/CT, system or at a small radiation oncology/radiotherapy facility that requires its radiation oncology/radiotherapy MPE to take responsibility for protocol optimization of a given imaging modality. The full curriculum framework can be found in the guidelines document, while the core KSC inventory and three specialty KSC inventories are given in Annex 1 of the same document. A candidate seeking recognition by the competent authorities as an MPE in a given specialty of medical physics should reach level 8 in the core KSC and the KSC specific to that particular specialty. In 2014 these recommendations were subsequently adopted as EFOMP Policy Statement 12.1 (Caruana et al. 2014).

10.7 CONCLUSION – FUTURE DIRECTIONS FOR THE MEDICAL PHYSICIST AND MEDICAL PHYSICS EXPERT IN THE EUROPEAN UNION IN THE IONIZING RADIATION AREA

The revised European BSS and "European Guidelines on the MPE" have provided a much-needed roadmap for the MP/MPE in the European Union. Educational and training programs up to the clinically qualified Medical Physicist level are now well established with high-level Masters courses and training programs throughout the EU. Yet the bridging of the gap between the clinically certified Medical Physicist and Medical Physics Expert levels is still mostly uncharted territory. The EFOMP is already working on the issue. We have received financial help from the EC to develop a program of courses to bridge the gap in the case of Diagnostic and Interventional Radiology (EUTEMPE-RX project, www.eutempe-rx.eu), and we have set up an EFOMP School for MPEs.

10.8 REFERENCES

Caruana CJ. 2011. The Tuning process and the Masters in Medical Physics in Europe - a personal vision. In: Tabakov S (Ed.), Medical Physics and Engineering - Education and Training. Italy: ICTP Press. ISBN 92-95003.

Caruana CJ, Christofides S, Hartmann GH. 2014. EFOMP Policy Statement 12.1: Recommendations on Medical Physics Education and Training in Europe 2014. Phys Med. 2014 Sep;30(6):598–603

Christofides S. et al. (2009). An Initial EFOMP Position on the Tuning Process for Masters Programs in Medical Physics in Europe. In DÃüssel O and Schlegel WC (Eds.): WC 2009, IFMBE Proceedings 25/XIII, 70–73.

Consolidated version of the Treaty establishing the European Atomic Energy Community. 2012. Official Journal C 327 Vol 55.

Council Directive. 2013. 2013/59/EURATOM of December, 2013 laying down basic safety standards for protection against the dangers arising from exposure to ionising radiation, and repealing Directives 89/618/Euratom, 90/641/Euratom, 96/29/Euratom, 97/43/Euratom and 2003/122/Euratom. Official Journal L 13 17/01/2014 pp. 0001–0073. http://eur-lex.europa.eu/legal-content/EN/TXT/HTML/?uri=OJ:L:2014:013:FULL (Accessed December 20, 2014).

Council Directive. 1996. 96/29/Euratom of May 13, 1996 laying down basic safety standards for the protection of the health of workers and the general public against the dangers arising from ionizing radiation. Official Journal L 159, 29/06/1996 pp. 0001–0114.

Council Directive. 1997. 97/43/Euratom of June 30, 1997 on health protection of individuals against the dangers of ionizing radiation in relation to medical exposure, and repealing Directive 84/466/Euratom. Official Journal L 180, 09/07/1997 pp. 0022–0027.

Council of EU. 2008. Recommendation 2008/C 111/01 on the establishment of the European Qualifications Framework for Lifelong Learning. Official Journal of the European Union 6.5.2008. http://eur-lex.europa.eu/legal-content/EN/ALL/?uri=CELEX:32008H0506(01) (Accessed December 20, 2014).

European Union. 2014a. Radiation Protection 174 European Guidelines on Medical Physics Expert http://ec.europa.eu/energy/nuclear/radiation_protection/publications_en.htm (Accessed December 20, 2014).

European Union. 2014b. Radiation Protection 175: Guidelines on Radiation Protection Education and Training of Medical Professionals in the European Union. http://ec.europa.eu/energy/nuclear/radiation_protection/publications_en.htm (Accessed December 20, 2014).

IAEA. 1996. International Basic Safety Standards for Protection Against Ionizing Radiation and for the Safety of Radiation Sources. IAEA Safety Series No. 115.

IAEA. 2009. Clinical Training of Medical Physicists Specializing in Radiation Oncology. Training Course Series, 37, International Atomic Energy Agency. http://www-pub.iaea.org/MTCD/publications/PDF/TCS-37_web.pdf (Accessed December 20, 2014).

IAEA. 2010. Clinical Training of Medical Physicists Specializing in Diagnostic Radiology. Training Course Series, 47, International Atomic Energy Agency. http://www-pub.iaea.org/MTCD/publications/PDF/TCS-47_web.pdf (Accessed December 20, 2014).

IAEA. 2011. Clinical Training of Medical Physicists Specializing in Nuclear Medicine. Training Course Series, 50, International Atomic Energy Agency. http://www-pub.iaea.org/MTCD/publications/PDF/TCS-50_web.pdf (Accessed December 20, 2014).

ICRP. 2007. The 2007 Recommendations of the International Commission on Radiological Protection. ICRP Publication 103. Ann. ICRP 37 (2–4).

ICRP. 2012. Publication 118, ICRP Statement on Tissue Reactions/Early and Late Effects of Radiation in Normal Tissues and Organs - Threshold Doses for Tissue Reactions in a Radiation Protection Context. Ann. ICRP 41(1/2).

CHAPTER 11

Regulatory Structure and Issues in the Middle East

Ibrahim Duhaini
Rafik Hariri University Hospital, Lebanon (Coordinator)

Huda Al-Naemi
Hamad Medical Corporation, Qatar

Shada Wadi-Ramahi
King Faisal Specialist Hospital and Research Center (KSA), Jordan

Nabaa Naji
Mustansiriya Medical College, Iraq

Hanan Awada Al-Dousari
Ministry of Health, Kuwait

Riad Shweikani
Atomic Energy Commission, Syria

Ali Al-Remeithi
Federal Authority for Nuclear Regulation, United Arab Emirates

CONTENTS

T HE USE OF RADIATION IN MEDICINE has been on the rise in many countries. Many diagnostic and therapeutic procedures may expose patients and staff to high radiation doses, which can be reduced to low levels to ensure safety and protection against the harmful effects of radiation exposures. The objective of this chapter is to highlight protocols pursued by some countries in the Middle East (Table 11.1) to ensure the following:

1. Implementing an effective radiation safety strategy.

2. Enforcing radiation safety practice for patients, staff, physicians, and visitors.

3. Providing regular radiation safety education to concerned staff.

4. Identifying opportunities to improve radiation safety performance.

5. Recognizing a regulatory framework for establishing, licensing, inspecting, and monitoring radiation facilities.

By adhering to the principles and doctrines of radiation safety set forth by some international organizations, the safety culture among radiation workers will be enhanced, and the productivity as well as performance of the protocols will be optimized.

Table 11.1: Homepages for regulatory programs and advisory and professional organizations discussed in this chapter

Country	Regulatory Body	Homepage
Lebanon	Lebanese Atomic Energy Commission (LAEC)	http://www.laec-cnrs.gov.lb/
United Arab Emirates	Federal Authority for Nuclear Regulation (FANR)	http://www.fanr.gov.ae/En
Jordan	Jordan Nuclear Regulatory Commission	http://www.jnrc.gov.jo/
Kingdom of Saudi Arabia	Saudi Arabian Atomic Regulatory Authority	http://www.kacare.gov.sa/en/
Iraq	Iraqi Radioactive Source Regulation Authority	http://irsra.gov.iq/
Kuwait	Radiation Protection Division (RPD) at the Ministry of Health	http://www.moh.gov.kw/
Qatar	Qatar Nuclear Regulatory Division at the Ministry of Environment	http://www.moe.gov.qa/English
Syria	Atomic Energy Commission of Syria (AECS)	http://www.aec.org.sy/

11.1 REGULATORY STRUCTURE AND RADIATION PROTECTION ISSUES IN THE FOLLOWING MIDDLE EASTERN COUNTRIES

11.1.1 Iraq

The Iraqi National Center for Radiation Protection controls and monitors radiation protection and safety of workers in the medical field and other fields where radiation sources have been used for a long time. This institution is very well equipped with modern devices for detecting and monitoring the exposure rate of the subjects and determining the area of radiation pollution. It is responsible for providing the monitoring tools for the workers in radiation fields to assess the amount of exposure periodically. It is also responsible for granting the licenses for use of radiation sources in governmental and private clinics. This task covers all the Iraqi provinces. This institution evaluates the amount of environmental pollution due to the improper disposal of radiation wastes. The media department of this center conducts awareness campaigns to educate the general public about the radiation sources, and their effects, benefits, and hazards.

This center works on the establishment of "The Early Radiation Alarm System" in all the Iraqi governorates, which will be used for early radiation detection and warning in the case of radiation accidents. This center is working in coordination with the Iraqi Ministry of Health and the Ministry of Environment. Besides the Iraqi national center of radiation protection, the Iraqi institution for controlling radiation resources works on evaluating the needs of teaching institutions, such as scientific colleges in the Iraqi universities, for using radiation sources for teaching purposes, i.e., scientific labs and for doing research. The tasks of this center include development of all the protection procedures in the scientific teaching institution before starting work with radiation-emitting sources (MHI 2015).

11.1.2   Jordan

11.1.2.1   Introduction

Jordan is located in the Middle East region on the northern part of the Arab Peninsula. Its population, according to the Department of Statistics, is over 6.5 million, as of November 30, 2014 (MHJ 2014). However, the actual number might vary and fluctuate due to the fluctuating number of expatriates and refugees that arrive in the country.

Radiation in Jordan is used in various sectors, such as industry, agriculture, research, and medicine. Almost every major public and private hospital has a radiology and nuclear medicine department, and many have stand-alone clinics that offer radiology or nuclear medicine services. Concerning cancer treatment, there are currently four radiation treatment centers in the country as of the end of 2014. One center is exclusively a Gamma-Knife® facility and one center is a private not-for-profit comprehensive cancer center, while the remaining two are general hospitals with radiation therapy departments.

The regulatory structure in the country is ever changing and evolving. In 2001, the Jordan Atomic Energy Commission (JAEC) was created and it took the role of regulatory and supervisory agency for the various applications of radiation and the radiation workers in the country. In addition, it also provides calibration services for survey meters and a TLD reading facility, and offers various training courses.

In 2007, the Jordan Nuclear Regulatory Commission (JNRC) was created and became the authority for regulation, licensure, and inspection covering all aspects of radiation application in the Kingdom.

"The main goals of the JNRC are to work, in coordination with relevant bodies, on achieving the following:

- Regulating and monitoring the use of nuclear energy and ionizing radiation.

- Protecting environment and human health and property from the hazards of radiation and related pollution.

- Ensuring the availability of conditions and requirements of general safety, radiation protection, and nuclear safety and security." (JNRC 2014).

In 2014, and under the re-organization and the restructuring of governmental institutions and authorities, the status of JNRC as an independent commission directly linked to the office of the Prime Minister changed as it became integrated under the Energy and Mineral Regulatory Commission (EMRC) (EMRC 2014).

11.1.2.2 Current Practice

JNRC has the following duties in terms of radiation usage in the Kingdom:
  a. Personal and site licensure.
  b. Granting approval for radiation protection officers.
  c. Granting approval for use of a radiation device.
  d. Granting approval for import/export of radiation materials.
  e. Border inspection (for radiation material entry/exit).
  f. Facility inspection.

A Site License is issued after inspecting the facility that will accommodate the radiation device. Proper shielding as well as appropriate radiation safety measures have to be provided before the license is issued. The JNRC conducts annual inspection of all sites to assure their continued compliance to radiation safety regulations throughout the lifetime of the device.

On the other hand, a Personal License is issued to all radiation workers after they have successfully completed a week-long general radiation safety course. The license is the official permit that allows the worker to work with and handle radiation. If the licensee is in good standing, the license is renewed every three years. Annual inspections to the facilities also include reviewing licensees' status and competence.

In recent years, the JNRC has put together task force groups to provide national guidelines for and standardize quality assurance for radiation oncology and to provide guidelines for incident-reporting systems.

11.1.2.3 JNRC Community Outreach

In any country, there are two major factors that might be causes for hindering major projects or proper implementation of radiation: Lack of proper qualifications of staff and general public fear of radiation. Recent years have seen extensive community outreach by top officials at the JNRC to provide the public, through multiple media (TV shows, news interviews, newspapers), with information on the safe use and implementation of radiation; in particular, answering questions on the pros and cons of Nuclear Energy versus renewable sources of energy.

The JNRC has also reached out to national professional societies, such as the medical physics association, and to public and private hospitals, to co-sponsor workshops with these stakeholders, aimed at increasing the level of competency of radiation workers.

11.1.3 Regional and International Agreements and Cooperation

Understanding well that radiation safety and protection is not a one-country responsibility, Jordan has established close ties with international and regional bodies. Cooperation on various levels exists between JAEC and JNRC on one side, and the International Atomic Energy Agency (IAEA) on the other, for the purpose of improving the performance and training of staff in the regulatory, bodies as well as building proper human capacity capable of safely handling radiation sources and generators. IAEA, through certain funds established with the Kingdom, provides 1-to-3-month fellowships for radiation workers in other countries for training and education.

In addition, Jordan also co-signed and is a member of the ARASIA group. The Arab States in Asia, a.k.a. ARASIA, is a "Co-operative Agreement for Arab States in Asia for Research, Development and Training Related to Nuclear Science and Technology" (ARASIA 2015). Ever since its inception, the ARASIA agreement has contributed to numerous agricultural, marine, and medical projects that utilize radiation isotopes or radiation generators. The projects initiated by ARASIA fall under two categories, research-oriented, and training and education. The latter projects mainly focus on training of medical radiation workers, not only on the safe and proper handling of radioisotopes and radiation generators, but also on specific clinical issues very relevant to patient care in nuclear medicine and radiation safety. More often than not, people forget that radiation protection and its regulation is not only for the workers but more so for the patients who represent the receiving end of radiation application in medicine.

11.1.4 Kuwait

Diagnostic Services have been evolving rapidly in the state of Kuwait. In particular, Kuwait has one of the most advanced clinical Nuclear Medicine Services in the Gulf region, with nine Nuclear Medicine Departments providing a wide range of services, including PET/CT and PET Radiopharmaceutical Production Centers. In 1964, the first rectilinear scanner was introduced in Kuwait. In the 1970s, the use of radioiodine 131I therapy was established for thyroid abnormalities. During the 1980s, the single-headed cameras were brought to the new departments. In the 1990s dual- and triple-headed cameras were introduced to some nuclear medicine departments in the country as well as the Strontium 89 (^{89}Sr) bone palliation therapy. In the new millennium, hybrid imaging systems and more radiotherapy procedures such as ^{90}Y Zevalin therapy, Radiosynovectomy, and Therasphere therapy all were introduced and practiced along with the in-house production of positron emitters from the cyclotron. In 2013, more than 35,000 patients underwent a nuclear medicine procedure. The Radiation Protection Department (RPD) was established by an Executive Decree of the Minister of Public Health in 1977. The main role of the RPD is protection against ionizing radiation and non-ionizing radiation and the issuance of licenses. The RPD has the following functions (MHK 2009):

1. To consider matters relating to radiation protection and to formulate the policy for its implementation.

2. To approve the licenses stipulated under this law and to cancel, modify, and suspend those licenses.

3. To draft the regulations and decrees referred to the law.

4. To make recommendations and proposals concerning legislation relating to radiation protection.

The Competent Authority is responsible for matters relating to the licensing, control, and inspection of radiation devices, radioactive substances, the premises where they are located, and of persons using them, in accordance with the provision of the law.

Anyone licensed to use or keep radiation devices or radioactive substances must notify the Competent Authority (MHK 2003):

1. In the case of loss of any radioactive substance or radiation device, within 24 hours of the loss;

2. In the case of any accident that may result in the exposure of any person to an ionizing radiation dose above the permissible limit stipulated in the requirements to be laid down by an executive decree of the Minister of Public Health, within 24 hours of the accident, with a detailed report on the accident and its causes.

The Competent Authority shall co-operate with the authorities concerned in taking the necessary measures to prevent the hazards of accidents and emergency situations that might lead to radiation exposure, and to avoid their repetition. The licenses shall implement the measures decided upon by the Competent Authority in this matter.

The Radiation Protection Commission may cancel the license referred to in Sections 2 and 3 of this law in the following cases (MHK 2005):

1. If it is found that the licensee submitted incorrect statements or resorted to illegal means as a result of which the license was issued.

2. If the licensee has violated the conditions or requirements stipulated in this law or in executive decrees issued under it.

3. If the licensee has violated the conditions stipulated in the license.

4. If the licensee dies or is affected by a disease rendering him incapable of work with ionizing radiation.

5. If it is found that there are exposure hazards for the licensee or his employees or third parties.

6. If public interest so demands.

The Commission may with immediate effect suspend a license for a period specified by it. It may also grant the licensee time to comply with the stipulated conditions and requirements or to take the appropriate measures before cancelling his license.

The Commission's decision about cancellation or suspension of a license shall be implemented by administrative action and the licensee may appeal to the Minister of Health against the Commission's decision on cancellation or suspension within a month of being notified thereof. The Minister shall give his decision on the appeal after obtaining the opinion of the Radiation Protection Commission, and his decision in this matter shall be final.

No person under the age of 18 years shall be employed in any work in which he may be exposed to ionizing radiation.

The licensee should establish a policy that encourages and provides a continuing professional development program, with the aim to improve staff skills, maintain familiarity with current practices, and foster a safety culture throughout the institution. Such training and development schemes can be accomplished through informal meetings of the nuclear medicine department, seminars, accredited continuing education programs, or other means.

11.1.5 Lebanon

11.1.5.1 Introduction

The first Legislative Decree No. 105/83 that deals with "Regulating the Use of and Protection against Ionizing Radiations" was established in 1983 by the Lebanese Government (LPAD 1983). Upon this law the following Regulatory Decisions have been inaugurated by the Council of Ministers in the Lebanese Government (LGN 2006):

1. Decision No. 28/1996: Granting the Lebanese Atomic Energy Commission established within the National Council for Scientific Research (CNRS) the authority to carry out a comprehensive survey and fulfill radiation control tasks.

2. Decision No. 30/1997: Affiliation of Lebanon with the international project of supporting the infrastructure of radiation protection, and delegation to the CNRS to prepare the draft text laws for workers in the radiation protection field.

3. Decision No. 14/1999: Granting the Lebanese Atomic Energy Commis-

sion (LAEC) the authority to control the ionizing radiation dose in imported scrap.

4. Decision No. 705/1/2005: Regulating applications for import of radiation devices and radioactive materials, and licensing of facilities dealing with such products.

5. Decision No. 55/2006: Acceptance of the amendment introduced by the Council of Governors — IAEA to the Small Quantities Protocol — SQP.

Thus, in 1996, the National Council for Scientific Research or Centre National de la Recherche Scientifique (CNRS) headed by Dr. Mouin Hamze, instituted LAEC with the full support of the IAEA for preparing the National Legal and Technical infrastructures allowing an effective implementation of a comprehensive Radiation Safety Scheme in the country.

11.1.5.2 International Agreements

Lebanon has signed and ratified the following multilateral and international legally binding treaties in the field of Safeguards, Safety, Security, and Liability (CNRS 2012):

A. Safeguard:

1. Treaty on the Non-Proliferation of Nuclear Weapons (NPT) (Signed 1970).

2. Agreement between Lebanon and the IAEA for the Application of Safeguards in Connection with the Treaty on the Non-Proliferation of Nuclear Weapons SQP (Signed in 1973 and its Amendments in 2007).

B. Safety

1. Convention of Nuclear Safety (Ratified 1996).

2. Convention on Assistance in the Case of Nuclear Accident or Radiological Emergency (Ratified 1996).

3. Convention on Early Notification of a Nuclear Accident (Ratified 1996).

4. Joint Convention on the Safety of Spent Fuel Management and on the Safety of Radioactive Waste Management (Signed 1997).

C. Security

1. Convention of Physical Protection of Nuclear Materials 1998.

2. Code of Conduct on Safety and Security of Radioactive Sources (Committed 2004) and its Guidance on the import and export of radioactive sources (Committed 2007).

D. Liability

1. Vienna Convention on Civil Liability for Nuclear Damage (Ratified 1963 and its Amendments in 1997).

2. Paris Convention on Third Party Liability in the Field of Nuclear Energy (Ratified 1968).

11.1.5.3 Regulatory Structure

The LAEC is closely cooperating with the IAEA for harmonizing its legal and regulatory framework in accordance with its international obligations and in compliance with the latest IAEA Basic Safety Standards and Guides. Also, cooperation with the IAEA covers technical aspects that aim to substantially enhance radiological safety and nuclear security in several key areas.

The large progress of the use of radiation sources in industries, research centers, and particularly in the medical sector, as well as the continuous national need for testing and certification laboratories, pressed the LAEC, in the last decade, to multiply its effort by increasing the number of its technical staff, to make important change in its organizational chart, and to develop its cooperation with the IAEA for extending its end-user-oriented laboratories, and enhance its skills in regulatory issues and radiation safety services.

According to its General Director, Dr. Bilal Nsouli, the LAEC now is a key institution in the national regulatory system and is playing a central role for fulfilling the technical aspects of the international obligations and commitments of Lebanon pertaining to safety, security, and safeguard. It has become a regional center of excellence in the use and development of analytical methods for applications in various key areas that have direct impact on the socioeconomic development of the country (LAEC 2006).

The two departments that are related to the medical fields in the structural organization of the LAEC are the following.

11.1.5.4 Department of Authorization and Inspection of Ionizing Radiation (DAIIR)

The LAEC issues certificates for authorizing the activities involving ionizing radiation (X-ray, gamma rays, alpha rays, beta rays, and neutrons). The activity may be, but is not restricted to, operation, import, export, transport, storage, disposal, amendment, interruption, and decommissioning. Any person or entity intending to carry out any activities specified above must notify the LAEC and apply the necessary application form relevant to the practice as a first step in the authorization process. The required information in the application forms depends on the potential exposure risk present in the practice, and it may vary from simple to more detailed information, which should cover the following items:

1. General information on the institution
2. Information on radiation worker personnel
3. Information on radioactive sources used
4. Information on the location and safety assessment
5. Protective tools used
6. Radiation protection program of the institution
7. Radiation emergency program of the institution
8. Waste and transport preparedness

9. An official undertaking should be signed by the legal counsel of the institution

The Department of Authorization and Inspection of Ionizing Radiation (DAIIR) adopts, during the scientific assessment of the application, the Basic Safety Standards and the Safety Series of the IAEA, in addition to the ICRP publications.

The validity period of the Certificate for Authorization (CFA) depends on the categorization of sources and the significant potential exposure of the practice: 2 years for nuclear medicine and radiotherapy, 5 years for dental radiology, 3 years for diagnostic radiology, and 2 to 5 years for industry.

A.1. Inspection

According to Article 3 of the Applicatory Decree 15512, the LAEC conduct different types of inspections:

1. Commissioning Inspection:

Definition: The commissioning inspection is the inspection conducted prior to starting the practice and prior to issuing the CFA.

Objective:

1. To verify the conditions set in the application for authorization.

2. To verify the compliance with the standards adopted.

3. Periodic Inspection:

Definition: The periodic inspection is the inspection conducted regularly after issuing the CFA.

Objectives:

1. To verify on a timely basis that the operator is managing safety in a proper manner.

2. To verify that relevant documents and instructions are valid and that the authorization conditions are being complied with.

Types:

1. Announced

2. Unannounced

3. Additional Inspections:

Carried out in abnormal events:

1. Safety-related incidences or accidents.

2. Significant changes in the practice.

Inspectors enter the place where the practice takes place; make measurements when needed; require, review, and receive information and documents relevant to the conduction of an inspection.

Periodicity: The frequency of the inspections differs according to the significant potential exposure:

3 inspections/year for nuclear medicine

2 inspections/year for radiotherapy

1 inspection/year for diagnostic radiology, blood irradiator, and industry

2 inspections/5 years for dental radiology

The operator is responsible for submitting the CFA issued from the LAEC to the Ministry of Public Health (MHL 2005). Based on this certificate, the

MPH issues the license signed by the Minister of Health according to Article 1 of the Applicatory Decree 15512.

A.2. Enforcement

The LAEC may take enforcement action for:

- Noncompliance discovered during inspection

- Noncompliance in transfer and trade

- Transfer to unauthorized person

- No notification of radioactive source transfer

- Noncompliance with the application decree requirements

The enforcement action depends on encountered risk to health and safety. It can vary gradually in terms of severity:

- Formal instructions to correct the infraction in a specific period of time.

- Ban on import, export, sale, and transfer.

- Reporting the violation to the Minister of Health when the operation is not safe, so that he may take the necessary actions according to Decree Law 105/83, which includes license withdrawal.

A.3. Regulations for Practices

The LAEC is responsible for issuing and implementing regulations for the safe use of ionizing radiation according to Article 5 of the Applicatory Decree 15512. The LAEC is responsible for regulating the following:

- The use of radiation sources to provide an appropriate standard of protection and safety for humans and environment without unduly limiting the benefits of using ionizing radiation in various practices

- The introduction and the conduct of any practice involving the use of radiation sources.

The objective of these regulations is to protect the health and safety of patients, workers, the public, and the environment, and to ensure the security of radioactive sources. Currently, the Lebanese regulations for radiation protection are present in a draft form, and not formally issued yet. These regulations are based on the IAEA Basic Safety Standards (BSS-115) (IAEA 2011) and ICRP Publication 103 (ICRP 2007). Moreover, these regulations specify the principles, requirements, and associated criteria for safety upon which its regulatory judgments, decisions, and actions will be based. Also, it covers all Lebanese practices dealing with ionizing radiation, such as Industry, Medical (nuclear medicine, radiology, and radiotherapy), Research, and Transport.

11.1.5.5 Department of Radiation Safety Support

B.1. Calibration of Radiation Measurement Instruments

The Secondary Standard Dosimetry Laboratory (SSDL) is the national laboratory in the field of calibration of ionizing radiation detectors. It was established in 2007 and it provides calibrations in terms of air kerma, personnel equivalent dose Hp(10), and ambient dose equivalent H*(10) in the field of radiation protection. Radiation survey detectors such as survey meters and dosimeters are usually used to assess any possible leak of radiation from relevant equipment or through different kinds of shielding. These detectors need to be calibrated systematically once per year in order to ensure the accuracy of their results within acceptable levels of uncertainty. The SSDL is equipped with an X-ray system with a 250KV X-ray tube and a ^{137}Cs irradiator, and standard dosimeters. The reference standards dosimeters are traceable to the IAEA Primary Standard Dosimeter Laboratory.

B.2. Individual Dose Management

The IMS (Individual Monitoring System) laboratory is monitoring 3500 workers dealing with regulated areas. The laboratory uses TLD (thermoluminescence dosimeter) for both whole body assessment and area monitoring purposes. The IMS laboratory's performance has been demonstrated through its participation in many proficiency tests, with highly satisfactory results especially at high doses. In addition, the development of quality control and quality assurance systems will lead the laboratory to generate high reliability results and make it well prepared for accreditation according to ISO 17025.

B.3. Quality Control of X-Ray Equipment

Quality control of X-ray equipment is an important step in the Quality Assurance Program. Later, this is a prerequisite for any license request dealing with practices involving X-ray equipment. The department covers the QC of such equipment, and delivers official certificates in this regard that are used for the license request as well as for the accreditation demand. The QC is performed on a yearly basis. The number of QC certificates delivered for X-ray facilities all around the country is 500 per year.

B.4. Workplace Monitoring

Workplace monitoring achieves and maintains an acceptable protection of the working environment from radiation hazards. Monitoring includes measurements and radiological assessment of the obtained results. A workplace monitoring program is implemented by the LAEC to ensure radiation safety for all relevant practices and in different kinds of facilities. The workplace monitoring is also a prerequisite safety measure for licensing and accreditation processes of any practice dealing with ionizing radiation. More than 400 workplace monitoring missions are carried out yearly by the department.

11.1.6 Qatar

11.1.6.1 Introduction

The Decree-law No. 31 (2002) assigned to the Supreme Council for the Environment and Natural Reserves (SCENR) and the National Health Authority (NHA) lets the authority to supervise the regulation and control of the use of radioactive material and sources, and protection against the associated hazards. NHA is responsible for personal and practice licensing in the medical field, while SCENR is responsible for all others.

The Emiri decree No. 16 (2009) is on "the Competences of the Ministries." According to article 13 of this decree, the competences of the Ministry of Environment include the "supervision of the handling of chemical and radioactive material and the disposal of waste," and the "follow-up of the civil uses of nuclear energy." Article 16 states that the competences of NHA are transferred to the Supreme Council of Health. This includes the responsibility for "issuing practice licenses to the applications of radiation for medical purposes" stated in Article 3 of the Decree-law no 31 (2002).

11.1.6.2 The Regulatory Body

The Emiri Decree No. 39 (2009) on the "Organizational Structure of the Ministry of Environment" assigns the regulatory functions to the Radiation and Chemicals Protection Department (RCPD), whose responsibilities include (SCENR 2015):

1. Assessment of licensing applications of establishments and individuals for working in the radiological field, and issuing licenses.

2. Preparation of regulations and requirements for work in the radiological field, including the dose limits, the management of radioactive waste, and the conditions for transport, storage, and work with radioactive material.

3. Inspection of facilities, operations, and practices involving radioactive sources or radiation emitting devices, in addition to inspection of places and equipment affected by radiation.

The organizational structure of the units within Ministry of Environment (MoE) that have a role in regulatory control of ionizing radiation is attributed to the Radiation and Chemicals Protection Department (RCPD). This department controls two other division namely the Safety and Risk Assessment Division and the Licensing Division.

With regard to the medical field, the Medical Licensing Committee within Supreme Council of Health (SCH) is in charge of issuing practice and personal licenses. This committee is chaired by the head of Occupational Health and Safety (OHS) in HMC, which is the largest user of ionizing radiation in the medical field in Qatar. The SCH relies on technical support from OHS, in conducting the Radiation Personal License Exams.

11.1.6.3   Regulations and Guidelines

The following set of regulations has been issued by SCENR in the Ministry of Environment:

- SCENR Decision No. 4 of 2003, "The Executive Regulations of the Decree-Law No. 31 of the Year 2002 Concerning Radiation Protection."

- SCENR Decision No. 1 of 2004, "Safe Transport of Radioactive Materials."

- SCENR Decision No. 2 of 2005, "Radiation Work and Radiation Doses."

- SCENR Decision No. 11 of 2005, "Radioactive Waste Management."

- SCENR Decision No. 2 of 2007, "Regulations for Radiation Protection Officers."

- SCENR Decision No. 3 of 2007, "Decontamination of Radioactive Material."

- SCENR Decision No. 4 of 2007, "Conditions for Obtaining Licenses in the Radiological Field."

- MoE Decision No. 45 of 2013, "Management of Radioactive Waste Resulting from Natural Radioactivity in the Oil and Gas Industry."

11.1.6.4   Authorization, Inspection, and Enforcement

A. Authorization Article 7 of the Decree Law No. 31 of 2002 states that SCENR is responsible for issuing personal licenses for individuals to perform radiological work, and institutional licenses, which include site, institution, and practice licenses. Article 3 assigns the responsibility of issuing practice licenses in the medical field to NHA. The Decree No. 16 (2009) assigns the licensing responsibilities of SCENR and NHA to MoE and SCH, respectively (NHA 2015).

The implementing regulations, Decree No. 4 (2003), specify details on the four license types, their requirements, conditions, issuing procedures, and renewal requirements. The license duration is specified in the licensing procedures. Article 12 of this decree allows for the registration of low-risk facilities and activities, but this seems not to be applied in the current licensing system.

Decree No. 4 (2007) establishes detailed requirements, including training and qualification requirements, for personnel licensing. Those licenses are issued by SCH for persons working in the medical field and by MoE for all others. For issuing site licenses, MoE requires information on the site location and its vicinity. An inspection of the site is performed by MoE prior to issuing the license. The site license is valid for five years and subject to renewal (IAEA 2007).

For issuing the institution license, MoE requires a radiation protection program and an emergency plan. MoE has adopted guidelines for the content of the radiation protection program, which equally apply to all facilities and activities. The program and the emergency plan are assessed by the safety and risk assessment division. In parallel an inspection to the facility is conducted by the monitoring and industrial inspection department. Only then the licensing division issues the license. Radiation sources can only be installed in the facility after it receives the institution license, which is valid for three years.

Practice licenses are issued by SCH in the medical field and by MoE in the other areas. MoE requires the applicant to submit personnel licenses and an institution license to support the application. An inspection is also conducted prior to issuing the license. The license validity is one year. In issuing practice licenses in the medical field, SCH additionally considers health requirements. The practice license duration in the medical field is one year (IAEA).

B. Inspection

The Emiri Decree No. 16 (2009) and the Decree No. 31, (2002) empower the Ministry of Environment to conduct inspections in order to ensure compliance with regulatory requirements in all sectors, including the medical facilities and activities for which the personnel and practice licenses are issued by SCH. MoE could carry out many types of inspections: pre-licensing, routine announced, routine unannounced and in response to abnormal events.

C. Enforcement

Enforcement actions are mainly requests to implement corrective actions based on non-compliances identified in inspections. Such requests are imposed by the licensing division.

The licensee is requested to notify, within a prescribed period of time, RCPD on the implantation of the corrective actions. A follow-up inspection is always conducted to verify the adequacy of the corrective actions.

11.1.7   Syria

11.1.7.1   Introduction

The legal framework in the field of radiation safety in Syria consists of the Atomic Energy Commission of Syria (AECS 2015):

1. Law No. 12, issued on April 5, 1976; and its amendments.
2. Prime Minister's Decree No. 6514, issued on February 8, 1997.
3. Legislative Decree No. 64, issued on August 3, 2005 by the President of the Syrian Arab Republic; and its amendments, which canceled all the previous provisions, including the Prime Minister's Decree No. 6514, issued on February 8, 1997.
4. Prime Minister's Decree No. 134, issued on January 17, 2007; General regulation for the implementation of the Legislative Decree No. 64, which replaced the AECS Decision No. 112/99 (February 3, 1999).

5. Director General Decision No. 623/2008, issued on May 22, 2008/Instructions on Licensing Radiation Practices.

6. Director General Decision No. 1385/2011, issued on December 26, 2011, for licensing RNRO inspectors.

11.1.7.2 Regulations and Guidelines

Legislative Decree No. 64 complies with the BSS and GS-R-1, and also covers the commitments of the Syrian Government in its letter to the IAEA Director General regarding the application of the Code of Conduct and the Security Council Decision No. 1541. Decree No. 64 nominates the AECS as:

- The regulatory body with respect to radiation protection and safety and security of radiation sources;

- the body responsible for emergency planning and coordination of radiological or nuclear accidents; and

- the competent authority responsible for issuing approval certificates for package design.

Legislative Decree No. 64 (2005) requests AECS to establish a Regulatory Organ, which is to be directly affiliated to the AECS Director General. AECS Board Decision No. 23/6/2006 (2006) designates the Radiological and Nuclear Regulatory Office (RNRO) as the regulatory organ.

By this legislative decree, the AECS is empowered to:

1. Prepare regulations to be issued by the Prime Minister.

2. Issue authorizations for medical and industrial applications of ionizing radiation.

3. Perform inspections on facilities and sites in which radiation sources are used or located, including in medical uses.

4. Impose enforcement actions.

5. Undertake measures to detect illicit trafficking.

6. Verify the absence of contamination exceeding permissible limits in the goods imported or crossing Syria.

7. Promote protection, safety and security culture among the public.

Moreover, this legislative decree prescribes sanctions in cases of non-compliance, as also stated above. The sanctions scheme takes into account the risk associated with the radiation sources. It also defines the civil liability for damage due to radiological or nuclear accidents.

11.1.7.3 Updates

New general regulations were issued by the Prime Minister in January 2007 with many chapters on safety of radiation sources, evaluation of radiation protection and safety and security of radiation sources, management responsibilities, radiation protection officers, occupational exposure, medical exposure, public exposure, exposure to NORM, and on the safe discharge of radioactive waste.

11.1.8 United Arab Emirates

11.1.8.1 Introduction

The Federal Authority for Nuclear Regulation (FANR) was formally established as the nuclear regulatory body in accordance with Federal Law by Decree No. 6 of 2009, which is also known as the UAE Nuclear Law. FANR is an independent organization with full legal competence, and financial and administrative independence. The Nuclear Law gave FANR licensing and regulatory powers over nuclear safety, nuclear security, radiation protection, and non-proliferation. FANR is committed to maintain the highest standards of transparency in performing its functions, in addition to cooperating with and advising relevant government agencies (FANR 2015).

FANR's Board of Management established the National Radiation Protection Committee as an advisory and consultative body to FANR. The Committee is headed by FANR's Director General and its membership includes representatives from different local and federal authorities. It is constituted to work with competent authorities to develop training programs as appropriate, and to promote radiation safety awareness.

11.1.8.2 Regulatory Framework

Although FANR is the "new" radiation regulator for all radiation practices, FANR developed a new regulatory framework to be able to regulate and control the users of radiation sources. This was achieved by establishing a new process of licensing, publications of advisory material, and regulatory guidance documents that are per practice — for example, dental, radiotherapy, and nuclear medicine. FANR's medical licensees are required to adhere to FANR's Regulation 24 (REG-24) titled "Basic Safety Standards for Facilities and Activities Involving Ionizing Radiation Other Than in Nuclear Facilities," which is largely based on the International Atomic Energy Agency (IAEA) BSS GSR Part 3. FANR also published a regulatory guide 07 (RG-07) describing the methods and/or criteria acceptable to FANR for meeting and implementing specific requirements in REG-24.

FANR has the authority to implement an inspection program in relation to any regulated activity, to assure itself that the operator complies with the applicable law and regulations and any conditions set out in the license requirements.

11.1.8.3 Issues and Challenges Related to Radiation Protection

Currently, the market of radiation safety training is not well developed in this country. Training courses should be tailored to cater to the need of each medical practice and the different level of knowledge associated with it. As a result, this directly affects and slows the qualification process of the Radiation Protection Officer (RPO). Another major issue is the limited number of quality assurance service providers available to cover the entire seven emirates of the UAE. There is lack of qualified medical physicists and little capability of personnel in hospitals to perform the quality control tests on their own. Dosimetry is also another challenge in the UAE, as there are no suitable internal dosimetry services for both direct and indirect dosimetry, or adequate numbers of local external dosimetry services. Another challenge is the shortage of sufficient transport companies for the transport of radiopharmaceuticals and radioisotopes.

11.1.8.4 Radiation Protection Infrastructure

FANR is working on developing and improving the radiation protection infrastructure in the UAE. Initiatives include establishing a National Training Strategy on Radiation Safety in cooperation with local stakeholders, launching the Secondary Standards Dosimetry Laboratory (SSDL), the approval process of Dosimetry and Training Services Providers, finalizing the criteria for FANR recognition of Qualified Experts and RPOs, and the continuous enhancement of the safety culture in the country.

11.2 MEFOMP ROLE IN ENHANCING RADIATION PROTECTION STANDARDS

As a regional organization of IOMP, the Middle East Federation of Organizations of Medical Physics (MEFOMP) includes official Medical Physics Societies and Associations in the Middle East region. One of the main roles of MEFOMP is enhancing the Radiation Safety Culture in the profession by the following (MEFOMP 2015):

1. To promote the co-operation and communication between medical physics organizations in the region.

2. To promote medical physics and related activities in the region.

3. To promote the advancement of the status and standard of practice of the medical physics profession.

4. To organize and/or sponsor international conferences, regional and other meetings and/or courses.

5. To collaborate or affiliate with other scientific organizations worldwide.

Finally, a collective work of translating ICRP Publication 105, "Radiological Protection in Medicine," into Arabic was completed by a Task Group constituted of many medical physicists and radiation safety professionals from the Middle East region. The Arab Atomic Energy Agency (AAEA) supported the work of the Task Group and printed out the Arabic Translation document, which was distributed free of charge to all Arabic members.

11.3 ACKNOWLEDGMENTS

The authors would like to thank Mouein Hamzeh and Dr. Bilal Nsouli from Lebanon; Dr. Huda Al Sulaiti and Ms. Aisha Al-Baker from Qatar; Dr. Ibrahim Othman and Dr. Hassan Kharita from Syria and Dr. Jamila Al-Suwaidi and Ms. Aayada Alshehhi from UAE.

11.4 REFERENCES

Atomic Energy Commission of Syria. 2015. www.aec.org.sy (Accessed August 2015).

Co-operative Agreement for Arab States in Asia for Research, Development and Training Related to Nuclear Science and Technology (ARASIA). http://web.aec.org.sy/arasia/ (Accessed August 2015).

Federal Authority for Nuclear Regulation United Arab Emirates. 2015. www.fanr.gov.ae (Accessed August 2015).

International Atomic Energy Agency. 2007. Report of the Radiation Safety Infrastructure Appraisal in Qatar, RaSIA 2/2007/QAT, IAEA.

International Atomic Energy Agency. 2013. Report of the Advisory Mission to Qatar, H. Suman, B. Djermouni, A. Hammou.

International Commission on Radiological Protection (ICRP). 2007. The 2007 Recommendations of the International Commission on Radiological Protection. Annals of the ICRP 37: (2–4), 2007.

International Atomic Energy Agency (IAEA). 2011. Radiation Protection and Safety of Radiation Sources: International Basic Safety Standards Interim Edition.

Lebanese Atomic Energy Commission (LAEC), 2006. Program, Tasks and License Requirements for Radiation Protection in Lebanon Brochure (Arabic). http://www.laec-cnrs.gov.lb (Accessed August 2015).

Middle East Federation of Organizations of Medical Physics. 2015. www.MEFOMP.org

Ministry of Health in Kuwait. http://www.moh.gov.kw (Accessed August 2015).

National Council for Scientific Research (CNRS). http://www.cnrs.edu.lb (Accessed August 2015).

National Health Authority (NHA), State of Qatar. 2002. Guidelines and Protocols for Radiation Licensing.

Radiation Protection Laws and Regulations, Supreme Council for the Environment and Natural Reserves (SCENR) in Qatar

The Iranian Ministry of Health and Medical Education. http://behdasht.gov.ir (Accessed August 2015).

The Jordanian Energy and Mineral Regulatory Commission. http://www.erc.gov.jo (Accessed December 2014).

The Jordanian Ministry of Health. http://www.moh.gov.jo (Accessed August 2015).

The Lebanese Ministry of Health Archived Resolution (Arabic) 2005.

CHAPTER 12

Regulatory Structures and Issues in North America

Ruth E. McBurney

Conference of Radiation Control Program Directors, Inc.

CONTENTS

A LTHOUGH THE REGULATORY STRUCTURE for control of radiation-emitting devices and radioactive material in North America varies in the federal agencies involved and the roles and responsibilities of state, provincial, territorial and local authorities, a network of radiation control programs has been established with authorizing legislation and regulation. This chapter will focus on the description of the framework for regulation of radiation in medical facilities in the United States and Canada, since the structure pertaining to Mexico is included in the chapter on South America.

In both the United States (U.S.) and Canada, radiation protection issues are considered throughout the life cycle of facilities using radiation-producing material, sources, and machines. In the licensing and inspection programs, facility design, operation, source security, and in some cases, decommissioning are all important aspects of radiation protection. Safety culture, in which radiation protection is a fundamental component of overall safety for workers and the public, is also promoted in the programs. To aid the reader, Table 12.1 provides the website homepage for the regulatory programs and advisory and professional organizations discussed in this chapter. More specific websites are provided in the references included within the discussion of specific programs.

Table 12.1: Homepages for regulatory programs and advisory and professional organizations discussed in this chapter

Program/Organization	Homepage
American Association of Physicists in Medicine	http://aapm.org/
American College of Radiology	http://www.acr.org/
American Society of Radiation Oncology	https://www.astro.org/
Canadian Nuclear Safety Commission	http://nuclearsafety.gc.ca/eng/
Conference of Radiation Control Program Directors	http://crcpd.org/
Health Canada	http://www.hc-sc.gc.ca/
Health Physics Society	http://www.hps.org/
Nationall Council on Radiation Protection and Measurements	http://www.ncrponline.org/
National Defense and Canadian Armed Forces	http://www.forces.gc.ca/en/
Occupational Safety and Health Administration	https://www.osha.gov/
Society of Nuclear Medicine and Molecular Imaging	http://www.snmmi.org/
U.S. Department of Transportation	http://www.dot.gov/
U.S. Environmental Protection Agency	http://www.epa.gov/
U.S. Food and Drug Administration	http://www.fda.gov/
U.S. Nuclear Regulatory Commission	http://www.nrc.gov/

12.1 REGULATORY STRUCTURE, AUTHORITIES, AND REQUIREMENTS IN THE UNITED STATES OF AMERICA

In the U.S., radiation protection standards and recommendations for the protection of occupational workers, members of the general public, patients, and the environment are complex and varied. The regulatory authority for regulation depends on the source of the radiation (radioactive material or radiation machines). Federal legislation has been adopted for the regulation of most sources of radioactive material (except for diffuse naturally occurring radioactive material); worker protection; the manufacture of radiation machines; and the use of mammographic X-ray machines. The processes for both federal and state regulations provide for stakeholder input. Following the publication of proposed regulations, regulators encourage comments from individuals including members of the public and professional members of the medical community. Also, professional organizations provide advice to regulators on pros and cons of regulations under consideration. Such organizations include but are not limited to the American Association of Physicists in Medicine, American College of Radiology, American Society of Radiation Oncology, Conference of Radiation Control Program Directors, Health Physics Society, National Council of Radiation Protection and Measurements, and Society of Nuclear Medicine and Molecular Imaging.

The following have been the principal milestones in the development of federal radiation protection legislation and regulation:

(a) The Atomic Energy Act of 1946 (The McMahon Bill) established the Atomic Energy Commission (AEC) to manage the U.S.'s atomic energy program (McMahon 1946). This bill primarily focused on the security and control of materials related to nuclear weapons development, and designated the new federal agency, AEC, to have jurisdiction over fissionable nuclear material. Health and safety standards were not a major consideration in this legislation (Jones 2005). A definition for byproduct material was established, which was not significantly changed until 2005.

(b) The Atomic Energy Act of 1954 added a regulatory program for radiation safety of workers and the public in facilities using radioactive materials covered under the Act.

(c) The 1959 Amendments to the Atomic Energy Act, Section 274, authorized the AEC to enter into Agreements with individual States to relinquish regulatory authority over byproduct and source material, and special nuclear material below the amount to create a critical mass (NRC 2013).

(d) Presidential Executive Order 31 and Public Law (P.L.) 86-373 created the Federal Radiation Council (FRC) to advise the President on radiation matters affecting health and provide guidance for all Federal agencies in the development of radiation standards (Jones 2005). In later years, the Environmental Protection Agency was given this responsibility.

(e) The National Council on Radiation Protection and Measurements (NCRP) was chartered by the U.S. Congress in 1964 (P.L. 88-376) to: 1) collect, analyze, develop, and disseminate information and recommendations about radiation protection and radiation measurements, quantities, and units; 2) provide a means for organizations concerned with radiation protection and measurement to cooperate for effective utilization of combined resources; 3) develop basic concepts about radiation protection, and quantities, units, and measurements and the application of these concepts; and 4) cooperate with the International Commission on Radiological Protection (ICRP), the FRC. The International Commission on Radiation Units and Measurements, and other national and international organizations, governmental and private, are concerned with these issues (NCRP 2015).

(f) The Radiation Control for Health and Safety Act (P.L. 90-602) of 1968 established standards for X-ray producing equipment. This Act focused on the growing health problem of exposure to ionizing and nonionizing radiation from radiation machines, and directed the Food and Drug Administration to establish a program to protect the public health from unnecessary emissions from machines that produce radiation (McBurney 2008).

(g) The Energy Reorganization Act of 1974 established the Nuclear Regulatory Commission (NRC) as an independent regulatory agency, and set up a separate agency, the Energy Research and Development Administration (ERDA) for policy planning, research, and development concerning sources of Energy, including nuclear energy. In 1977, the Department of energy was established to take over this role (NRC 2013).

(h) The Mammography Quality Standards Act of 1992 was enacted to improve the quality of mammography exams (FDA 2004).

(i) The Energy Policy Act of 2005 expanded the definition of byproduct material to include discrete sources of radium and accelerator-produced radioactive material and security requirements based on the International Atomic Energy Agency (IAEA) Code of Conduct for higher-activity sources (GPO 2005).

12.2   REGULATION OF RADIOACTIVE MATERIAL

By law and under regulations of the NRC, certain radioactive materials are defined and regulated under the Atomic Energy Act, as amended. These include source material, special nuclear material, and byproduct material. Most radioactive materials used in medical settings are byproduct material (Type 1, 3, and 4).

The Atomic Energy Act, as revised in 1978 and in 2005 by the Energy Policy Act (EPAct), defines byproduct material in Section 11e.(1) as radioactive material (except special nuclear material) yielded in or made radioactive by exposure to the radiation incident to the process of producing or using special nuclear material (GPO 2005).

The definition of byproduct material in Section 11e.(2) is the tailings or wastes produced by the extraction or concentration of uranium or thorium from any ore processed primarily for its source material content.

The definition in Section 11e.(3) is any discrete source of radium-226 that is produced, extracted, or converted after extraction, before, on, or after the date of enactment of the EPAct for use for a commercial, medical, or research activity; or any material that has been made radioactive by use of a particle accelerator and is produced, extracted, or converted after extraction, before, on, or after the date of enactment of the EPAct for use for a commercial, medical, or research activity.

The definition in Section 11e.(4) is any discrete source of naturally occurring radioactive material, other than source material, that the NRC, in consultation with the Administrator of the Environmental Protection Agency (EPA), the Secretary of the Department of Energy (DOE), the Secretary of the Department of Homeland Security (DHS), and the head of any other appropriate Federal agency, determines would pose a threat similar to the threat posed by a discrete source of radium-226 to the public health and safety or the common defense and security; and is extracted or converted after extraction before, on, or after the date of enactment of the EPAct for use in a commercial, medical, or research activity (NRC 2006).

The NRC's Office of Nuclear Material Safety and Safeguards (NMSS) regulates activities that provide for the safe and secure production of nuclear fuel used in commercial nuclear reactors; the safe storage, transportation, and disposal of high-level radioactive waste and spent nuclear fuel; and the transportation of radioactive materials regulated under the Atomic Energy Act of 1954. The agency's four regional offices (Region I — Northeast, Region II — Southeast, Region III — Midwest, and Region IV — West/Southwest) implement the NRC's materials program in the states for which they are responsible. In addition, the NMSS develops and oversees the regulatory framework for the safe and secure use of nuclear materials; medical, industrial, and academic applications; uranium recovery activities, low-level radioactive waste sites; and the decommissioning of previously operating nuclear facilities and power plants.

Materials regulation is also supported by independent advice from the Advisory Committee on the Medical Uses of Isotopes (ACMUI). ACMUI advises the NRC on policy and technical issues that arise in the regulation of the medical uses of radioactive material in diagnosis and therapy. The ACMUI membership includes healthcare professionals from various disciplines, who comment on changes to NRC regulations and guidance; evaluate certain nonroutine uses of radioactive material; provide technical assistance in licensing, inspection, and enforcement cases; and bring key issues to the attention of the Commission for appropriate action. This Committee, along with other advisory committees to the Commission, are structured to provide a forum where experts representing many technical perspectives can provide independent advice that is factored into the NRC's decision-making process (NRC 2014a).

12.2.1 Agreement State Program

Section 274 of the Act provides a statutory basis under which the NRC relinquishes to the States portions of its regulatory authority to license and regulate byproduct materials (radioisotopes); source materials (uranium and thorium); and certain quantities of special nuclear materials. The mechanism for the transfer of the NRC's authority to a State is an agreement signed by the Governor of the State and the Chairman of the Commission, in accordance with section 274b of the Act (Jones 2005, NRC 2013).

NRC assistance to States entering into Agreements includes review of requests from States for 274b Agreements, or amendments to existing Agreements, meetings with States to discuss and resolve NRC review comments, and recommendations for Commission approval of proposed 274b Agreements. Additionally, the NRC conducts training courses and workshops; evaluates technical licensing and inspection issues from Agreement States; evaluates State rule changes; participates in activities conducted by the Conference of Radiation Control Program Directors, Inc.; and provides early and substantive involvement of the States in NRC rulemaking and other regulatory efforts. The NRC also coordinates with Agreement States the reporting of event information and responses to allegations reported to the NRC involving Agreement States. Today, 37 States have entered into Agreements with the NRC, and others are being evaluated. Of the more than 20,000 active source, byproduct, and special nuclear materials licenses in place in the United States, about a quarter are administered by the NRC, while the rest are administered by the 37 Agreement States (NRC 2014b).

12.2.2 Applicable Regulations to Uses of Radioactive Material

Radioactive material regulations of the NRC are from Title 10 of the Code of Federal Regulations. Those used by Agreement States must be compatible with, and in some cases, such as basic radiation safety standards, be identical to, those of NRC (NRC 2014c).

To maintain consistency in the regulation of radioactive materials throughout the Agreement State Programs, the Conference of Radiation Control Program Directors (CRCPD) develops Suggested State Regulations for Control of Radiation (SSRs), a set of peer-reviewed model regulations that can be used by radiation control programs when adopted through administrative, and in some cases legislative, procedures (McBurney 2008).

Committees of members with expertise and interest in a specific regulatory area are assigned to address sections of the regulations, along with federal resource persons and medical and health physicists, and in some cases, physicians, to assist as advisors in the development of the model standards. The SSRs are an important service to state radiation control programs in that they provide a consistent national approach to regulations. The SSRs are updated as needed by changes in federal standards, new technologies, and changes in risk evaluations. For applicable radioactive material regulations, the SSR parts must have federal concurrence by the NRC to be accepted in Agreement State regulations (McBurney 2008).

The current primary radiation protection standards are in Part 20 of Title 10 in the Code of Federal Regulations (NRC 2014c). The regulations establish standards of protection for occupational workers and members of the public from ionizing radiation resulting from licensed activities. The standards, last updated in 1991, adopted the basic tenets of the ICRP system of radiation dose limitation described in ICRP Publication 26 (ICRP 1977), and incorporated the principles of justification, optimization, and limitation. The ICRP Publication 26 also provided for the summation of internal and external exposures.

The majority of the ICRP Publication 26 recommendations were adopted by the National Council on Radiation Protection and Measurements (NCRP) Report No. 91 (NCRP 1987), "Recommendations on Limits for Exposure to

Ionizing Radiation." Therefore, the majority of the NCRP Report No. 91 recommendations were adopted in the 10 CFR Part 20 amendments of 1991.

The NRC has issued an Advance Notice of Proposed Rulemaking (ANPR) regarding potential changes to 10 CFR Part 20 to obtain input from stakeholders on the development of a draft regulatory basis that would support potential changes to the NRC's current radiation protection regulations. The goal of this effort is to achieve greater alignment between the NRC's radiation protection regulations and the 2007 recommendations of ICRP in the ICRP Publication 103 (ICRP 2007). Through this ANPR, the NRC has identified specific questions and issues with respect to a possible revision of the NRC's radiation protection requirements (NRC 2014d).

Table 12.2: Primary regulations of the NRC applicable to medical facilities and corresponding suggested state regulations for the control of radiation (NRC 2014c, CRCPD 2014a)

Title 10, Code of Federal Regulations Citation	Title of Part	Suggested State Regulations Part Designation	Title of Part
10 CFR Part 19	Notices, Instructions, and Reports to Workers: Inspection and Investigations	Part J	Notices, Instructions and Reports to Workers; Inspections
10 CFR Part 20	Standards for Protection Against Radiation	Part D	Standards for Protection Against Radiation
10 CFR Part 30	Rules of General Applicability to Domestic Licensing of Byproduct Material	Part C	Licensing of Radioactive Material
10 CFR Part 35	Medical Use of Byproduct Material	Part G	Use of Radionuclides in the Healing Arts
10 CFR Part 37	Physical Protection of Category 1 and Category 2 Quantities of Radioactive Material	Part V	Physical Protection of Category 1 and Category 2 Quantities of Radioactive Material
10 CFR Part 71	Transportation and Packaging of Radioactive Material	Part T	Transportation and Packaging of Radioactive Material

In addition to the basic standards in Parts 20 and 19, and the licensing requirements in Part 30 of 10 CFR, medical licensees are regulated under

Part 35, Medical Use of Byproduct Material (NRC 2014e). This part contains regulations on supervision and directives required for diagnostic and therapeutic procedures using radioactive material; training requirements for Radiation Safety Officers, authorized users (physicians) for the various medical uses, authorized medical physicists, and nuclear pharmacists; technical requirements for use of calibration and survey equipment; safety precautions and requirements specific to each type of use; calibration requirements for therapy devices (brachytherapy, remote afterloaders, teletherapy units, and gamma stereotactic radiosurgery units); reports and recordkeeping requirements, including reporting and notification of a medical event, dose to an embryo/fetus or a nursing child, and leaking source; and enforcement. Licenses of broad scope for research and use is covered under 10 CFR Part 33, the corresponding provisions of which are included in Part C of the SSR's.

The primary regulations applicable to medical facilities in the United States and the corresponding part of the Suggested State Regulations are shown in Table 12.2.

12.3   REGULATION OF RADIATION-GENERATING DEVICES

Unlike the regulation of most radioactive materials, for the most part, no federal agency regulates the uses of radiation machines, and it is the responsibility of the individual States. Under the Radiation Control for Health and Safety Act of 1968, the U.S. Food and Drug Administration (FDA) was charged with administration and enforcement of the performance standards prescribed for electronic products under Section 38 of the Public Health Service Act, as amended (42 USC 263f). Under that act, manufacturers of electronic products must certify as to the compliance of those products with applicable federal performance standards. However, the use of the products was not covered under federal law and was left to the States to regulate. The Public Health Service Act did recognize the fact that States also established standards on products, since the Act requires that a State performance standard for which there is a Federal standard established by the FDA must be identical to the Federal standard (McBurney 2008).

Most States and Territories in the U.S. have been authorized through State legislation to protect occupational and public health and safety and the environment through a regulatory program for certain sources of radiation that is consistent with federal and other States' systems and that permits development and use of sources of radiation while protecting public health and safety. All States with such enabling legislation regulate radiation machines in facilities that are not under exclusive federal jurisdiction, such as military and Veterans Administration hospitals. In addition, some local jurisdictions, such as New York City, have established radiation control programs within a local agency.

States normally use a licensing and/or registration process followed by routine inspections to assure compliance with established worker protection, patient protection, machine performance, and quality assurance standards. The regulatory program usually includes all facilities that use radiation machines, including X-ray machines and accelerators used for human and veterinary diagnosis and treatment, including those accelerators and cyclotrons used for radioisotope production, and industrial, academic, and research facilities using X-ray-machines and accelerators (McBurney 2008).

CRCPD develops model regulations for the use of radiation machines through the SSR process, including federal concurrence, for use by State, local and territorial radiation control programs (CRCPD 2014a). Basic radiation protection standards as set out in 10 CFR Part 20 and Part D of the SSRs are also applied to registered or licensed machine-based radiation facilities in the States.

Several of the SSRs are focused on the registration and licensure and use standards for radiation machines in the healing arts. These include:

Part A: General Provisions
Part B: Registration of Radiation Machines, Facilities, and Services
Part F: X-Rays in the Healing Arts
Part H: Radiation Safety Requirements for Analytical X Rays
Part I: Radiation Safety Requirements for Particle Accelerators
Part X: Medical Therapy

In addition to providing model State regulations, CRCPD, in cooperation with the FDA, conducts the Nationwide Evaluation of X-Ray Trends (NEXT). Under the NEXT program, CRCPD and the FDA have captured exposure data from a nationally representative sample of U.S. medical and dental facilities for over 30 years on a variety of radiographic procedures. Specific protocols and phantoms are developed, and training is provided to the state inspectors conducting the surveys. NEXT surveys have been completed on mammographic, dental, pediatric chest, adult chest, dental, abdomen/lumbar spine, fluoroscopy, computed tomography, and cardiac catheterization procedures. The survey protocol, data summaries, and vetted survey results for most of the NEXT surveys conducted are available from CRCPD (CRCPD 2014b).

12.4 MAMMOGRAPHY QUALITY STANDARDS

The one area in which the FDA has been authorized to regulate the use of radiation machines is mammography, under the Mammography Quality Standards Act of 1992. Under that law, the FDA certifies and inspects facilities for compliance with quality standards for machine parameters and quality assurance, as well as training and experience requirements for physicians, technologists, and medical physicists.

Congress enacted MQSA to ensure that all women have access to quality mammography for the detection of breast cancer in its earliest, most treatable stages. The Act refers to the MQSA as amended by the Mammography Quality Standards Reauthorization Acts of 1998 and 2004 (MQSRA) (FDA 2004).

Congress tasked the FDA with developing and implementing MQSA regulations. Interim regulations, issued in December 1993, became effective on October 1, 1994. The FDA began enforcing the accreditation and certification provisions of the Act on that date, and began annual inspections of mammography facilities in January 1995. On October 28, 1997, the FDA issued more comprehensive final regulations, which became effective on April 28, 1999. Reauthorizations of MQSA came with new requirements, which were incorporated into the regulations.

Under MQSA regulations, the FDA approves accreditation bodies for accreditation of mammography facilities. To date, the American College of Radiology and three state agencies have been approved to accredit mammography facilities. States can also be approved as State Certifying Agencies (SCAs) for delegation of FDA's certification authority in those states. To date, four state agencies have been approved as SCAs (FDA 2004).

12.5 OTHER FEDERAL AGENCIES HAVING A ROLE IN RADIATION PROTECTION IN MEDICAL FACILITIES

12.5.1 United States Environmental Protection Agency

The EPA is responsible for issuing general radiation guidance to federal agencies. They are in the form of the following:

- Federal Guidance Policy Recommendations, which are policy statements signed by the President and usually reflected in federal and state regulations for radiation protection of workers and the general public

- Technical Reports, which help standardize radiation dose and risk assessment methodologies.

The latest guidance report, proposed Federal Guidance Report No. 14, Radiation Protection Guidance for Diagnostic and Interventional X-Ray Procedures, was prepared by the Interagency Working Group on Medical Radiation to address the significant increase in the use of digital imaging technology and high-dose procedures such as computed tomography (CT scans). This report provides federal facilities that use diagnostic and interventional X-ray equipment with recommendations for keeping patient doses as low as reasonably achievable without compromising the quality of patient care. It is an update of Federal Guidance Report No. 9, which was issued in 1976 (EPA 2013).

12.5.2 Department of Labor

The Occupational Safety and Health Administration in the U.S. Department of Labor establishes workplace radiation safety standards for workers. Their standards are applicable in facilities that are not regulated by the NRC or an Agreement State. Unlike the standards of NRC and the Agreement States,

OSHA still relies on ICRP 2 as the main foundation for its occupational health and safety regulations (Jones 2005).

12.5.3 Department of Transportation

The transportation of radioactive material is governed by the U.S. Department of Transportation. Their regulations for radioactive material have been implemented through the Code of Federal Regulations. Regulations specific to transportation of radioactive material have also been adopted by the NRC in 10 CFR Part 71, Packaging and Transportation of Radioactive Material (NRC 2014f), and by Agreement States under similar regulations. The Suggested State Regulations for transportation are found in Part T of the SSR's (CRCPD 2014a).

12.6 QUALIFICATIONS FOR MEDICAL PHYSICISTS IN THE UNITED STATES

Medical physicists are required for certain medical uses of radioactive material and radiation machines in U.S. regulations and in state regulations. The qualifications vary somewhat, based on the laws, regulatory authority, duties, and use of radiation involved.

12.6.1 Medical Physicists in Facilities Using Radioactive Material

For the use of radioactive material in the healing arts, the requirements are found in 10 CFR 35 (NRC 2014c). As defined in that part, authorized medical physicists are required to be named on the licenses for most therapeutic uses of radiation sources, such as brachytherapy and external beam therapy. Authorized medical physicists must:

1) Be certified by a specialty board recognized by NRC or an Agreement State; or

2) Hold a Master's or Doctor's degree in physics, medical physics, or other physical science, engineering, or applied mathematics from an accredited college or university; and have completed one year of full-time training in medical physics or an additional year of work experience under an individual qualified to be an authorized medical physicist.

More details of the qualification requirements can be found in 10 CFR 35.51 (NRC 2014c).

12.6.2 Medical Physicists in Mammography Facilities

For mammography facilities, the regulations under MQSA require that the facility have a medical physicist for conducting surveys of the equipment and providing oversight of the facility quality assurance program. The qualifications for a medical physicist in mammography facilities includes:

1) Be State licensed or approved or have certification in an appropriate specialty area of one of the FDA-approved certifying bodies; and

2) a) Have a Master's degree or higher in a physical science from an accredited institution with no less than 20 semester hours or equivalent of college undergraduate or graduate level physics;

b) Have 20 hours of documented training in surveys of mammography facilities; and

c) Have experience of conducting surveys of at least 10 mammography units; and

3) Maintain continuing education and training requirements.

Alternative initial qualifications are also included in the rules for medical physicists that were practicing prior to April 1999 (grandfather clause). More detail on the requirements can be found in the FDA regulations in 21 CFR Part 900, Mammography (FDA 2004).

12.6.3 Medical Physicists in Diagnostic and Therapeutic Radiation Machine Facilities

Requirements for qualified medical physicists (QMP) are also included in Part X of the Suggested State Regulations, and proposed standards for qualified medical physicists are included in Part F of the SSRs for diagnostic machines that require the physicist expertise for quality assurance surveys of computed tomography (CT) and fluoroscopic units.

In Part X, the services of a Qualified Medical Physicist is required in facilities having therapeutic radiation machines with energies of 500 kV and above. The Qualified Medical Physicist is responsible for:

1) Full calibration(s) and protection surveys required;

2) Supervision and review of dosimetry;

3) Beam data acquisition and transfer for computerized dosimetry, and supervision of its use;

4) Quality assurance, including quality assurance check review;

5) Consultation with the authorized user in treatment planning, as needed; and

6) Performing calculations/assessments regarding misadministrations.

The qualifications for a Qualified Medical Physicist (QMP) under Part X for therapy include:

1) Registration with the appropriate state agency; and

2) Certification by the American Board of Radiology in an appropriate area that includes radiological physics or therapeutic radiological physics, or by the American Board of Medical Physics in Radiation Oncology Physics, or by the Canadian College of Medical Physics; or

3) Holding a Master's or doctorate degree in physics, medical physics, other physical science, engineering, or applied mathematics from an accredited college or university, and having completed one year of full-time training in medical physics and an additional year of full-time work experience in the

use of external beam therapy under the supervision of a Qualified Medical Physicist (CRCPD 2014a).

In proposed changes to SSR Part F for diagnostic X-ray facilities, a qualified medical physicist or a qualified expert is required for certain quality assurance, machine testing, and surveys required on X-ray equipment. The definition of a qualified medical physicist and qualified expert in this Part are as follows:

"Qualified medical physicist (QMP)" means an individual who meets each of the following credentials:

1) Has earned a master's and/or doctoral degree in physics, medical physics, biophysics, radiological physics, medical health physics, or equivalent disciplines from an accredited college or university; and

2) Has been granted certification in the specific subfield(s) of medical physics with its associated medical health physics aspects by an appropriate national certifying body and abides by the certifying body's requirements for continuing education.

"Qualified Expert (QE)" means an individual who is granted professional privileges based on education and experience to provide clinical services in diagnostic medical physics by the appropriate State radiation control agency (CRCPD 2014c).

For federal facilities, the EPA's proposed Federal Guidance No. 14, Radiation Protection Guidance for Diagnostic and Interventional X-Ray Procedures, identifies the need for a Qualified Physicist to perform some of the duties involved in certain facilities that carry out diagnostic and interventional X-ray procedures. Their definition of a Qualified Physicist is:

"An individual who is competent to practice independently in the relevant medical subfield of medical physics or health physics. In general, a health physicist or medical physicist with appropriate training and experience regarding the medical use of X-rays is considered a qualified physicist. Ideally, persons should have certification from the American Board of Health Physics, the American Board of Medical Physics, the American Board of Radiology, or the American Board of Industrial Hygiene, to be considered a qualified expert in these respective fields. For the purposes of this document, the relevant subfield of medical physics is diagnostic radiological physics. Certification, continuing education, and experience are factors toward demonstrating that an individual is a qualified physicist. Individual federal agencies may develop their own criteria for determining when a physicist is a 'Qualified Physicist' as defined in this document" (from EPA-402R-10003 Draft Proposal, Federal Guidance Report No. 14, 2012) (EPA 2013).

Some states (4 currently) license medical physicists, and many others (27 currently) register medical physicists. The requirements for licensure and registration vary among the states in general expertise required and qualifications for the type of medical physics services to be provided (medical therapy, diagnostic x-ray or nuclear medicine, and medical health physics).

CRCPD maintains a registry of board-certified medical physicists. This registry provides independent verification to radiation control program staff and employers, of written documentation presented with applications for employment or to the radiation control program upon application for a radioactive material license or machine license or registration.

Data on certifications and specialties have been provided to the registry by the following certifying boards:

American Board of Radiology
American Board of Medical Physics
American Board of Health Physics
American Board of Science in Nuclear Medicine
Canadian College of Physicists in Medicine

12.7 REGULATORY STRUCTURE, AUTHORITIES, AND REQUIREMENTS IN CANADA

In Canada, responsibilities are divided between Federal and Provincial/Territorial Authorities. For example, the federal government has primary responsibility for natural resources (including uranium mining) and the development, production, and use of nuclear energy. Healthcare (including radiation doses to patients) is within the scope of each of the provincial governments. Given the breadth and depth of the nuclear industry in Canada, and the various levels of governments having different areas of responsibility, some aspects of radiation protection regulation in Canada can be complex. Canada has a cooperative system for the regulation of ionizing radiation protection covering federal, provincial, territorial, and military jurisdictions. A Federal/Provincial/Territorial Radiation Protection Committee (FPTRPC) exists to aid in cooperation between the various agencies. Their mandate encompasses regulation and guidance on all aspects of radiation protection: federal and provincial; naturally occurring radioactive material (NORM) and anthropogenic; ionizing and nonionizing (Clement 2008).

12.7.1 Canadian Nuclear Safety Commission

The Canadian Nuclear Safety Commission (CNSC) is the federal nuclear regulator whose mandate includes radiation protection regulation of most occupational and public exposures. The CNSC does not regulate medical (patient) exposures, some aspects of NORM, or military applications. Provincial authorities are the primary regulators with respect to doses to patients and occupational doses arising from X-rays. Health Canada plays a role in X-ray device certification, development of national guidance (e.g., on radon) and direct regulation of certain federal facilities. NORM is regulated provincially, with varying regulatory mechanisms across the provinces and territories. Radiation protection regulation for National Defence and the Canadian Armed Forces is performed by the Director General Nuclear Safety. The Canadian Nuclear Safety Commission (CNSC) regulates the use of nuclear energy and materials to protect health, safety, security, and the environment and to respect Canada's international commitments on the peaceful use of nuclear energy. Created in 1946 as the Atomic Energy Control Board, the name of the agency changed in 2000 with the enactment of the Nuclear Safety and Control Act (NSCA). Since that time, the CNSC has worked to ensure that the production of nuclear energy does not pose an unreasonable risk to the public and the environment (Clement 2008).

Radiation protection aspects of nuclear regulation in Canada are consistent with the current recommendations of the International Commission on Radiological Protection, whose fundamentals are embodied in publication ICRP-60. The CNSC participates in a number of international activities related to radiation protection and other aspects of nuclear regulation, in order to contribute to the harmonization of international nuclear safety and security regulatory standards and to ensure that the CNSC's activities are consistent with international best practices. These activities involve organizations such as the International Nuclear Regulators Association, the International Atomic Energy Agency, the Nuclear Energy Agency of the Organization for Economic Co-operation and Development, the International Commission on Radiological Protection, and the United Nations Scientific Committee on the Effects of Atomic Radiation. The CNSC is also actively involved in the bilateral exchange of regulatory information and collaboration with foreign nuclear regulators (Clement 2008).

In addition to the regulation of the types of radioactive material used in medical facilities, CNSC also certifies, licenses, and regulates the use of linear accelerators operating at 1 MeV or above. These facilities are considered to be Class II facilities, under the Nuclear Safety and Control Act. The regulations for Class II Nuclear Facilities and Prescribed Equipment Regulations (SOR/2000-205) can be found at the following site: http://laws-lois.justice.gc.ca/eng/regulations/SOR-2000-205/FullText.html.

Regulations of CNSC focus on occupational worker requirements and protection of the general public. Medical radiation, patient protection, medical events, and quality assurance issues are not included in those regulations.

12.7.2 Certification of Radiation Safety Officers

In Canada, Radiation Safety officers in Class II facilities (that include medical radioactive material and accelerators) must be certified by the Commission or an authorized designated officer. Regulations regarding certification requirements are found in Section 15.02 of the Class II Nuclear Facilities and Prescribed Equipment Regulations. Sections 15.03 to 15.12 of the Class II Nuclear Facilities and Prescribed Equipment Regulations list several other RSO requirements for every licensee operating a Class II facility or holding a Class II servicing license.

For medical facilities, the following qualifications are required for consideration for certification:

12.7.2.1 Education

The candidate seeking certification must have at least a Bachelor's degree in engineering or science from a recognized university. Alternate education qualification will be reviewed for acceptability by the CNSC on a case-by-case basis.

12.7.2.2 Minimum Experience

The candidate must have a minimum of two years' experience working in a medical facility that works directly with Class II nuclear facilities, or two to five years of relevant experience in one or more of the following fields:

1) Health or medical radiation physics
2) Radiation protection
3) Class II prescribed equipment
4) Other types of nuclear facilities

12.7.2.3 Level of Knowledge

The candidate must demonstrate an appropriate level of knowledge covering:

1) The relevant provisions of the NSCA
2) The relevant sections of the following regulations:
a) General Nuclear Safety and Control Regulations
b) Radiation Protection Regulations
c) Class II Nuclear Facilities and Prescribed Equipment Regulations
d) Nuclear Substance and Radiation Devices Regulations
e) Packaging and Transport of Nuclear Substances Regulations (if applicable).
3) The operational activities that are licensed by the CNSC, and for which the candidate will be the RSO
4) Any operational requirement from the CNSC, as may be listed in license conditions
5) Radiation physics
6) Principles of radiation safety
7) The radiation protection program of the facility — more specifically:
a) policies and procedures of the organization with respect to radiation safety;
b) details of the construction of the facility, including shielding, safety systems, interlocks, and prescribed equipment specifications;
c) the responsibilities and authority of senior management;
d) the responsibilities of people working under the radiation protection program (CNSC 2014).

12.7.3   Health Canada

Health Canada is the federal department responsible for helping Canadians maintain and improve their health, while respecting individual choices and circumstances. The Radiation Protection Bureau's mandate is to promote and protect the health of Canadians by assessing and managing the risks posed by radiation exposure in living, working, and recreational environments. Health Canada's Radiation Protection Bureau is responsible for reducing the health and safety risks associated with different types of radiation. To protect Canadians from these effects, Health Canada:

- conducts research into the biological effects of environmental and occupational radiation;

- develops better methods for internal radiation dosimetry and its measurement;

- provides radiation safety inspections of federally regulated facilities containing radiation-emitting devices, the devices themselves, as well as training on the proper operation of the devices;

- develops regulations, guidelines, standards, and safety codes pertaining to radiation-emitting devices;

- provides radiation advice and collaborates with government departments and agencies, industry, and the general public (Clement 2008).

Health Canada also provides a similar role to the U.S. Food and Drug Administration in the regulation of manufacturers of radiation machines. The use of the machines of less than 1 MeV in energy is regulated at the provincial and territorial levels for the protection of radiation workers, patients, and members of the public.

12.7.4   The Director General Nuclear Safety

The Director General Nuclear Safety (DGNS) is accountable for the development, co-ordination, and assurance of the implementation of a comprehensive nuclear safety program for the Department of National Defence and the Canadian Forces (DND/CF). This responsibility encompasses the many radioactive materials and other sources of ionizing radiation in use within the DND/CF with a view to assuring overall design, development and operational safety. DGNS is responsible for auditing compliance with the nuclear safety program, which includes technical safety analyses of the adequacy of design and behavior of equipment and activities initiated by or including DND/CF personnel. (Clement 2008).

12.7.5 The Role of Provincial and Territorial Authorities

Responsibility for workplace health and safety is under the jurisdiction of the provinces and territories, typically through their Worker Compensation Boards or Departments of Health or Labour. The exception is where this is explicitly a federal domain, such as in federal departments, agencies and corporations, the armed forces, national research organizations, and those industries involved in interprovincial land transportation, air and maritime services, and telecommunications. The Canadian Nuclear Safety Commission is the federal agency responsible for the control of nuclear substances and facilities, as well as the resulting radiation exposure, except within the Canadian armed forces. Naturally occurring radioactive materials (NORM), however, are not regulated by the CNSC except when these materials are being transported or imported/exported. NORM remains the responsibility of the provinces and territories, except where this is explicitly in a federal setting.

Radiation-emitting devices, other than those using nuclear substances, are subject to federal requirements at the point of sale or importation regarding standards for design, construction, and functioning. Otherwise, the installation and use of those in the jurisdiction is again provincial or territorial, except for those devices installed and used in federal facilities or in federally regulated industries. The Canada Labour Code prevails in federal jurisdictions, and for radiation protection, the standards developed by Health Canada, as specified in its series of radiation protection safety codes, are applicable.

At the provincial and territorial levels, radiation protection is administered either through designated radiation protection programs or as part of the broader duties of occupational health and safety officers from the Workers Compensation Board or similar organizations. For the provinces, radiation protection programs are based in various ministries or agencies. Some of the programs have regulatory authorities, while for others their functions are restricted to advisory and service roles and supporting the regulatory programs of other environmental, occupational, and public health bodies.

One key area of responsibility for the provinces and territories is the delivery of healthcare, which is a major user of medical X-ray equipment. Protection of patients is a trade-off in the optimization of exposure to achieve an acceptable level of diagnostic information. The introduction of computerized imaging modalities is leading to larger doses to the population. The advent of digital imaging requires renewed attention to the means for controlling exposures. Protection of healthcare workers is important as this group accounts for the largest number of occupationally exposed persons, at around 70% of all persons currently monitored routinely for occupational radiation exposure in Canada (Clement 2008).

12.7.6 Role of the Federal/Provincial/Territorial Radiation Protection Committee

The mission of the Federal/Provincial/Territorial Radiation Protection Committee (FPTRPC) is to advance the development and harmonization of practices and standards for radiation protection within Federal, Provincial, and Territorial jurisdictions. The FPTRPC is an intergovernmental Committee established to support Federal, Provincial, and Territorial radiation protection agencies in their respective mandates, in part, by providing a national focus for government radiation protection agencies and harmonizing radiation protection programs, standards, and guidelines (Clement 2008).

Through the use of the FPTRPC and safety codes and guidance provided by Health Canada, the provinces are able to maintain consistency in the regulation of radiation under their jurisdiction. The provincial regulations for X-ray facilities are based primarily on the Safety Code of Health Canada. For hospitals and other large facilities, they are located in Safety Code 35: Safety Procedures for the Installation, Use and Control of X-ray Equipment in Large Medical Radiological Facilities. Safety Code 33, "Radiation Protection in Mammography," is used for mammographic facilities and the Safety Code for Small Medical Radiological Facilities is used for other applicable medical institutions (Health Canada 2014). All of the Safety Codes for use of X-ray equipment can be found on Health Canada's website at: http://www.hc-sc.gc.ca/ewh-semt/pubs/radiation/index-eng.php#codes

12.8 RADIATION SAFETY OFFICER/MEDICAL PHYSICIST REQUIREMENTS FOR X-RAY FACILITIES

Like Type II radiation facilities regulated under CNSC, there must be a Medical Physicist or Radiation Safety Officer to act as an advisor on all radiation protection aspects of facility design, construction, and operations. The medical physicist/radiation safety officer must:

1) possess qualifications required by any applicable federal, provincial, or territorial regulations or statutes and be certified according to a recognized standard, such as for medical physicists, the Canadian College of Physicists in Medicine; and

2) acquire re-qualification or refresher training according to any applicable federal, provincial, or territorial regulations or statutes and according to a recognized standard, such as for medical physicists, the Canadian College of Physicists in Medicine (Health Canada 2014).

12.9 CURRENT REGULATORY ISSUES REGARDING MEDICAL PHYSICISTS IN NORTH AMERICA

1) Roles and responsibilities of medical physicists and medical health physicists: As described under the requirements for medical physicists in both the

U.S. and Canada, the delineation of duties that require expertise in medical physics and medical health physics or for which either could perform the duties is not clearly defined. In addition to the laws and regulations of the various agencies involved in the U.S. and in Canada, other accrediting bodies also have medical physics requirements. For example, the Joint Commission, which certifies facilities that are acceptable under Medicare, has developed standards changes focusing on CT, nuclear medicine, positron emission tomography (PET), and MRI services. For those involving ionizing radiation, a medical physicist (as defined by the standard) will be required for several of the duties involved.

2) Training and experience for qualified medical physicists: The above descriptions of the regulatory framework in the U.S. and Canada also point out the variety and complexity of training and experience requirements for medical physicists. The agencies and standard-setting bodies involved in establishing the standards continue to work together to gain greater consistency in these standards.

3) Medical Event Reporting: In the U.S., medical events as described in Section 35.3045 of 10 CFR for radioactive material and in the Suggested State Regulations Part X and F for machine-based radiation (defined as a misadministration), are required to be reported and notified to the appropriate regulatory authority. Medical events that must be reported are any events, except for an event that results from patient intervention, in which the administration of byproduct material or radiation results in:

a) a dose differing from the prescribed dose by an amount stated in the standard;

b) an administration of a dose or dosage to the wrong individual or human research subject;

c) an administration of a dose or dosage delivered by the wrong mode of treatment;

d) for radioactive material use, a dose that exceeds 0.05 Sv effective dose equivalent, 0.5 Sv to an organ or tissue, or 0.5 Sv shallow dose equivalent to the skin from:

i. an administration of a wrong radioactive drug containing byproduct material;

ii. an administration of a radioactive drug containing byproduct material by the wrong route of administration; or

iii. a leaking source (NRC 2014c, CRCPD 2014a).

Other events that require reporting to the NRC or Agreement States are also found in 10 CFR 35.3045. Reported events are placed into the Nuclear Material Events Database (NMED).

For machine-based radiation, the CRCPD has established a database of medical events that have been reported to the state radiation control programs and defined the parameters. The American Association of Physicists in Medicine is assisting in the analysis of these events for trend analysis and root causes. In addition, the American Society for Radiation Oncology (ASTRO) has also established a voluntary event reporting system for therapy facilities.

12.10 CONCLUSION

Although regulatory structure, responsibilities, and the regulations developed by the agencies involved vary among the jurisdictions and legal authorities in North America, the agencies in the U.S. and those in Canada have established relationships with the other agencies and radiation professional organizations in their boundaries to effect greater consistency in the regulation of medical radiation. Both countries have established multi-agency and organizational committees to address common issues and develop regulatory solutions, and will continue to do so in the future.

12.11 REFERENCES

Canadian Nuclear Safety Commission. 2014. Personnel Certification: Radiation Safety Officers REGDOC-2.2.3. CNSC. http://nuclearsafety.gc.ca/eng/acts-and-regulations/regulatory-documents/published/html/regdoc2-2-3/index.cfm (Accessed January 12, 2015).

Clement, C.H. 2008. Ionizing Radiation Protection Regulation in Canada: The Role of the Federal Provincial Territorial Radiation Protection Committee, Proceedings of the 2008 IRPA Congress, Buenos Aires, Argentina. IAEA/INES.

Conference of Radiation Control Program Directors. 2014a. Suggested State Regulations for Control of Radiation. CRCPD. https://www.crcpd.org/SSRCRs/default.aspx (Accessed January 12, 2015).

Conference of Radiation Control Program Directors. 2014b. What is NEXT? https://www.crcpd.org/Pubs/NEXT.aspx (Accessed January 19, 2015).

Conference of Radiation Control Program Directors. 2014c. SR-F Committee's proposed changes to SSR Part F. CRCPD. (unpublished)

Health Canada. 2014. Medical X-Ray Safety Guidelines. http://www.hc-sc.gc.ca/ewh-semt/radiation/clini/xray/index-eng.php (Accessed January 12, 2015).

International Commission on Radiation Protection. 1977. Recommendations of the ICRP. ICRP Publication 26. Ann. ICRP 1 (3).

International Commission on Radiation Protection. 2007. The 2007 Recommendations of the International Commission on Radiological Protection. ICRP Publication 103. Ann. ICRP 37 (2–4).

Jones C. G. 2005. A Review of the History of U.S. Radiation Protection Regulations, Recommendations, and Standards. Health Phys. 88:6, 2005.

McBurney, RE. 2008. Regulation of Radiation Generating Devices: Challenges and Approaches for State Radiation Control Programs. Proceedings of the 2008 HPS Midyear Meeting on Radiation Generating Devices, Oakland, CA.

McMahon Bill (Atomic Energy Act of 1946). Senate Special Committee on Atomic Energy, Atomic Energy Act of 1946. Hearings on S. 1717. Washington DC: January 22–April 4, 1946.

National Council on Radiation Protection and Measurements. 1987. Recommendations on Limits for Exposure to Ionizing Radiation. Bethesda, MD: NCRP Report No. 91.

National Council on Radiation Protection and Measurements. 2015. About NCRP. Bethesda, MD. http://www.ncrponline.org/AboutNCRP/About_NCRP.html (Accessed January 12, 2015).

Public Law 109-58. 109th Congress. 2005. Energy Policy Act of 2005. http://www.gpo.gov/fdsys/pkg/PLAW-109publ58/pdf/PLAW-109publ58.pdf (Accessed January 12, 2015).

U.S. Environmental Protection Agency. 2013. Radiation Protection Guidance for Diagnostic and Interventional X-Ray Procedures: Federal Guidance Report No. 14 (Proposed). EPA. http://www.epa.gov/radiation/federal/fgr-14.html (Accessed January 12, 2015).

U.S. Food and Drug Administration. 2004. Mammography Quality Standards Act. FDA. http://www.fda.gov/Radiation-EmittingProducts/MammographyQualityStandardsActandProgram/Regulations/ucm110823.htm (Accessed January 12, 2015).

U.S. Nuclear Regulatory Commission. 2006. Title 10, Code of Federal Regulations, Part 30. NRC. http://www.nrc.gov/reading-rm/doc-collections/cfr/part030/part030-0004.html (Accessed January 12, 2015).

U.S. Nuclear Regulatory Commission. 2013. Nuclear Regulatory Legislation, NUREG-0980, Vol. 1, No. 10. NRC.

U.S. Nuclear Regulatory Commission. 2014a. Advisory Activities. Advisory Committee on the Medical Use of Isotopes. NRC http://www.nrc.gov/about-nrc/regulatory/advisory/acmui.html (Accessed January 12, 2015).

U.S. Nuclear Regulatory Commission. 2014b. Agreement State Program. NRC. http://www.nrc.gov/about-nrc/state-tribal/agreement-states.html (Accessed January 12, 2015).

U.S. Nuclear Regulatory Commission. 2014c. NRC. Title 10, Code of Federal Regulations. http://www.nrc.gov/reading-rm/doc-collections/cfr/ (Accessed January 12, 2015).

U.S. Nuclear Regulatory Commission. 2014d. NRC. Advance Notice of Proposed Rulemaking. Federal Register 2014-17252. July 2014. http://www.regulations.gov/#!documentDetail;D=NRC-2009-0279-0067 (Accessed January 12, 2015).

U.S. Nuclear Regulatory Commission. 2014e. Title 10, Code of Federal Regulations, Part 35. http://www.nrc.gov/reading-rm/doc-collections/cfr/part035/ (Accessed January 19, 2015).

U.S. Nuclear Regulatory Commission. 2014f. Title 10, Code of Federal Regulations Part 71. http://www.nrc.gov/reading-rm/doc-collections/cfr/part071/ (Accessed January 19, 2015).

CHAPTER 13

Regulatory Structures and Issues in Latin America

Simone Kodlulovich Renha
National Nuclear Energy Commission, Rio de Janeiro, Brazil, Asociación Latinoamericana de Física Médica (ALFIM)

Lidia Vasconcellos de Sá
National Nuclear Energy Commission, Rio de Janeiro, Brazil, Asociación Latinoamericana de Física Médica (ALFIM)

Ileana Fleitas Estévez
PAHO Pan-American Health Organization, Havana, Cuba, Asociación Latinoamericana de Física Médica (ALFIM)

CONTENTS

LATIN AMERICA is a region of great social, economic, and cultural diversity; therefore, different patterns of technological development can be found. A wide variety of radiological sources and practices can be observed. However, most countries have implemented practices applying radiation for medical, industrial, agricultural, and research purposes.

Epidemiological data show that non-communicable diseases (NCDs) are the leading cause of preventable and premature death and illness in the region. Three of every four deaths occur due to NCDs and 34% of all deaths are considered to be premature (30 to 69 years). Four NCDs account for the greatest burden of disease in the region: cardiovascular diseases, cancer, diabetes, and chronic respiratory diseases (PAHO 2013). Diagnostic imaging and radiation therapy services are essential for early detection and effective treatment of these diseases, and are critical for improving quality of life. Diagnostic imaging services also cover a wide range of clinical applications in prenatal care, and maternal and child health. In addition, they play an important role in the early diagnosis of communicable diseases, such as pneumonia, HIV, and TB. Medical radiation technologies have seen an important evolution in recent decades. Each day new procedures and equipment are introduced in Latin American health services, but not always accompanied by an adequate infrastructure and health workforce. In this context, there is a need to maximize the benefits of medical radiation technologies and minimize the health risks for patients and staff. Regulatory framework plays an essential role in this objective, and in allowing health authorities to work together.

13.1 GENERAL DESCRIPTION

The region of Latin America and the Caribbean comprises 33 countries, including Antigua and Barbuda, Argentina, Bahamas, Barbados, Belize, Bolivia, Brazil, Chile, Colombia, Costa Rica, Cuba, Dominica, Dominic Republican, Ecuador, El Salvador, Guatemala, Guyana, Granada, Haiti, Honduras, Jamaica, Mexico, Nicaragua, Panama, Paraguay, Peru, Saint Kitts and Nevis, Trinidad and Tobago, Saint Lucia, Saint Vincent and Grenadines, Suriname, Uruguay, and Venezuela. More than 590 million people live in the region, which represents 8.5% of the world population in a total area of about 20 million km2 (UN 2011). The human development index (HDI), as a composite statistic of life expectancy, education level, and income, ranges from first-place Chile (37th in the world) to last-place Haiti (168th in the world). Some countries have a high level of industrialization, mainly in the petroleum industry, as with Brazil and Venezuela. Argentina, Brazil, and Mexico, have nuclear power plants, while Chile, Peru, Colombia, and Jamaica are among those operating research reactors. Uruguay and Mexico, amount others, have cyclotrons to produce radionuclides for medical purposes.

With a gross regional income of 9,393 USD per capita (2013) and an annual gross domestic product (GDP) growth of 2.9% (2012–2013) according to PAHO (Pan-American Health Organization), national or total expenditures in healthcare and related goods and services (public and private) in Latin America and the Caribbean (LAC) represent 6.7% of the region's GDP and an average per-capita expenditure of USD 661 (PAHO 2014).

The public/private mix of health expenditure in the region is 52/48. Most of these countries provide a middle level of assistance: Antigua and Barbuda, Argentina, Colombia, Cuba, Montserrat, St. Vincent and the Grenadines, Anguilla, Chile, Costa Rica, Dominica, Mexico, Panama, Peru, and Uruguay, with the investment from $8,000 to 20,000 PPP (Per Capita Total Expenditure on Health) in the region's health system. Countries such as: Bolivia, Guyana, Honduras, Nicaragua, Paraguay, El Salvador, Guatemala and Haiti provide a low health assistance level with less than $8,000 in investments. Only Barbados, Bahamas, Aruba, and Trinidad and Tobago have a high level of assistance (PAHO 2012). Moreover, following the growth of life expectancy to around 74 years, certainly the health services available to the population are expected to grow at the same rate. In this scenario of investment, modern medical technology has reached most countries in the region, but in a heterogeneous manner and often without an adequate infrastructure (WHO 2014).

As a result, there is no uniformity in laws and standards regulating the use of radioactive sources or ionizing radiation-emitting equipment. The complexity of legal systems can result in standards and regulations not updated to current technologies or to international recommendations. Furthermore, legislation and standards generally depend on approval from more than one regulatory authority, such as the Ministry of Health or Nuclear Authorities.

Taking into account the rapid increase in the number of health centers and equipment and also their complexity, the deficit of well-trained professionals has been exacerbated. The number of accredited institutions that provide education and training in the region is much lower than required and mainly concentrated in Argentina, Brazil, and Cuba. The lack of adequately trained health workers, particularly medical physicists, can endanger patient safety. A properly established governmental, legal, and regulatory framework for security and safety should be provided to deal with risks in all exposure situations.

13.2   REGULATORY FRAMEWORK

There is a hierarchy of responsibilities within the framework, from governments to regulatory bodies to the organizations responsible for, and the persons engaged in, activities involving radiation exposure (Table 13.1). The government is responsible for the national legal system of such legislation, regulations, standards, and measures as may be necessary to fulfil all national and international obligations effectively, and for the establishment of an independent regulatory body. In Latin America, about 20 countries have a formal regulatory authority in place to address radiation protection and safety, but do not always have sufficient technical capability or resources to fulfil their functions (PAHO 2007). These authorities may be located in the ministry of health or in another ministry or even in another governmental organization. Coordination between them, although desirable, is not always adequate. Quality and safety of healthcare services is always a responsibility of the health authorities. Regulatory bodies with competences for radiation protection and safety should work with the health authorities for an effective and sustainable regulatory program for medical exposures. Currently, health authorities and medical professional societies are insufficiently involved in producing or adopting justification guidelines. Radiation doses to the patients are not optimized and Diagnostic Reference Levels are not established in Latin American countries. Supporting regulatory services, such as education and training centers, and a metrological infrastructure for calibration and dosimetry, are also lacking in many countries.

Table 13.1: Homepages for regulatory bodies discussed in this chapter

Country	Regulatory Body	Homepage
ARGENTINA	National Regulatory Authority (ARN)	www.arn.gob.ar
	Ministry of Health	http://www.msal.gov.ar
ANTIGUA AND BARBUDA	Ministry of Health, Social Transformation, and Consumer Affairs	http://www.antigua.gov.ag
BAHAMAS	Ministry of Health	http://www.bahamas.gov.bs/health
BARBADOS	Ministry of Health	http://www.health.gov.bb/
	Ministry of Energy and the Environment, Environmental Protection Department	www.energy.gov.bb
BELIZE	Ministry of Health	http://health.gov.bz/
	Environmental Department	http://www.doe.gov.bz/
BOLIVIA	Bolivian Institute of Technology and Nuclear Science (IBTEN), Ministry of Sustainable Development	http://www.ibten.gob.bo/
BRAZIL	National Commission of Nuclear Energy (CNEN)	http://www.cnen.gov.br
	Ministry of Health, National Health Sanitary Agency	www.anvisa.gov.br
CHILE	Chilean Nuclear Energy Commission (CCHEN)	http://www.cchen.cl
	Sanitary Authority Health Ministry	http://web.minsal.cl
COLOMBIA	Ministry of Mines and Energy	http://www.minminas.gov.co
	Ministry of Health	http://www.minsalud.gov.co
COSTA RICA	Ministry of Health	http://www.ministeriodesalud.go.cr
CUBA	National Center for Nuclear Safety, Ministry of Science, Technology, and Environment	http://www.medioambiente.cu/oregulatoria/cnsn
	Regulatory Central Group, Ministry of Public Health	http://www.sld.cu

Continued on next page

Table 13.1 — *Continued from previous page*

Country	Regulatory Body	Homepage
DOMINICA	No Regulatory Authority	
DOMINICAN REPUBLIC	National Energy Commission (CNE)	http://www.cne.gov.do
EL SALVADOR	Radiation Advisory and Regulatory Unit, General Health Direction, Ministry of Health	http://www.salud.gob.sv
ECUADOR	Secretariat for Control and Nuclear Applications, Ministry of Electricity and Renewable Energy	http://www.energia.gob.ec
GRENADA	Ministry of Health	http://www.gov.gd/ministries/health.html
GUATEMALA	Department of Protection and Radiological Security (DPSR), Ministry of Mines and Energy	http://www.mem.gob.gt
GUYANA	Ministry of Health	http://www.health.gov.gy/
HAITI	National Authority for Radiological Safety, Ministry of Health (ANSR)	http://mspp.gouv.ht/ http://mspp.gouv.ht/site/downloads/Loi_Organique.pdf
HONDURAS	Secretary of Natural Resources, Environment, and Mining (SERNA)	http://www.serna.gob.hn
JAMAICA	Radiation Safety Authority (RSA) in the Bureau of Standards Jamaica (BSJ)	http://www.bsj.org.jm/
MEXICO	National Commission of Nuclear Security and Safeguards	http://www.cnsns.gob.mx
	Federal Commission for the Protection against Sanitary Risk (COFEPRIS)	http://www.cofepris.gob.mx
NICARAGUA	National Commission of Atomic Energy.	http://www.minsa.gob.ni/
PANAMA	General Directorate for Health, Ministry of Health	http://www.minsa.gob.pa

Continued on next page

Table 13.1 — *Continued from previous page*

Country	Regulatory Body	Homepage
PARAGUAY	Radiological and Nuclear Regulatory Authority	http://www.escritosdederecho.com/2014/06/ley-5169-del-8-de-mayo-de-2014.html
PERU	Peruvian Institute of Nuclear Energy, Ministry of Energy and Mines	http://www.ipen.gob.pe
SAINT LUCIA	No regulatory body	
ST. VINCENT AND THE GRENADINES	No regulatory body	
ST. KITTS AND NEVIS	No regulatory body	
SURINAME	No regulatory body	
TRINIDAD AND TOBAGO	Ministry of Health	http://www.health.gov.tt/
URUGUAY	National Regulatory Authority for Radiation Protection	http://www.arnr.gub.uy
VENEZUELA	Direction of Radiological Health, Ministry of Health	www.mpps.gob.ve

Note: *This table reflects the current status of the regulatory authorities for medical practices in Latin America. Many countries are changing or updating their regulatory framework in order to comply with international recommendations. Some regulatory authorities were created or have changed their scope recently. The websites change continuously, and many of them are in construction or updating.*

Regarding the Caribbean countries, most of them are in need of establishing a comprehensive legal and regulatory framework for radiation safety and security of radioactive sources and radiation-emitting equipment, although some of these countries have pieces of legislation or regulations addressing some aspects of radiation safety. Jamaica, as well as Trinidad and Tobago, have made recent progress in establishing a legal and regulatory framework and setting up their regulatory bodies. All countries have diagnostic X-ray machines while some have radioactive sources in medical diagnosis and/or treatment units and linear accelerators. Many of these facilities are not licensed by a legally empowered regulatory body. Being conscious of this situation, international organizations such as the IAEA and PAHO have promoted some collaborative initiatives aimed at creating regional capabilities in Caribbean countries to improve access to safe and secured radiation technologies for applications in health and industry. On June 11–15, 2012, they organized the

Caribbean Community Regional Workshop on Regulatory Infrastructure for the Control of Radioactive Sources, in Kingston, Jamaica (CARICOM, 2012). All delegations recognized the need for establishing their national infrastructure for radiation safety and security, and that the political commitment exists. A regional approach for establishing the legal and regulatory framework was also discussed as it would facilitate harmonization and compensate for the lack of expertise, but might be too slow to satisfy the urgent need of some countries.

A general overview of the framework of Latin American and Caribbean countries addressing Radiotherapy (RT), Nuclear Medicine (NM) and Diagnostic Radiology (DR) can be found further on in this chapter.

13.2.1   Argentina

Argentina folows Law No 24.804, 2/4/97, the National Law for Nuclear Activity. Its regulatory authorities are the Autoridad Reguladora Nuclear (ARN), Nuclear Regulatory Authority, (www.arn.gob.ar) and the Ministerio de Salud, Ministry of Health, (http://www.msal.gov.ar).

Medical uses of radiation are regulated by ARN for RT and NM services, and by the Ministry of Health for DR. ARN is an autonomous agency within the jurisdiction of the President; it is responsible for promulgating regulation and standards. For DR, the competent authority is the National Health Authority, for developing and implementing laws, regulations and guidelines. The licensing process includes classification of installations based on complexity and radiological risks, granting Authorizations (Class I and II) or Registrations (Class III). The process includes Building Permit, Commissioning, Operation and Decommissioning Licenses, Specific Individual Licenses, and Authorizations. Regulatory activities include documentation review, inspections, and audits. The Basic Safety Regulation for Radiation Protection (AR 10.1.1) and 62 standards for practices are established in the country. The National Atomic Energy Commission (CNEA) is the State Argentine Agency responsible for advising the National Executive Government in the definition of nuclear policies, promoting research and development and peaceful applications such as exploration of raw materials, environmental remediation, construction of nuclear power plants and cyclotrons as well as radionuclides production and uranium enrichment.

13.2.2   Antigua and Barbuda

At present, there is no existing law or bill related to radiation sources, nor is there any regulation relating to the medical uses of radiation. The Ministry of Health has some general regulations for health-related activities, but nothing specific for radiation services. Antigua and Barbuda is presently working towards developing a Cancer Treatment Center and requires development of laws and regulations on safety and security of radiation sources as soon as possible.

13.2.3 Bahamas

In the Bahamas, there is no comprehensive legislative and statutory framework establishing national regulatory authority or governing the use and possession of radiation sources currently in place.

The Statute Law has specific chapters referring to ionizing radiation (protection of workers, custom management, environmental health services, import and export of prohibited and restricted substances). Chapter 232, Enabling Act, empowers the Minister of Health to make regulations for the control and prevention of radiation hazards and the disposal of radioactive or otherwise hazardous wastes. The Act is not enforced. In the absence of a legal framework, a national system for authorization, inspection, and enforcement has not been established.

The Hospital and Health Care Facilities (General) Regulations, 2000 establishes requirements to be met by all licensed healthcare facilities. Regulations require that each clinic establishes a preventive maintenance program to ensure that all equipment is maintained in accordance with specifications of the manufacturers, and requirements for radioactive waste management and post areas to prevent inadvertent entry and exposure. This legislation contains a section dedicated to radiation oncology clinics.

13.2.4 Barbados

In Barbados, legislation and rules concerning radiation include the Radiation Protection Act, Chapter 353A, 1971; the Customs Act 1963 as amended 2010; Miscellaneous Controls (Importation and Exportation of Goods, Prohibition, Radioactive Materials) Regulations, 2004; Customs (List of Prohibited and Restricted Imports and Exports) Order, 2009, paragraphs 3, 5(2), Second Schedule; and Local Hospital Radiation Rules 2004.

No regulatory body exists for the authorization, inspection, and enforcement of related activities. The importation license is granted by the Environmental Protection Department (permission given through the Miscellaneous Customs Act). There is limited legislation to govern the use of ionizing radiation or the management of radioactive sources in radiation medicine services. The Ministry of Health Council regulates health professionals through relevant registration legislation.

13.2.5 Belize

In 1992, Belize instituted the Environmental Protection Act (EPA), Chapter 328 of the Laws of Belize. The Department of Environment (DOE) is a regulatory body within the Ministry of Fisheries, Forestry and Sustainable Development responsible for monitoring the implementation of the EPA and regulations and to take the necessary actions to enforce its provision. Part I of the EPA refers to radioactive sources as a hazardous substance that may be able to harm/affect human beings, other living creatures, plants, micro-organisms, property of the environment, including but not limited to the substances specified in the schedule. Part I, Subpart A, # 6 of the EPA refers to high-level radioactive wastes or other high-level radioactive matter. Subpart B, # 14 refers to other radioactive matter not included in Part I Subpart A.

The Ministry of Health is in charge of regulating health technologies. There is no specific legislation addressing the use of ionizing radiation or the management of radioactive sources in radiation medicine services.

13.2.6 Bolivia

Bolivia has enacted the Decree Law No. 19172, 19/09/1982, the Law on Protection and Radiation Safety.

Supreme Decree No. 24483, 22/12/2005 refers to Regulations of the Law 19172, in which IBTEN was recognized as the National Competent Authority.

The Regulatory Authorities are the Instituto Boliviano de Ciencia y Tecnología Nuclear (IBTEN), Ministerio de Desarrollo Sostenible, Bolivian Institute of Nuclear Science and Technology (IBTEN), Ministry of Sustainable Development.

As the National Competent Authority (ANC), IBTEN is responsible for all programs for the control of ionizing radiation and quality management in radiation sources applications, fulfilling the requirements of the Law on Protection and Radiation Safety. The regulatory framework for the medical uses of radiation includes requirements for transport, movement, import, and export of radioactive sources and X-ray generating equipment. There are also specific regulations for the licensing of radiological medical services (RD, NM and RT) as well as for the individual licensing of health workers professionally exposed to ionizing radiation. Information may be found at: http://www.ibten.gob.bo/portal/index.php?opt=front &mod=contenido &id=131&pid=72.

13.2.7 Brazil

Brazil's Law No 4.118, 27/8/1962: Lei de Atividades Nucleares in the National Law for Nuclear Activity.

Law No9.782,1/26/1999 created the Agência Nacional de Vigilância Sanitária, National Health Sanitary Agency.

The Regulatory Authority is the Comissão Nacional de Energia Nuclear (CNEN). The National Commission of Nuclear Energy is an autonomous agency at the Ministry of Science, Technology and Innovation that regulates RT and NM (www.cnen.gov.br). The Ministério da Saúde, Agência National de Vigilância Sanitária, Ministry of Health, National Health Sanitary Agency, regulates DR (www.anvisa.gov.br).

CNEN supports the government in the formulation of the National Nuclear Energy Policy, performs research, and promotes and provides services in nuclear technology and its applications for peaceful purposes. CNEN establishes regulations, carries out the licensing and authorization process, including inspections in nuclear and radioactive facilities and grants licenses for site approval, construction, operation, and decommissioning. The medical facilities are classified in three groups based on complexity and radiological risks. The more complex facilities require site approval, building license, license commissioning, operation and decommissioning license. Specific Individual Licenses are mandatory in NM and RT. Standards are established for nuclear facilities, security, radiation protection, nuclear materials and minerals control, radiotherapy, nuclear medicine, radioactive waste, transport, decommissioning and personal certification.

ANVISA is an independent agency of the Ministry of Health responsible for products and services that may affect the health of the population. In the medical field, ANVISA regulates and inspects DR installations, besides ensuring the sanitary conditions of all health services. For Diagnostic Radiology, the Minister of Health does not have an individual professional certification process established in a nationwide and legal manner. Although not officially recognized, CNEN and ANVISA maintain permanent collaboration in the development of complementary occupational and medical exposures standards and carrying out joint inspections.

13.2.8 Chile

In Chile, Law No 16.319, 16/4/1964: created the Comisión Nacional de Energía Nuclear, currently Comisión Chilena de Energía Nuclear (CCHEN), Chilean Nuclear Energy Commission. The Regulatory Authority is the Comisión Nacional de Energía Nuclear (CCHEN), an autonomous agency at the Ministry of Energy, which regulates RT and NM (http://www.cchen.cl). The Autoridad Sanitaria Ministerio de Salud, Sanitary Authority Health Ministry, regulates DR (http://web.minsal.cl).

CCHEN is the competent authority to evaluate and control all radioactive facilities. It is responsible for proposing laws and regulations to the government for the nuclear and radiological fields. The regulatory activities include: health, industry, mining, agriculture and food, addressing practices related to the production, acquisition, transfer, transport, and import and export of radiation sources. CCHEN regulates, authorizes, and oversees nuclear and radioactive sources according to the risk installation category. CCHEN grants authorization for construction, transport, import, and export of radioactive sources. Standards are established for building and operation licenses, export and import authorizations, transport licenses and personal authorizations.

13.2.9 Colombia

Columbia's Law No 90874, 11/8/2014 created the Ministerio de Minas y Energía, the Ministry of Mines and Energy. Law No9031, 12/4/1990 created the Ministerio de Salud, the Ministry of Health.

Its Regulatory Authority is the Ministerio de Minas y Energía, Ministry of Mines and Energy, which regulates RT (Cobalt, Brachytherapy) and NM (http://www.minminas.gov.co/). The Ministerio de Salud or Ministry of Health, regulates DR and RT (Linear Accelerator) (http://www.minsalud.gov.co/).

The Ministry of Mines and Energy is the regulatory authority responsible for promoting national policy and regulations on nuclear and radioactive applications. The regulatory standards include: regulation of radiation protection and safety, radioactive sources categorization, license management of radioactive materials, authorization and inspections for use, import, and transport authorization, radioactive waste management, and licensing to provide personal dosimetry service. There is also personal certification required to handle radioactive sources in place. Despite Law No 90874 that gives the Ministry of Mines the regulatory responsibility, the Law No9031 of Ministry of Health establishes requirements to be accomplished by the user.

13.2.10 Costa Rica

Law No 5395, 30/10/1973, Ley General de Salud is Costa Rica's General Health Law. Law No 4383, 8/18/1969, created the Comisión de Energía Atomica, the Atomic Energy Commission. Its Regulatory Authority is the Ministerio de Salud, Ministry of Health, which oversees Radiation Protection Regulation, 18/12/1994 (http://www.ministeriodesalud.go.cr).

The Ministry of Health is the competent authority to regulate and control all radioactive facilities, including the medical uses of radiation. It is responsible for interpreting laws and regulations about safety requirements for the nuclear and radiological fields. The regulatory activities are related to the production, acquisition, transfer, transport, import, and export of sources in the health and industry fields. It also regulates, authorizes, and oversees nuclear and radioactive sources, including waste management; the installations are classified in four risk categories. A certification process has been established for operators and radiation protection officers. General requirements for protection of workers and patients are included in Act 24037-S, which regulates all practices, but there are no specific regulations for RT, NM, and RD. According to Law No 4383, the Comisión de Energía Atomica promotes peaceful uses of ionizing radiation, its applications, development, and research. Its main role is to promote international cooperation in the peaceful uses of atomic energy.

13.2.11 Cuba

In Cuba, the relevant legislation includes Law Decree No207, 17/02/2000: On the Use of Nuclear Energy; Joint Resolution CITMA-MINSAP, 2002: Basic Safety Standards for Radiation Safety; Resolution 25/98 CITMA: Authorization of practices associated with the use of ionizing radiation; Resolution 437/2000 MINSAP: Regulation for authorization of practices associated with medical and dental radio diagnosis. The Regulatory Authority is the Centro Nacional de Seguridad Nuclear (CNSN), Ministerio de Ciencia, Tecnología y Medio Ambiente (http://www.medioambiente.cu/ oregulatoria/cnsn), National Center for Nuclear Safety, Ministry of Science, Technology and Environment; also involved are the Grupo Central Regulatorio (GCR), Ministerio de Salud Pública; Regulatory Central Group, Ministry of Public Health (http://www.sld.cu/).

The Ministry of Science, Technology and Environment (CITMA) executes regulation and control through CNSN, which is an independent body created in 1991 and composed of the national center and territorial delegations. Among others, they regulate radiation safety for radiotherapy and nuclear medicine services, granting authorization for the use of radiation sources, operating licenses, registrations, permissions, and certification of personnel. Practices are categorized according to the associated risks. Laws, regulations, and guides are based on the IAEA Security and Safety Standards Series. Some examples are Resolution 6/2004-CITMA, "Recognition of Competence for Radiation Safety Services;" Resolution 35/2003-CITMA, "Regulation for the safe management of radioactive wastes," as well as the Resolution 25/98 — CITMA "Authorization of practices associated with the use of ionizing radiation." The GCR from MINSAP regulates, in coordination with CITMA, X-ray diagnostic and dental services, and grants operating licenses or permits and certification of personnel.

13.2.12 Dominica

Relevant legislation in Dominica has included the Environmental Health Services Act of 1997, the Solid Waste Management Act, Schedule on Hazardous Waste 2000, the Occupational Health and Safety Act, the Accident and Notification Act, and the Standards Act.

There is no specific regulatory infrastructure related to radiation sources, or regulations for the medical uses of radiation. The Dominica Bureau of Standards acts as a Regulatory Agency, but has no specific mandates on radiation safety.

13.2.13 Dominican Republic

Law Decree No244-95, 10/13/1995: Reglamento de Protección Radiológica is the Dominican Republic's Radiological Protection Regulation. Law No496-06, 12/28/2006: delegates to the Comisión Nacional de Energía (CNE), or

National Energy Commission, the attributions of the National Commission for Nuclear Affairs. The regulatory authority is the Comisión Nacional de Energía (CNE), the National Energy Commission (http://www.cne.gov.do/).

CNE regulates the medical uses of radiation in the country. Departments of the Radiation Protection and Physical Security of Radioactive Sources were created in the CNE organizational structure to regulate the activities involving the use of radioactive substances and ionizing-radiation-generating devices, and to promote the applications of nuclear technology. The regulatory policy (Decree No244-95) applies to the production, purchase, import, export, use, and possession of ionizing radiation sources for industrial, medical, veterinary, agricultural, research, and teaching purposes, as well as to the transfer, transport, and storage of sources, and radioactive waste management.

13.2.14   El Salvador

In El Salvador, Law Decree No 41, 3/15/2002, Reglamento Especial de Protección y Seguridad Radiológica, contains Special Regulations for Radiological Protection and Radiation Safety.

The Regulatory Authority is the Unidad Reguladora y Asesora de Radiaciones (UNRA), Dirección General de Salud del Ministerio de Salud (MinSal), i.e., the Radiation Advisory and Regulatory Unit, General Health Direction, Ministry of Health (http://www.salud.gob.sv/).

UNRA controls and supervises the practices involving radiation devices and radioactive sources in medicine, industry, agriculture, research, and teaching. It is responsible for authorization, inspection, and control. The licensing process is classified into three risk-based categories. Ten standards have been published at http://www.salud.gob.sv/temas/politicas-de-salud/dir-reg-y-leg-en-salud/unra.html#RN, and are available on the homepage of the regulatory authority: Radiologic Protection and Safety, Diagnostic and Interventional Radiology, Quality Control for DR and RT, NM, Industrial Radiography X-ray and Gammagraphy, Management of Radioactive Waste and Transport.

13.2.15   Ecuador

In 2008, Ecuador's Law Decree 978, 4/8/2008 created the Ministerio de Electricidad y Energía Renovable, i.e., the Ministry of Electricity and Renewable Energy. Law Decree no3640, 8/8/1979: Reglamento de Protección Radiologica is the Regulation for Radiation Protection.

The Regulatory Authority is Subsecretaria de control y aplicaciones nucleares (SCAN), Ministerio de Electricidad y Energía Renovable — the Secretariat for Control and Nuclear Applications, Ministry of Electricity and Renewable Energy (http://www.energia.gob.ec/s).

Since 2008, the Government has incorporated the Ecuadorian Atomic Energy Commission into the Ministry of Electricity and Renewable Energy. The Secretariat for Control and Nuclear Applications (SCAN), under this

Ministry, is responsible for controls and regulates the peaceful use of ionizing radiation, including medical uses. Its main activities are the production, purchase, transportation, importation, exportation, use, and handling of radioactive radioisotopes and ionizing radiation-generating machines. Furthermore, SCAN advises the government, public agencies, and private sector on nuclear energy applications. SCAN issues licenses to individuals and institutions, issues importation permits, acts as National Liaison Office with IAEA, and provides technical services such as personal dosimetry and calibration, among others. All the regulatory activities are based on the Regulation of Radiation Safety. There are no specific regulatory standards applicable to each practice involving ionizing radiation in the country.

13.2.16 Grenada

The Waste Management Act # 16 of 2001 identifies waste from radioactive sources as hazardous waste. In Grenada the Ministry of Health is the designated or acting regulatory body for the authorization, inspection, and enforcement of activities related to radiation sources and radiation-based health technologies. There is no national inventory of radioactive sources and uses, nor a specific regulatory framework related to radiation protection in medical uses.

13.2.17 Guatemala

In Guatemala, Law Decree No11-86, 1/10/1986: Control, uso de radioisótopos y las radiaciones ionizantes, aplicaciones, law for control, use, and application of radioisotopes and ionizing radiation. The Regulatory Authority is the Departamento de Protección y Seguridad Radiológica (DPSR) Ministerio de Minas y Energía — the Ministry of Mines and Energy (http://www.mem.gob.gt/).

DPSR is responsible for overseeing all activities related to the control, use, and application of radioisotopes and ionizing radiation in medicine, including installation and operation, production, use, transport, import, and export, and performs inspections. DPSR establishes regulations and standards addressing the minimum requirements, permits, authorizations, or licenses for individuals or entities. Law No055-2001 establishes four classes of risk-related practices and licencing process, including personal permits and requirements for specific practices. Moreover, Law No559-1998 establishes minimum requirements for safe waste management.

13.2.18 Guyana

Guyana's Health Facilities Licensing Act 2007 and its Regulations 2008 contain specific requirements for imaging and radiation therapy facilities. The Program of Standards and Technical Services, Ministry of Health, issues licenses, performs inspections, and develops standards and regulations. Although not specific to radiation medical applications, the Health Facilities Licensing Act 2007 includes quality, safety, and personnel requirements.

13.2.19 Haiti

By Presidential Decree in 2005 Haili created the Autorite National de Surete Radiologique (ANSR), i.e., the National Authority for Radiological Safety.

ANSR develops practice guidelines and regulations, controls and monitors radioactive facilities and grants, and amends and suspends permits or licenses to operators. Currently, activities related to inspection and authorization, inventory, survey, as well as monitoring of exposed workers, including medical uses of radiation, are carried out by the Unit of Dosimetry and Radiation Protection of the Ministry of Health, which is awaiting full staffing of ANSR by the Government.

13.2.20 Honduras

In Hounduras Law Decree 198-2009, 11/15/2009, Ley de Seguridad Nuclear y Protección Radiológica, is the Law for Nuclear Security and Radiation Protection. The Regulatory Authority is the Secretaria de Recursos Naturales, Ambiente e Minas (SERNA) — Secretary of Natural Resources, Environment and Mining (SERNA) (http://www.serna.gob.hn/).

According to Decree 198-2009, the National Regulatory Authority ARN is part of the Ministry of Natural Resources and Environment (SERNA) and is responsible for the regulation and strengthening of radiation protection in medical, industrial, and research applications. Radioactive facilities are classified into four risk categories. SERNA grants Operating and Construction Authorizations, Final Closing, Transport, Export and Import and individual permissions. The National Autonomous University of Honduras (UNAH) hosts the first dosimetry laboratory in the country. There are two duly approved regulations: the Regulation of Radiation Protection and Licensing for medical, industrial, and research facilities and for Transport of radioactive material. General Regulations of the Law are in the process of being approved.

13.2.21 Jamaica

Cabinet Decision No. 01/11 of 10 January 2011, creating Jamaica's Radiation Safety Infrastructure, gave approval for the establishment of a Radiation Safety Authority (RSA) in the Bureau of Standards Jamaica (BSJ). The BSJ is a public body under the Ministry of Industry, Investment, and Commerce, and is directed by a 14-member board (Standards Council of Jamaica). Cabinet Decision # 01/11 extended the authority of the BSJ to regulate the importation, storage, usage, transportation, and disposal of radiation sources. Based on this law, the BSJ established the Regulatory Authority for Radiation Protection under the direct supervision of the Director of Engineering. BSJ and the Ministry of Health have initiated technical coordination for the development of a regulatory framework for medical uses of radiation.

13.2.22 Mexico

Mexico has two laws regulating the uses of radiological material: Constitutional Law-Article 27, 2/4/1985: Ley Reglamentaria del Artículo 27 Constitucional en Materia Nuclear, Regulatory Constitutional Law of Article 27 in nuclear matters and Ley General de Salud, 6/14/1999, General Health Law (Article 17 bis).

The Regulatory Authority is the Comisión Nacional de Seguridad Nuclear y Salvaguardias (CNSNS), National Commission of Nuclear Security and Safeguards (http://www.cnsns.gob.mx). The Secretária de Salud or Health Secretary, administers the Comisión Federal para la Protección contra Riesgos Sanitarios (COFEPRIS) (http://portal.salud.gob.mx/, http://www.cofepris.gob.mx/).

The CNSNS has the responsibility to control the implementation and enforcement of nuclear, radiological, and physical security and safeguards for the operation of nuclear and radioactive facilities; to authorize local site, design, construction, operation, modification, and decommissioning; transport and management of radioactive waste; to issue permits and licenses; to provide advice on security measures for nuclear, radiological and physical safeguards stipulated by abnormal or emergency conditions; to establish and manage the national system of registration and control of nuclear materials and fuels; to control extraction and processing of radioactive minerals facilities; to control import and export of nuclear materials and fuel; to perform audits and inspections; and to impose coercive measures.

COFEPRIS, the Health Secretary for Sanitary Risks, has the responsibility to regulate, control, and promote sanitary issues. This regulatory body controls licensing for possession, sale, distribution, transport, and use of radiation sources and radioactive materials for medical purposes, as well as waste removal, dismantling them, and the management of their waste, plus the importation and exportation of sources. The requirements established in the Sanitary Regulation are complementary to CNSNS by referencing their standards.

13.2.23 Nicaragua

Nicaragua's relevant legislation is Law No156, 3/23/1993: Ley sobre Radiaciones Ionizantes or the Law on Ionizing Radiation. The Regulatory Authority is the Ministerio de Salud (MINSA), Ministry of Health (http://www.minsa.gob.ni/), and there is Comisión Nacional de Energía Atómica (CONEA), the National Commission of Atomic Energy.

Law No. 156 provides the regulatory framework for the control of practices involving the use of ionizing radiation, and basic requirements for authorization, inspection, enforcement, and monitoring of radiation sources. Decree No. 24-93 established CONEA as the Competent Authority and defines the scope of their duties and responsibilities, but the Ministry of Health has found an intermediate point in the President to designate CONEA and the Director General of Health regulation. According to the Decree 25-2006, the Directorate General of Health Regulation of the Ministry of Health is responsible for the regulation of all sources of ionizing radiation and practices. The regulatory authority has established standards for radiotherapy, nuclear medicine, diagnostic radiology, and industry.

The Ministry of Health is responsible for ensuring the implementation of the Act and its Regulations, as CONEA is responsible for conducting regulatory measures required by the Act and its Regulations. CONEA has the responsibility of regulating the use, control, and application of radioisotopes and ionizing radiation, and possession, import, export, transfer, and transport of radioactive material. There is a close relationship between the two authorities. CONEA is chaired by a representative of the Ministry of Health, who is also the chairman of the Ministerial Decree. CONEA is also integrated by delegates of the Labor Ministry, the Nicaraguan Institute of Social Security and Welfare, and the National Autonomous University of Nicaragua.

13.2.24 Panama

Radiological issues in Panama fall under Law. No66, 11/10/1947: Codigo Sanitario, Sanitary Code, and Law Decree No. 770, 8/16/2012: Regulación en Protección Radiológica, Radiological Protection Regulation. The Regulatory Authority is Dirección General de Salud, Ministerio de Salud, i.e., General Directorate for Health, the Ministry of Health Resolution No. 471, 06/19/2009, (http://www.minsa.gob.pa).

The General Directorate for Health is responsible for the regulatory control of ionizing radiation sources in medical practices at the national level. The granting of the different types of authorizations (construction, operation, transport, etc.) is based on a system of inspections for the assessment of radiation protection and safety conditions. Resolution No. 471 establishes the procedures and forms for performing such radiological inspections. Health Authorities are also responsible for personnel authorization. Other relevant publications are: 2600SEG270 Radiological Protection Manual from the Panama Canal Authority, and Resolution 02 (02/02/2013) that recognizes the profession of Radiation Protection Officer.

13.2.25 Paraguay

Law Decree No10754/2000, 10/6/2000: Reglamento Nacional de Seguridad para la Protección contra las Radiaciones Ionizantes y para la Seguridad de las Fuentes de Radiación, contains Panama's National Safety Regulations for

Protection against Ionizing Radiation and Safety of Radiation Sources. Law No 5169, 5/8/2014: created the Radiological and Nuclear Regulatory Authority.

The regulatory authority is the Autoridad Reguladora Radiológica y Nuclear(ARRN), i.e., the Radiological and Nuclear Regulatory Authority (http://www.escritosdederecho. com/2014/06/ley-5169-del-8-de-mayo-de-2014).

Previous to Law No. 5169, the National Atomic Energy Commission (CNEA, Law 1081/65) and the Ministry of Public Health and Social Welfare (MSPyBS, Law 836/80) were responsible for regulatory activities in all ionizing radiation applications. The implementation of Law No. 5169 create the ARRN and extends to all events or actions that generate a certain or likely potential exposure to ionizing radiation from activities or practices in the medical, industrial, agricultural, research, education, and nuclear fields. This law also applies to the generation of radioactive waste, disused sources, and transport of radioactive waste.

13.2.26 Peru

Law Decree No. 21094, 02/04/1977: Ley Orgánica del Sector Energía y Minas, Peru's Organic Law of the Energy and Mining Sector, created the Instituto Peruano de Energía Nuclear (IPEN), the Peruvian Institute of Nuclear Energy. Law Decree No. 21875, 07/05/1977: Ley Orgánica del Instituto Peruano de Energía Nuclear is the Organic Law of the Peruvian Institute of Nuclear Energy.

The regulatory authority is the Instituto Peruano de Energía Nuclear, Ministerio de Energía y Minas, the Peruvian Institute of Nuclear Energy, Ministry of Energy and Mines (http://www.ipen.gob.pe/), (http://www.minem.gob.pe/).

IPEN regulates, promotes, controls, and develops nuclear energy activities based on Law 28028 — Use of Ionizing Radiation Sources, and its regulations. Through its Technical Office (Oficina Tecnica de la Autoridad Nacional, OTAN) IPEN executes regulation and control activities by granting permits (registrations, licenses, and specific authorizations) and executing inspections. The regulated sources include radioactive and nuclear material and ionizing-radiation-generating equipment, including medical applications.

13.2.27 Saint Lucia

To date there is no law, ordinance, or bill related to radiation sources in Saint Lucia. There is no regulatory body to oversee the uses of radioactive sources or related issues. A legal framework needs to be established for medical uses of radiation.

13.2.28   St. Vincent and the Grenadines

In this island nation the Environmental Health Services Act No. 14 of 1991 makes provisions for the conservation and maintenance of the environment in the interest of health generally, and particular, in relation to places frequented by the public. Public Health Act No. 9 of 1977 makes provisions for securing and maintaining health.

There are no specific laws, ordinances, bills, or regulations related to radiation sources. There is no regulatory body to oversee the uses of radioactive sources or issues relating to same. A legal framework needs to be established for the medical uses of radiation.

13.2.29   St. Kitts and Nevis

The National Conservation and Environmental Protection Act No. 5 of 1987 was succeeded by St. Kitts and Nevis' Act 21/2001. The Public Health Act No. 22 of 1969 is the main legislative instrument for managing environmental health issues in St. Kitts and Nevis.

There are no specific laws, ordinances, bills, nor regulations related to radiation sources. Section 3 of the Public Health Act empowers the Minister of Health to make provisions for all matters relating to the promotion or preservation of the health of the people. According to Section 10, the Minister may make regulations to give effect to the Act, including regulations for prevention of water pollution and waste disposal. The office of the Chief Medical Officer from the Ministry of Health in conjunction with the Ministry of Technology regulates health technologies. There is a lack of regulations regarding disposal of radioactive sources, safety, and security. A legal framework needs to be established for the medical uses of radiation.

13.2.30   Suriname

Suriname's Decree E 35, of May 25, 1983 deals with the Parate Execution which enables the Safety Inspector to temporarily terminate labor in case of immediate life-threatening situations. The Industrial Accidents Act (OSH Act), September 8, 1947 (GB 1947 No. 142) lastly amended by SB 1980 No. 116, is a framework act on safety and industrial hygiene in enterprises. Safety Regulation No. 8 (Ionizing Radiations Decree) of May 30, 1981 (SB 1981 No.73), aims to prevent industrial accidents and occupational diseases.

Labor Inspectorate, Ministry of Labor, is the acting regulatory body for the authorization, inspection, and enforcement of activities related to radiation sources. The Ministry of Health regulates health technologies in general by issuance of a permit, but there are not specific regulations for radiation-based health technologies. There is a need for a competent and designated authority to establish an effective radiation safety and security regulatory infrastructure.

13.2.31 Trinidad and Tobago

For Trinidad and Tobago, the Ministry of Health Occupational Safety and Health Management Policy (OSH Policy) of February 2012 is aimed at protecting the health, safety, and welfare of workers, visitors, and users of its facilities and those of the Regional Health Authorities.

The Radiation Protection Policy (RPP) of the Ministry of Health, currently in draft form, outlines recommendations for national radiation safety, including Ionizing Radiation Regulations (IRR) and Ionizing Radiation Medical Exposure Regulations (IRMER) based on the International Basic Safety Standards (BSS).

13.2.32 Uruguay

Uruguay's relevant legislation is Law No19.056, 1/4/2013: Protección y Seguridad Radiológica de Personas, Bienes y Medio Ambiente, i.e., Radiation Protection and Safety of People, Goods and Environment. The Regulatory Authority is the Autoridad Reguladora Nacional en Radioprotección, Ministerio de Industria, Energía y Minería — National Regulatory Authority for Radiation Protection, created by Law 17.930, 12/19/2005, Ministry of Industry, Energy and Mines (http://www.arnr.gub.uy/). The National Decree 270 of Law 19.056, September 22, 2014, specifies the responsibilities of the National Regulatory Authority.

The Reglamento Básico de Protección y Seguridad Radiológica, UY 100 (Basic Regulation for Radiation Protection and Safety) as well as 22 other specific regulatory standards for different practices, are currently in force. On July 29, 2014, a review of the UY 100 was published, taking into account the Fundamental Safety Principles from IAEA and the International Basic Safety Standards, GSR Part 3. The standards address the license process in all medical practices for five associated-risk grades, including transport, personal permits, and authorizations. The regulatory authority also provides individual dosimetry service. The National Regulatory Authority in Radiation Safety is the only such body, and in charge the control of all nuclear and radioactive applications in the country, including medical practices.

13.2.33 Venezuela

Venezuela's Law Decree No. 3.263 Resolution 401, 11/20/2004: Norma sanitaria para la autorización y el control de las radiaciones ionizantes en medicina, odontología y veterinaria, is the Sanitary Standard for authorization and control of ionizing radiations in medicine, odontology and veterinary practice. The Regulatory authority is the Dirección de Salud Radiológica, Ministerio del Poder Popular para la Salud (MPPS), Direction of Radiological Health, Ministry of Health (www.mpps.gob.ve/), along with the Dirección de Energía Atómica, Ministerio del Poder Popular para la Energía

Eléctrica (MPPEE), the Direction of Atomic Energy, Ministry of Energy (http://www.mppee.gob.ve/).

The Ministry of Health establishes the sanitary standards for medical, odontology, and veterinary practices. Resolution 401 (11/04/2006) defines requirements for authorization, registration and operational permits for NM, RT, and DR services. The Atomic Energy Direction is responsible for authorizations and permits for other than medical applications. These two regulatory authorities delegate the metrological and dosimetric aspects to the Venezuelan Institute of Scientific Research (IVIC).

13.3 RADIATION PROTECTION OF PATIENTS AND WORKERS IN MEDICAL FACILITIES

Radiation protection means that the government, through a national regulatory body, provides regulations, standards, rules, and resources to regulate and control facilities, equipment, and radioactive materials to protect the worker, the patient, and the environment. According to GSR Part 3 (IAEA, 2014a), the regulatory body shall require specialized health professionals and that they fulfil the requirements for education, training, and competence in the relevant specialty. Furthermore, the regulatory body shall ensure that the authorization for medical exposures be performed by radiological medical practitioners, medical physicists, medical radiation technologists, and any other health professionals with specific duties. However, the compiled data, as well as the experience acquired in studies developed in Latin American countries in the last decade, showed that the regulatory authority and all the actors involved in medical practice should gather their efforts in order to accomplish the international recommendations.

Constant technological advances in diagnostic imaging and radiation therapy may have an impact on the population's radiation dose throughout the world is very difficult to predict. Although some developments have led to more sensitive and effective detection systems, the ease by which new technologies capture images could give rise to unnecessary radiation exposure of patients and workers.

13.3.1 Dosimetry of Patients

International recommendations advise that Registrants and Licensees shall ensure that dosimetry of patients is performed and documented by or under the supervision of a medical physicist, using calibrated dosimeters and based on internationally or nationally accepted protocols (IAEA, 2014b). Therapeutic radiological procedures are classified as "high doses procedures." Absorbed doses to the planning target volume and to relevant tissues or organs for each patient treated with external beam therapy, brachytherapy, and/or radiopharmaceuticals should be obtained. In medical imaging, "relatively high doses" would include doses from image-guided interventional procedures, computed tomography, and nuclear medicine, where typical doses to patients for common procedures must be estimated. In these procedures it is recommended to carry out an independent verification by a different, independent medical physicist, using different dosimetry equipment. In checking for compliance, the regulatory body needs to be aware of the limitations on local resources.

For Latin American countries to ensure the quality of procedures, as to carry out this independent verification, it is essential to build adequate capacity and competences. There is a lack of trained health professionals, especially medical physicists, as identified earlier (Kodlulovich, S. and de SÃą, L.V. 2013). A regulatory body needs to be aware of these limitations on local resources. Other methods, such as verification using a second set of dosimetry equipment, or by postal thermoluminescence dosimetry, are considered acceptable in checking for compliance.

As medical technology is generally imported in most Latin American countries, maintenance and calibration services continue to be a concern. The IAEA/WHO Secondary Standard Dosimetry Laboratory (SSDL) Network includes only 11 centers in the region: in Argentina, Brazil, Chile, Colombia, Cuba, Ecuador, Guatemala, Mexico, Peru, Uruguay, and Venezuela (IAEA 2014). The lack of standard dosimetry laboratories influenced country capacities to ensure the quality of service providers. The situation is also critical for carrying out the internal dosimetry of patients, because of the failure to standardize methodology and lack of financial resources.

13.3.2 Dosimetry for Occupationally Exposed Workers

IAEA recommends that a regulatory body establish and enforce requirements to ensure that protection and safety is optimized, and to enforce compliance with dose limits for occupational exposure. Monitoring and recording occupational exposures in planned exposure situations should be provided by government or by other institutions accredited by an official board (IAEA 2014). In general, occupational external dosimetry programs for RT services are in place in most countries; however, for DR and NM services, those programs are not always mandatory, and consequently, they are not established yet. Regarding internal occupational dosimetry, the number of laboratories is not sufficient and legislation is not addressing this issue. IAEA has made efforts to implement an internal dosimetry network and currently there are about 15 laboratories working to develop standard methodologies. The application of optically stimulated luminescence dosimeters (OSL) in occupational external dosimetry has increased, especially in Argentina, Colombia, Costa Rica, and Peru. Dosimetry for the lens of the eyes is implemented only in Argentina and Colombia as a regular program; it is expected that this technology should be implemented in all regions over the short term.

13.4   EDUCATION AND TRAINING

The regulatory body shall ensure the application of the requirements for education, training, qualification, and competence in protection and safety of all persons with responsibilities for medical exposure (IAEA 2014). In Latin America, approximately 35 institutions provide training in medical physics, concentrated in Argentina, Brazil, and Cuba. However, other countries like Venezuela, Peru, Chile, and Colombia have postgraduate programs for medical physicists well established. There is a lack of an accreditation process for the institutions; in a few countries the formal accreditation process is conducted by the Ministry of Education. However, a specific syllabus for medical physics courses or a minimum curriculum has not been established by the authorities. This is fundamental in order to guarantee that these courses meet the minimum standards requirements.

Currently, the main demand is for accreditation of medical physics clinical programs, which is fundamental to the development of competence. Only a few hospitals and clinics are accredited to provide this training; most of these professionals do not have the opportunity to get practical training in hospitals. Even in Brazil, where there are at least 10 accredited hospitals, the work schedule is not standardized and the annual number of vacancies is not enough. The number of medical physicists certified in each country indicates that there is not a formal process for professional regulation, or even a certification scheme established in each country.

The lack of educational programs directly affects the regulatory requirements, especially where the presence of a medical physicist is mandatory. The government is responsible for establishing the regulatory framework, and as part of their duties to ensure, if necessary, education, training, and technical services. However, these requirements have not been fulfilled by the national authorities. There are only from 2.8% (Brazil) to 38% (Cuba) of the number of these professionals necessary working in the Nuclear Medicine field and 0.03% (Colombia) to 0.8% (Cuba) in Diagnostic Radiology. Only 30% of the countries have clinical residence programs and about 28% have graduate courses. Despite the complexity of the new technologies and the requirements for reporting typical doses, the medical physicist profession is not recognized as essential in NM and DR. The region should establish appropriate structure and syllabuses for Medical Physicist based on international recommendation to achieve the defined required professional competences.

13.5   CONCLUSIONS

In general, the technical capacity of regulatory authorities in Latin America is limited, especially for authorization and inspection. Although many efforts have been made resulting in significant improvements in medical applications of ionizing radiation, there are many more actions to be performed in order to implement GSR requirements in most countries. The division of regulatory responsibilities imposes many difficulties because of duplication of processes, delays in authorizations, mistakes in technical documentation due to different requirements, different fees, and other reasons. Again, coordination among regulatory authorities and health authorities is imperative for effective control. On the other hand, the broad use of different sources and procedures in medicine, their technical complexity, the high radiation doses in some treatments, as well as the increased number of patients, impose the necessity for a graded regulatory system, according to the risks associated. However, in the licencing process, most checklists are just a list of topics to be controlled with similar prioritization, although the risks could be very different. There is a need to implement quality assurance (QA) programs in order to improve clinical outcomes and ensure patient safety. The impact of various accidents in radiation therapy services at the global and regional level, as well as the increasing complexity of treatments, have increased the awareness of health and regulatory authorities about the importance of QA programs. However, this is not the case for imaging services where QA programs are still very scarce. The low number of well-trained professionals, especially medical physicists, imposes additional risks for patients. Development of educational programs and clinical residencies, as well as advocacy for the recognition of the medical physicist profession, should be considered as priorities for Latin America.

13.6 ACKNOWLEDGMENTS

The authors kindly thank the National Regulatory Authorities of Latin American countries for their information support.

13.7 REFERENCES

CARICOM, 2012. Regional Workshop on Regulatory Infrastructure for the Control of Radioactive Sources, organized by IAEA in cooperation with PAHO with support from the EU - Kingston, Jamaica, June 11–15, 2012 Meeting report.

International Atomic Energy Agency - IAEA. 2014a. Radiation Protection and Safety of Radiation Sources: International Basic Safety Standards - GSR Part 3, Vienna, Austria.

International Atomic Energy Agency - IAEA. 2014b. Secondary Standards Dosimetry Laboratories, http://www-naweb.iaea.org/nahu/dmrp/SSDL (accessed December 15, 2014).

Kodlulovich, S., de Sá, L.V. 2013. Education and Training in Latin America: Current Status and Challenges, Medical Physics International Journal, v.1, n1, pp 30–34.

PAHO. 2007. Pan American Health Organization - Health in the Americas - 2007 Edition, Washington DC, USA.

PAHO. 2012. Pan American Health Organization - PAHO Health Economics and Financing (HEF): Health Care Expenditure and Financing in Latin America and the Caribbean, Washington DC, USA.

PAHO. 2013. Pan American Health Organization - Plan of action for the prevention and control of non-communicable diseases, 2012–2015, CD52/7, Washington DC, USA.

UN. 2011. United Nations Population Fund Urbanization processes and disaster risks: a regional vision of Latin America. http://www.unisdr.org (accessed December 16, 2014).

WHO, 2014. World Health Organization, World Health Statistics, http://www.who.int (accessed December 15, 2014).

VIII

Role of the International Organizations in Worldwide Medical Radiation Protection

Role and Activities of the International Atomic Energy Agency in Radiation Protection in Medical Imaging and Radiation Oncology

Ahmed Meghzifene

International Atomic Energy Agency, Vienna, Austria

Ola Holmberg

International Atomic Energy Agency, Vienna, Austria

CONTENTS

T HE INTERNATIONAL ATOMIC ENERGY AGENCY (IAEA) is the world's platform for scientific and technical cooperation in the peaceful use of nuclear technology. It was established by the United Nations as an independent organization in 1957.

14.1   THE INTERNATIONAL ATOMIC ENERGY AGENCY

14.1.1   The IAEA and Its Mandate

The IAEA works to foster the role of nuclear science and technology in sustainable development. This involves both advancing and employing knowledge to address pressing worldwide challenges — ensuring access to food, water, and energy, fighting poverty and disease, and adapting to climate change. The IAEA works to maximize the safe operation of nuclear facilities that generate power, support industry, deliver healthcare, and serve research. The IAEA has five major departments, but the work described in this chapter is mostly carried out within three departments: the Departments of Nuclear Sciences and Applications, the Department of Nuclear Safety and Security, and the Department of Technical Cooperation. In the fields of medical physics and radiation safety, the development work, including coordination of research in various areas, and the preparation of safety standards, guidelines, and training material, is done by the first two departments; the IAEA facilitates the transfer of nuclear technology to Member States and implementation of safety standards through the Department of Technical Cooperation.

To provide for the application of the Standards and transfer of know-how, the IAEA uses five main tools: coordinating research, promoting education and training, providing assistance, fostering information exchange, and rendering services to its Member States. All these tools are used in the IAEA programs on medical physics and radiological protection of patients described below.

The statutory function regarding the establishment of safety standards and providing for their application constitutes a core work of the IAEA, the first part being expressed in the International Basic Safety Standards (BSS). The Board of Governors first approved radiation protection and safety measures in March 1960 and the first basic safety standards in June 1962. Since then, revisions of these standards were published, in 1967, 1982, 1996, and, most recently, in 2014 (IAEA 2014c). Such standards provide the basic requirements that must be satisfied to ensure safety for particular activities or application areas. The Agency's safety standards have a clearly defined pedigree: The United Nations Scientific Committee on the Effects of Atomic Radiation (UNSCEAR), a body set up by the United Nations in 1955. UNSCEAR compiles, assesses, and disseminates information on the health effects of radiation and on levels of radiation exposure due to different sources (UNSCEAR 2010a); this information is taken into account in developing the standards. In addition, account is taken of the recommendations of the International Commission on Radiological Protection (ICRP). Recommendations on how to comply with the requirements of the BSS relating to medical exposure are provided in the Safety Guide on Radiation Protection in Medical Exposure (IAEA 2002a, IAEA 2012), now in the process of updating. The updated Safety Guide is anticipated to be jointly co-sponsored by the International Labor Organization (ILO), the Pan American Health Organization (PAHO), and the World Health Organization (WHO). They describe strategies to involve organizations outside the regulatory framework, such as professional bodies, whose co-operation is essential to ensure compliance with the BSS requirements for medical exposures. Examples where this is necessary include the establishment of guidance levels for diagnostic medical exposures, acceptance testing processes for radiation equipment, calibration of radiotherapy units, and reporting of accidental medical exposure.

14.1.2 Role of the IAEA in Radiation Protection in Medicine and Medical Physics

The radiation exposure of patients is by far the largest type of exposure to the world's population from man-made radiation sources. It has been estimated that the number of medical procedures using ionizing radiation grew from 1.7 billion in 1980 to around 4 billion in 2007 (UNSCEAR 2010a). Too little or too much dose is problematic, and the risk of any given procedure ranges from negligible to potentially fatal. Radiation protection of patients must deal with the issues of not having dose limits, exposing sensitive subgroups, and using doses that could have deterministic effects. Furthermore, the number of occupationally exposed persons is much greater in medicine than in any other source or practice. UNSCEAR has estimated there are over 2.5 million monitored workers in medicine compared to 0.8 in industry and 0.3 as a result of military uses. Additionally, radiation accidents involving medical uses have accounted far more deaths and early acute health effects than any other type of radiation or nuclear accident, including accidents at nuclear facilities (UNSCEAR 2010b). Radiation protection and safety in medicine is of highest importance, considering these factors. In addition to the IAEA having a responsibility to prepare international safety standards related to radiation protection and safety in medical uses of ionizing radiation, as well as in relation to radiation protection and safety in other radiation and nuclear uses, the IAEA also has the responsibility to provide for the application of these standards.

There is no doubt that the application of ionizing radiation and radioactive substances in diagnostic, interventional, and therapeutic applications in medicine is beneficial for hundreds of millions of people each year. However, employing radiation in medicine has to involve a careful balance of the benefits of enhancing human health and welfare, and the risks related to the radiation exposure of people. In its program and activities related to the radiation protection of patients, the IAEA aims to help reduce unnecessary and unintended exposures in practice. Unnecessary exposures of patients can arise from procedures that are not justified for a specified objective, from the application of medical radiation procedures to individuals whose condition does not warrant such interventions, and from medical exposures that are not appropriately optimized for the situation in which they are being used. Unintended exposure of patients and can also arise from unsafe design or inappropriate use of medical radiation technology.

In order to form a coherent strategy for strengthening radiation protection of patients globally, an international conference on this topic was held in 2001 in Malaga, Spain. On the basis of the conference, an action plan for international work in this area was established. Among the important issues included in this plan, there are many actions that the IAEA has promoted over the last few years, such as the long-term tracking of individual patient exposures (the SmartCard/SmartRadTrack project), the international campaign on strengthening justification of medical exposure in diagnostic imaging: AAA — Awareness (effective communication about risk), Appropriateness (up-to-date referral guidelines), and Audit (clinical audit of justification), Retrospective Evaluation of Lens Injuries and Dose (RELID) in interventional cardiology, and reporting of safety related events in medical applications through the web-based reporting systems, Safety in Radiation Oncology (SAFRON), and Safety in Radiological Procedures (SAFRAD). All of these activities, and more, are detailed on the RPOP (radiation protection of patients) website, which was also developed under the international action plan for the radiation protection of patients, and which currently receives more than 1 million hits per month (IAEA 2013c).

At the end of 2012, a second international conference was organized to focus efforts in this area for the next decade and to maximize the positive impact of future international work in radiation protection in medicine. With the WHO as co-sponsor, and the Government of Germany, through the Federal Ministry for the Environment, Nature Conservation, and Nuclear Safety as host, the IAEA organized the International Conference on Radiation Protection in Medicine — Setting the Scene for the Next Decade. The conference was held in Bonn, 3–7 December, 2012 and aimed, in particular, to:

- indicate gaps in current approaches to radiation protection in medicine,

- identify tools for improving radiation protection in medicine,

- review advances, challenges, and opportunities in the field of radiation protection in medicine, and

- assess the impact of the International Action Plan for the Radiation Protection of Patients, in order to prepare new international recommendations, taking into account newer developments.

The conference was attended by more than 500 participants and observers from 77 countries and 16 organizations, and resulted in the so-called Bonn Call-for-Action (IAEA 2013a), a joint position statement between the IAEA and the WHO, which will continue to guide stakeholders over the next years on the efforts that are necessary to strengthen radiation protection in medicine.

In the field of medical physics, the IAEA's main activities focus on the Quality Assurance (QA) aspects of the use of radiation in medicine and education and training, to ensure safety and effectiveness, and deal with the science and technology involved in this area. Medical physicists play a key role in the implementation and optimization of protection and safety in radiation oncology and medical imaging. However, there is a shortage of these professionals, especially in low- and middle-income countries, mainly due to insufficient education and structured clinical training in hospitals, and lack of understanding of the roles and responsibilities of medical physicists working in a clinical environment (Meghzifere 2012). This shortage of Clinically Qualified Medical Physicists (CQMPs) has a negative effect on many projects that aim at establishing or upgrading radiotherapy or imaging departments. In some countries, modern and complex equipment is acquired without the corresponding investments in human resources, leading to underuse or misuse of the equipment. In response to these findings, the IAEA engaged in worldwide consultations with professional societies and stakeholders to achieve a consensus on the roles and responsibilities of clinically qualified medical physicists and their education and clinical training requirements. An international consensus was achieved on this this subject and the harmonized guidelines were published in the IAEA Human Health Series No. 25 in 2013 (IAEA 2013d). This publication aims at defining appropriately and unequivocally the roles and responsibilities of a CQMP in specialties of medical physics related to the use of ionizing radiation, such as radiation therapy, nuclear medicine, and diagnostic and interventional radiology. Important non-ionizing radiation imaging specialties, such as magnetic resonance and ultrasound, are also considered for completeness. On the basis of these tasks, this publication provides recommended minimum requirements for the academic education and clinical training of CQMPs, including recommendations for their accreditation, certification, and registration, along with continuing professional development. The goal is to establish criteria that support the harmonization of education and clinical training worldwide, as well as to promote the recognition of medical physics as a profession.

14.2 ESTABLISHING SAFETY STANDARDS AND PROVIDING FOR THEIR APPLICATION

The IAEA safety standards reflect an international consensus on what constitutes a high level of safety for protecting people and the environment from harmful effects of ionizing radiation, and they are established in consultation and cooperation with the competent organs of the United Nations and with specialized agencies concerned in these subject matters. The IAEA safety standards consist of three levels: (1) Safety Fundamentals, which present the fundamental safety objectives and principles of protection and safety; (2) Safety Requirements, which establish the requirements that must be met to ensure protection of people and the environment; and (3) a set of supporting Safety Guides, which provides recommendations and guidance on how to comply with the Safety Requirements.

14.2.1 The International Basic Safety Standards

One of the most widely recognized set of IAEA safety standards is the International Basic Safety Standards (BSS) (IAEA 2014c). This was developed by a joint secretariat consisting of eight international organs: the European Commission (EC), Food and Agricultural Organization of the United Nations (FAO), the International Labor Organization (ILO), the Nuclear Energy Agency of the Organisation for Economic Co-operation and Development (OECD NEA), the Pan American Health Organization (PAHO), the United Nations Environment Program (UNEP), and the World Health Organization (WHO), together with the IAEA. The BSS is a Safety Requirements document that applies to all facilities and activities that give rise to radiation risks. The BSS was first published in 1962.

The requirements of the BSS are based around the three types of exposure situations: planned exposure situations, existing exposure situations, and emergency exposure situations. Medical uses of ionizing radiation are a planned exposure situation. This includes situations when the radiological procedures do not go as planned, when unintended and accidental medical exposures occur.

Medical exposures, as outlined in the BSS, differ from occupational and public exposures in that persons (primarily patients) are deliberately, directly, and knowingly exposed to radiation for their benefit. In medical exposures applying a "dose limit" is inappropriate as it may limit the benefit for the patient; consequently, only two of the radiation protection principles apply: justification and optimization. Justification plays the role of gatekeeper, as it will determine whether the exposure will take place or not. If it is to take place, the radiological procedure has to be performed in such a way that the radiation protection and safety is optimized.

The IAEA statute makes the safety standards binding on the IAEA in relation to its own operations and also on States in relation to IAEA-assisted operations. While it is not mandatory, many countries adopt the IAEA safety standards as the basis of their legal and regulatory framework in the area of radiation protection.

14.2.2 Application of the Radiation Protection Requirements

The application of the justification principle to medical exposures requires a special approach, using three levels. As a primary justification of medical exposures, it is accepted that the proper use of radiation in medicine does more good than harm. At the second level there is a need for generic justification, to be carried out by the health authority in conjunction with appropriate professional bodies, of a given radiological procedure. This applies to the justification of new technologies and techniques as they evolve. For the third level of justification, the application of the radiological procedure to a given individual has to be considered. The specific objectives of the exposure, the clinical circumstances and the characteristics of the individual involved, have to be taken into account. National or international referral guidelines, developed by professional bodies together with health authorities, need to be used. Over the last number of years, the IAEA has had several international Technical Meetings to help increase the availability and use of referral guidelines globally, in many instances in co-operation with the WHO, the American College of Radiologists (ACR), the Royal College of Radiology (RCR), and the International Society of Radiology (ISR), as well as other international organizations and professional bodies.

The application of the requirements for optimization of radiation protection and safety to the medical exposure of patients also requires a special approach. Too low a radiation dose could be as bad as too high a radiation dose, in that the consequence could be that a cancer is not cured or the images taken are not of suitable diagnostic quality. It is of paramount importance that the medical exposure leads to the required clinical outcome. In diagnostic imaging and image-guided interventional procedures, diagnostic reference levels (DRLs) are a tool used in optimization of radiation protection and safety. Periodic assessments are to be performed of typical patient doses or, for radiopharmaceuticals, activities administered in a medical radiation facility. If comparison with established diagnostic reference levels shows that the typical patient doses or activities are either unusually high or unusually low, a local review is to be initiated to ascertain whether protection and safety have been optimized and whether any corrective action is required. The IAEA conducts training courses for health professionals in the utilization of DRLs. Other tools used in optimization of protection and safety include, inter alia, design and operational considerations and programs of quality assurance.

14.2.3   Roles and Responsibilities of Different Stakeholders

The roles and responsibilities of the government with regard to radiation protection and safety include establishing an effective legal and regulatory framework for protection and safety for all exposure situations; establishing legislation that meets specified requirements; establishing an independent regulatory body with the necessary legal authority, competence, and resources; establishing requirements for education and training in protection and safety; and ensuring that arrangements are in place for the provision of technical services (including radiation monitoring services and standards dosimetry laboratories), and education and training services.

All medical facilities must be authorized by the health authority to ensure that the facility meets the applicable requirements for quality of medical services. When the medical facility uses ionizing radiation, authorization for medical practice and healthcare should be granted by the health authority only if radiation safety requirements are met.

Professional bodies represent the collective expertise of the given health profession and specialty and, as such, they also have an important role in contributing to radiation protection and safety in medical uses of ionizing radiation. This includes setting standards for education, training, qualifications, and competence for a given specialty, and setting technical standards and giving guidance on practice.

In medical uses of ionizing radiation, the prime responsibility for radiation protection and safety rests with the person or organization responsible for the medical radiation facility — normally referred to as the registrant or licensee. However medical uses of ionizing radiation involve a multidisciplinary team led by a health professional who often is not the registrant or licensee of the authorized medical radiation facility. Because of the medical setting in which such exposures occur, primary responsibility for radiation protection and safety for patients lies with the health professional responsible for the radiological procedure, who is referred to in the BSS as the "radiological medical practitioner."

The medical radiation technologist is usually the interface between the radiological medical practitioner and the patient, and his/her skill and care in the choice of techniques and parameters determines to a large extent the practical realization of the optimization of radiation protection and safety for a given patient's exposure in many modalities.

The medical physicist provides specialist expertise with respect to radiation protection of the patient. The medical physicist has responsibilities in the implementation of the optimization of radiation protection and safety in medical exposures, including source calibration, clinical dosimetry, image quality, and patient dose assessment, and physical aspects of the quality assurance program, including medical radiological equipment acceptance and commissioning.

For a medical radiation facility, the radiation protection officer oversees the application of requirements for occupational and public radiation protection, and may provide general radiation protection advice to the registrant or licensee. The medical physicist is usually requested to fulfil the radiation protection function in hospitals.

There are also other health professionals with responsibilities for radiation protection of the patient. These include, for example, radiopharmacists, radiochemists, dosimetrists, and biomedical or clinical engineers.

These and other guidance can also be found in greater details in an IAEA Safety Guide on Radiation Protection and Safety in Medical Uses of Ionizing Radiation, which is currently being developed (IAEA 2012).

14.3 SUPPORTING THE DEVELOPMENT AND IMPLEMENTATION OF QUALITY ASSURANCE IN MEDICAL APPLICATIONS

The medical use of radiation requires implementation of an appropriate QA program to ensure that the technology is used in a safe and effective manner. It consists of procedures that ensure a consistent and effective delivery of the treatment or imaging procedure, while ensuring safety of patients and keeping minimal exposure to personnel and the public. It includes both clinical, medical physics and radiation protection, and safety aspects. The IAEA assists its Member States in establishing and implementing national QA programs. This IAEA assistance has two complementary components: (i) development of harmonized guidelines for the use of radiation-based technologies in the medical field, taking into account clinical and medical radiation physics and radiation protection and safety aspects; (ii) support to Member States that can benefit from the IAEA Technical Cooperation (TC) program aimed at establishing or upgrading an infrastructure in an integrated manner. In addition, the IAEA support also enables setting up a national infrastructure for calibration of dosimeters and external dosimetry audit programs.

The development of harmonized guidelines by the IAEA is done through consultants' meetings and often includes representatives of professional organizations such as the International Organization for Medical Physics (IOMP), the American Association of Physicists in Medicine (AAPM), or the European SocieTy for Radiotherapy & Oncology (ESTRO). Many IAEA publications have an important impact on the professional practice of medical physics; an example of a worldwide impact of an IAEA publication is the International Code of Practice for radiotherapy dosimetry (IAEA 2000), which is implemented in many hospitals around the world.

14.3.1 Providing Traceability and Quality Audit Services

Acknowledging the need for accurate dosimetry in radiation oncology, medical imaging, and radiation protection, the IAEA provides dosimetry calibration services to Member States through its joint IAEA/WHO Network of

Secondary Standards Dosimetry Laboratories (SSDLs). Through the calibration services, the IAEA supports its Member States to establish a link to the international measurement system through its reference dosimetry system within the framework of the Mutual Recognition Arrangement of the International Committee for Weights and Measures (BIPM 1999). In addition to calibration services, the IAEA provides dosimetry verification services both for SSDLs and for end-user institutions engaged in radiotherapy and radiation protection. These verification services aim at supporting the Member States to check the integrity of their dosimetry standards and implementation of dosimetry protocols in radiotherapy hospitals. The primary beneficiaries of these activities are hospital patients undergoing medical procedures involving radiation, and radiation workers and the general public that benefit from improved dosimetry practices. The IAEA's dosimetry services are provided cost-free mainly to those countries that are not members of the "meter convention" and do not have access to a recognized calibration service. The IAEA calibration and measurement capabilities are published in the international database of the BIPM (BIPM 2002), following a peer-review process conducted by relevant regional and international metrology bodies. The IAEA calibration and measurement capabilities are supported through international dosimetry comparisons published in Metrologia (for example, Tanaka et al. 2014). The Thermoluminescent dosimetry (TLD) service for hospitals aims at ensuring proper calibration of radiotherapy beams and checks approximately 600 clinical beams per year. Since the service started in 1969, it has checked a total of more than 4500 radiotherapy beams in approximately 2000 centers. Follow-up actions on results outside the acceptance limit support the radiotherapy centers in resolving the deviation, thus preventing further mistreatment of patients. The TLD program is implemented through a close collaboration between the IAEA and WHO (Pan American Health Organization, PAHO, in Latin America). A comprehensive QA program supports the IAEA dose quality audit program. An important element of that QA program relies on reference irradiations for the IAEA that TLD sets by primary standards for dosimetry laboratories and reference hospitals, acting as an external quality control of the IAEA service. The results of the TLD program for radiotherapy hospitals were published in a special issue of the SSDL Newsletter No. 58 (IAEA 2010d). The long-term impact on the improvement of the results with time is significant, as highlighted in Figure 14.1, which shows the trends of the fraction of the results that are within the 5% acceptance limit.

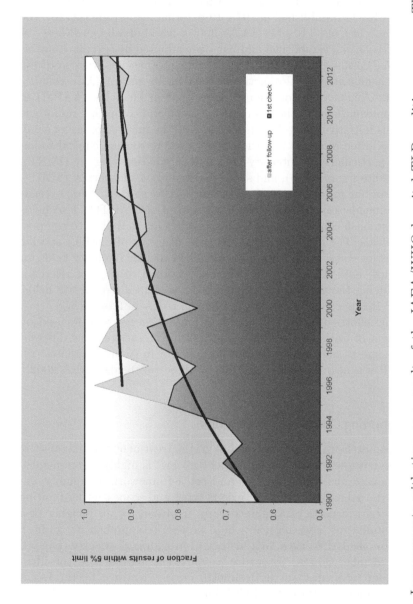

Figure 14.1 Improvements with time as a result of the IAEA/WHO hospital TLD auditing program. The dashed black line shows the trend as a result of the first postal TLD check. The solid black line shows the improvement as a result of follow-up when the first exposure is outside of the 5% acceptance criterion.

14.3.2   Harmonization of Radiation Dosimetry

One of the main tasks of the IAEA in the field of medical physics is the development of Codes of Practice (CoP), or protocols, for the dosimetry of beams used in radiotherapy. The first publication was a "Manual of Dosimetry in Radiotherapy" (TRS-110) published in 1970. In external beam radiotherapy, the IAEA's most recent CoP, TRS-398 (IAEA 2000), based on absorbed-dose-to-water standards, is well-established and has been adopted by several countries as their national dosimetry CoP. More recently, an international working group on small and composite field dosimetry was set up by the IAEA in collaboration with the American Association of Physicists in Medicine (AAPM) to develop recommendations for the dosimetry of small and composite radiotherapy fields. This initiative was taken in response to the increasing use of small and composite, often dynamic radiation fields used in routine clinical radiation therapy, particularly in intensity-modulated radiation therapy (IMRT), stereotactic radiosurgery (SRS), and stereotactic body radiotherapy (SBRT). In 2008, the working group proposed a formality (Alfonso et al. 2008) extending the recommendations of IAEA TRS-398 and of AAPM TG-51 to fields that cannot establish conventional reference conditions, as well as to composite radiation fields. The formality introduces the concepts of machine-specific reference fields for static small fields and plan-class specific reference fields for composite fields, both deviating from the conventional concept of reference fields and bridging the gap with smaller fields and clinical composite fields. The new CoP is expected to be published by the IAEA in late 2015.

Most of the efforts of the IAEA in radiation dosimetry were focused on the development of CoPs for external beam therapy. To address the different aspects of dosimetry and QA in brachytherapy and in X-ray diagnostic radiology, a CoP was also developed for each of these two fields (IAEA 2002b, IAEA 2007b).

14.3.3   Supporting Comprehensive Audits

As part of a comprehensive approach to QA in the treatment and diagnosis of patients, an independent external audit (peer review) is highly recommended to ensure adequate quality of practice and set a framework for quality improvement. External audits can be of various types and scopes, either limiting the review to critical parts of a radiotherapy treatment or an imaging process (partial audits) or assessing the whole process (comprehensive audits). In radiotherapy, the IAEA has a long history of providing dosimetry (partial) audits to its Member States. To ensure a consistent approach to the identification of sources of mistakes in dosimetry errors identified by the IAEA through its TLD audit service, the IAEA has developed and published a set of procedures for experts undertaking missions to radiotherapy hospitals in Member States for on-site review of dosimetry equipment, data, techniques, measurements, and training of local staff (IAEA 2007c). This methodology is limited to dosimetry and medical radiation physics aspects of the radiotherapy process without considering other steps such as dose prescription, simulation, planning, etc. However, it was soon realized that many centers would also benefit from a systematic review of the entire radiotherapy process. This was the foundation for developing the concept of "Comprehensive Audits of Radiotherapy Practices: A Tool for Quality Improvement, Quality Assurance Team for Radiation Oncology (QUATRO), " which was published in 2007 (IAEA 2007a). The objective of QUATRO is to review and evaluate the quality of all of the components of the practice of radiotherapy at an institution, including its professional competence, with a view to quality improvement. A multidisciplinary team, comprising a radiation oncologist, a medical physicist, and a radiotherapy technologist, carries out the audit. Until now, the IAEA has conducted about 70 missions worldwide.

The IAEA has expanded the concept of QUATRO to two other disciplines of medical applications with ionizing radiation, namely nuclear medicine and diagnostic radiology. For nuclear medicine, the IAEA guidelines focus on quality management audits and were published in 2009 (IAEA 2009b), for radiology, the IAEA guidelines published in 2010 (IAEA 2010b) are very similar to those used in radiation oncology. Although these auditing guidelines might differ in the details of their content, they all share the same basic characteristics and they are performed by multidisciplinary teams of experts, experienced in the corresponding area of radiation medicine. In order to assist the auditors during the audit, and at the same time, to facilitate the objectivity of the review process, standard detailed questionnaires and audit report forms have been developed and included in the IAEA guidelines. The auditing process, as described by the IAEA, takes place in a completely voluntarily manner, and what a facility should expect, as an output of a comprehensive clinical audit, is a systematic review of the current practice and identification of areas of improvement.

14.4 SUPPORTING IMPLEMENTATION GLOBALLY

14.4.1 Providing Guidance

An essential component of the IAEA work in the field of medical physics and radiation protection is to develop internationally harmonized guidelines. The guidelines are developed through technical and consultants' meetings attended by experienced medical physicists or radiation protection specialists selected by the IAEA or professional societies. Through these meetings, the IAEA aims at achieving a global consensus among all participants to ensure that most of the relevant concerns are factored into the final recommendations. The draft guidelines are subsequently submitted to another group of consultants for critical review. The final recommendations are usually published as IAEA Human Health Series/Reports or Safety Reports. Examples of such guidelines are given in (IAEA 2014b).

The IAEA website on radiation protection of patients (http://rpop.iaea.org) was released in September 2006, and has become the top website in the world in the field of radiation protection of patients, and on medical radiation protection, overall, with more than 10 million hits per year. It appears on the first page of search engines such as Google when relevant search terms are used, such as radiation protection in pregnancy, mammography, fluoroscopy, CT, radiology, nuclear medicine, radiotherapy, cardiology, children, dental radiology, DEXA, PET/CT, interventional radiology, gastroenterology, urology, orthopedic surgery, etc. There are specific pages for radiation protection of children and pregnant women, as well as specific pages for patients.

14.4.2 Providing Training and Education

Supporting education and training of professionals working in medical imaging and radiation oncology is a high-priority area for the IAEA programs in medical physics and radiation protection of patients. The IAEA activities include the development of training material such as handbooks (IAEA 2005, IAEA 2014a), an online training resource for nuclear medicine professionals (IAEA 2009a), clinical training guides (IAEA 2010a, IAEA 2011a, IAEA 2011b), and websites (IAEA 2010c, IAEA 2013c). In addition to the development of education material, the IAEA organizes training courses and workshops, mainly through the technical cooperation program. For example, during the past 5 years, the IAEA has organized around 20 training courses and workshops and 4 international workshops jointly with the International Centre for Theoretical Physics (ICTP). To ensure that the training programs proposed through these training events respond to the needs of the participating countries, the topics are decided in consultations with counterparts participating in relevant TC projects. The training courses aim at building knowledge in a specific area, whereas workshop, focus on implementation of a specific technique, such as the use of in-vivo dosimetry for checking treatment delivery or implementation of a quality control program for gamma cameras. In addition to capacity building, this approach helps establish local expertise and strengthens networking by bringing together experienced professionals and technicians from developing and developed countries. In addition, the IAEA supports continuous professional development through a fellowship mechanism. Under this training scheme, fellows are supported to go abroad for comprehensive training in a suitable institution for periods ranging from a few months to one or two years. Scientific visits, with duration of up to two weeks, are awarded to staff in managerial positions in order to broaden the scientific or managerial qualifications of specialists in developing countries.

In addition to education and training activities in radiation protection in medical imaging and radiation oncology, the IAEA has recently initiated the development of a specific training package for medical physicists in support of nuclear or radiological emergency situations, in collaboration with the IOMP. Although medical physicists have a solid knowledge of radiation dosimetry and dose reconstruction, and a fairly good understanding of radiation biology, only a few of them have actually been involved in supporting the response to a nuclear or radiological emergency situation. Such a training package should be reviewed and agreed upon by emergency preparedness centers to ensure that the role of medical physicists is well understood and accepted by hospitals, nuclear power plant managers, and civil defense teams. The first international workshop to train trainers in this area was held in Fukushima, Japan in June 2015.

14.4.3 Providing Technical Assistance

The IAEA technical cooperation (TC) program is the main mechanism through which the IAEA delivers technical assistance to its Member States (IAEA 1998). Through the program, the IAEA helps Member States to build, strengthen, and maintain capacities in the safe, peaceful, and secure use of nuclear technology in support of sustainable socioeconomic development. TC projects provide expertise in fields where nuclear techniques offer advantages over other approaches, or where nuclear techniques can usefully supplement conventional means. All Member States are eligible for support, although in practice technical cooperation activities tend to focus on the needs and priorities of less-developed countries. The TC program is unique in the United Nations system, as it combines specialized technical and development competencies. The program focuses on applying nuclear technology to improve human health, support agriculture and rural development, advance water resource management, address environmental challenges, and help sustainable energy development, including the use of nuclear power for electricity. The program also focuses heavily on supporting nuclear safety and security. The TC program operates in four geographic regions: Africa, Asia and the Pacific, Europe, and Latin America. Within each region, it helps Member States address their specific needs, taking into consideration existing capacities and different operational conditions. The IAEA's TC program supports human resource capacity building activities, networking, knowledge sharing, and partnership facilitation, as well as the procurement of equipment. Human resource capacity building is provided through expert missions and meetings, fellowships (for trainees of new technologies or trainees of medical physics expertise), scientific visits (for more senior staff), and special training courses focusing on the safe and effective use of peaceful applications of nuclear energy and nuclear technology. In addition, the IAEA organizes regional and interregional workshops, and supports national workshops. An example of IAEA support in medical physics through IAEA regional TC projects in Africa are RAF/6/044, "Strengthening Medical Physics in Support of Cancer Management," and RAF/6/038, "Promoting Regional and National Quality Assurance Programs for Medical Physics in Nuclear Medicine." Under these 2 projects, more than 20 regional workshops were held in Africa on various fields of medical physics during the past 5 years.

14.4.4 Facilitating Coordinated Research

The IAEA promotes and provides support for research and development on practical applications in health, including radiation safety aspects, and fosters the exchange of scientific and technical information and exchange of scientists. The IAEA support for research is provided through its Coordinated Research Activities (CRAs), which have been designed to contribute to the fulfilment of this mandate by stimulating and coordinating the undertaking of research by institutes in IAEA Member States in selected fields. The IAEA's Coordinated Research Activities create fertile ground for bringing together scientists from developing and developed countries to meet, focus on well-defined areas of research and exchange knowledge, experience, and ideas for their mutual benefit. Most of the Coordinated Research Activities are carried out under its Coordinated Research Projects (CRPs), which bring together an average of 15 scientific institutes from developing and developed countries to concentrate on problems of common interest. Proposals for participation in the CRA should be prepared by institutes in IAEA Member States and submitted directly to the Research Contracts Administration Section. Once the proposal has been positively evaluated, the IAEA may offer institutes in developing countries a research, technical, or doctoral contract, and offer institutes in developed countries a research agreement or technical contract, which, if under the auspices of a CRP, includes participation at periodic Research Coordination Meetings (RCMs). Research is completed in the institutes' countries, with the Chief Scientific Investigator for each contract/agreement invited to the periodic RCMs for face-to-face discussions and exchanges of information, and to facilitate the building of professional bonds.

14.4.5 Facilitating Knowledge Exchange

IAEA Conferences and symposia are considered to be major meetings and are designed to support the exchange of information between experts and specialists from various countries. In addition, the IAEA organizes smaller meetings aiming at reviewing a specific technology or modality, and often endup with recommendations to the IAEA and to the scientific community.

In early 2009, the IAEA initiated the Information System on Occupational Exposure in Medicine, Industry, and Research, referred to as the ISEMIR project. The catalyst for the ISEMIR project was the experience of the Information System on Occupational Exposure (ISOE) of nuclear power plant operators around the world, where having a database that contained detailed information on operational occupational doses across many nuclear power plants enabled the comparison and benchmarking of doses for specific occupations, functions, and tasks. This in turn enabled the assessment of the impact of various radiation protection actions. As the ISOE database became populated with data covering many years, dose trends were also able to be analyzed. If such an approach was successful for nuclear power plant workers, perhaps a similar approach could be utilized in the non-nuclear domain — i.e., medicine, industry, and research. The ISEMIR Working Group on Interventional Cardiology (WGIC) met for the first time in February 2009. The mandate for WGIC was to gain a worldwide overview of occupational exposures and radiation protection of staff in IC; to identify both good practices and shortcomings, and hence define actions to be implemented for assisting each of the regulatory bodies, medical physicists, medical staff, technicians and nurses, dosimetry service providers, and X-ray machine suppliers, in improving occupational radiation protection; to propose recommendations for harmonizing monitoring procedures; and to set up a system for regularly collecting and analyzing occupational doses for individuals in IC and for dissemination of this information to improve occupational radiation protection. This has led to the design and development of the ISEMIR international database (IAEA 2013b). The purpose of the ISEMIR database is not to assess compliance with occupational dose limits, but rather to be an active tool for assessing the level of, and hence guiding, implementation of the radiation protection principle of optimization of protection at a given IC facility.

The IAEA has launched a voluntary reporting system called Safety in Radiological Procedures (SAFRAD), in which patients who are submitted to defined trigger levels or events in fluoroscopically guided diagnostic and interventional procedures are included in an international database (IAEA 2009c). The primary objective of the system is educational. It is believed that the process of entering information into SAFRAD might itself lead to increased focus on safety and quality of service. The data furnished will remain accessible to the participant, who also will periodically have access to analyzed results. The IAEA will publish overall summary reports of SAFRAD data from time to time. The IAEA will not supply identifiable SAFRAD data to any governmental authority or other third party.

The IAEA has also developed a reporting and learning system for voluntary reporting of safety-significant events in radiotherapy, called Safety in Radiation Oncology (SAFRON) (IAEA 2013e). The objectives of the SAFRON project are to implement a global safety reporting and learning system that encompasses retrospective reporting and prospective risk analysis within a learning environment that will improve the safe planning and delivery of radiotherapy. The integration of prospective risk analysis together with retrospective reporting enables the system to be proactive, which is of value when considering the rapid development of new medical technology.

14.5 HOW TO PARTICIPATE IN IAEA ACTIVITIES

Some of the opportunities for individuals and organizations to participate in the activities of the IAEA are through the IAEA Technical Cooperation program. At all times, there are national, regional, and international IAEA technical cooperation projects ongoing, through which there can be requests for fellowships and scientific visits, and for participation in meetings, workshops, and trainings. These requests should be channeled through the National Liaison Officer of the applicant's country. The National Liaison Officer is designated by the relevant government authority and recognized by the IAEA as the primary contact person between the IAEA and the Member State on matters relating to the IAEA technical cooperation program.

Scientific visits are intended to broaden the scientific or managerial qualifications of specialists in developing countries and for a duration not exceeding two weeks. Fellowships are normally awarded for periods of up to one year or two years, and are available to university graduates mainly through project-oriented on-the-job training. Candidates are usually selected on the basis of educational and professional qualifications, the needs of the Member State concerned, and the language proficiency of the nominee.

The IAEA organizes meetings, both under its technical cooperation program, and under its technical departments' regular programs. The purpose of the meetings is often to work out a set of recommendations and advice on particular aspects of individual projects, or of specific programmatic areas. Participation in meetings is upon invitation, or nomination by a Member State or invited organization.

The IAEA also organizes interregional and regional training courses and supports national training courses in cooperation with its Member States. Invitation letters announcing these courses are issued by the IAEA to relevant Member States. Application forms for training course participants must be submitted to the Agency through governmental channels, and priority is given to requests associated with projects of direct benefit to the Member States.

14.6   REFERENCES

A. Meghzifene. 2012. A call for recognition of the medical physics profession. The Lancet 379, no. 9825 (April): 1464–1465. http://www.thelancet.com/journals/lancet/article/PIIS0140-6736(11)60311-5/fulltext.

International Atomic Energy Agency (IAEA). 1998. IAEA Technical Cooperation. http://www.iaea.org/technicalcooperation/Home/index.html. (Accessed January 23, 2015.)

International Atomic Energy Agency (IAEA). 2000. Absorbed Dose Determination in External Beam Radiotherapy: An International Code of Practice for Dosimetry Based on Standards of Absorbed Dose to Water. Technical Reports Series 398.

International Atomic Energy Agency (IAEA). 2002a. Radiological Protection for Medical Exposure to Ionizing Radiation Safety Guide. IAEA Safety Standards Series RS-G-1.5. http://www-pub.iaea.org/books/IAEABooks/6290/Radiological-Protection-for-

Medical-Exposure-to-Ionizing-Radiation-Safety-Guide. (Accessed January 23, 2015.)

International Atomic Energy Agency (IAEA). 2002b. Calibration of photon and beta ray sources used in brachytherapy: Guidelines on standardized procedures at Secondary Standards Dosimetry Laboratories (SSDLs) and hospitals. IAEA-TECDOC-1274. http://www-pub.iaea.org/books/IAEABooks/6474/Calibration-of-Photon-and-Beta-Ray-Sources-Used-in-Brachytherapy. (Accessed January 23, 2015.)

International Atomic Energy Agency (IAEA). 2005. Radiation Oncology Physics: A Handbook for Teachers and Students. http://www-pub.iaea.org/books/IAEABooks/7086/Radiation-Oncology-Physics-A-Handbook-for-Teachers-and-Students. (Accessed December 5, 2014.)

International Atomic Energy Agency (IAEA). 2007a. Comprehensive Audits of Radiotherapy Practices: A Tool for Quality Improvement Quality, Assurance Team for Radiation Oncology (QUATRO). http://www-pub.iaea.org/books/IAEABooks/7680/Comprehensive-Audits-of-Radiotherapy-Practices-A-Tool-for-Quality-Improvement-Quality-Assurance-Team-for-Radiation-Oncology-QUATRO. (Accessed January 23, 2015.)

International Atomic Energy Agency (IAEA). 2007b. Dosimetry in Diagnostic Radiology: An International Code of Practice. IAEA TRS-457. http://www-pub.iaea.org/books/IAEABooks/7638/Dosimetry-in-Diagnostic-Radiology-An-International-Code-of-Practice. (Accessed January 23, 2015.)

International Atomic Energy Agency (IAEA). 2007c. On-site Visits to Radiotherapy Centres: Medical Physics Procedures. IAEA-TECDOC-1543. http://www-pub.iaea.org/books/IAEABooks/7681/On-site-Visits-to-Radiotherapy-Centres-Medical-Physics-Procedures-Quality-Assurance-Team-for-Radiation-Oncology-QUATRO. (Accessed January 23, 2015.)

International Atomic Energy Agency (IAEA). 2009a Distance Assisted Training for Nuclear Medicine Professionals. http://nucleus.iaea.org/HHW/NuclearMedicine/DATOL/. (Accessed December 4, 2014.)

International Atomic Energy Agency (IAEA). 2009b. Quality Management Audits in Nuclear Medicine Practices. http://www-pub.iaea.org/books/IAEABooks/7948/Quality-Management-Audits-in-Nuclear-Medicine-Practices. (Accessed January 23, 2015.)

International Atomic Energy Agency (IAEA). 2009c. SAFRAD https://rpop.iaea.org/safrad/. (Accessed January 23, 2015.)

International Atomic Energy Agency (IAEA). 2010a. Clinical Training of Medical Physicists Specializing in Radiation Oncology. Training Course Series 37. Also available in French, Spanish and Russian. http://www-pub.iaea.org/books/iaeabooks/8222/Clinical-Training-of-Medical-Physicists-Specializing-in-Radiation-Oncology. (Accessed January 23, 2015.)

International Atomic Energy Agency (IAEA). 2010b. Comprehensive Clinical Audits of Diagnostic Radiology Practices: A Tool for Quality Improvement Quality Assurance Audit for Diagnostic Radiology Improvement and Learning (QUAADRIL). IAEA Human Health Series 4. http://www-pub.iaea.org/books/IAEABooks/8187/Comprehensive-Clinical-Audits-of-Diagnostic-Radiology-Practices-A-Tool-for-Quality-Improvement-Quality-Assurance-Audit-for-Diagnostic-Radiology-Improvement-and-Learning-QUAADRIL. (Accessed January 23, 2015.)

International Atomic Energy Agency (IAEA). 2010c. IAEA Human Health Campus. http://nucleus.iaea.org/HHW/MedicalPhysics/index.html. (Accessed January 23, 2015.)

International Atomic Energy Agency (IAEA). 2010d. SSDL Newsletter No. 58. http://www-pub.iaea.org/MTCD/publications/PDF/Newsletters/SSDL-NL-58.pdf. (Accessed December 5, 2014.)

International Atomic Energy Agency (IAEA). 2011a. Clinical Training of Medical Physicists Specializing in Diagnostic Radiology. Training Course Series 47. Also available in French and Spanish. http://www-pub.iaea.org/books/IAEABooks/8574/Clinical-Training-of-Medical-Physicists-Specializing-in-Diagnostic-Radiology. (Accessed January 23, 2015.)

International Atomic Energy Agency (IAEA). 2011b. Clinical Training of Medical Physicists Specializing in Nuclear Medicine. Training Course Series 50. Also available in French and Spanish. http://www-pub.iaea.org/books/IAEABooks/8656/Clinical-Training-of-Medical-Physicists-Specializing-in-Nuclear-Medicine. (Accessed January 23, 2015.)

International Atomic Energy Agency (IAEA). 2012. Radiation Protection and Safety in Medical Uses of Ionizing Radiation (DRAFT SAFETY GUIDE DS399). Vienna: IAEA.

International Atomic Energy Agency (IAEA). 2013a. Bonn Call-for-Action. https://rpop.iaea.org/RPOP/RPoP/Content/Documents/Whitepapers/conference/bonn-call-for-action-statement.pdf. (Accessed November 26, 2014.)

International Atomic Energy Agency (IAEA). 2013b. ISEMIR https://rpop.iaea.org/RPOP/RPoP/Modules/login/isemir-ic.htm. (Accessed January 23, 2015.)

International Atomic Energy Agency (IAEA). 2013c. Radiation Protection of Patients. https://rpop.iaea.org/RPoP/RPoP/Content/index.htm. (Accessed November 26, 2014.)

International Atomic Energy Agency (IAEA). 2013d. Roles and Responsibilities, and Education and Training Requirements for Clinically Qualified Medical Physicists. IAEA Human Health Series 25. http://www-pub.iaea.org/books/IAEABooks/10437/Roles-and-Responsibilities-and-Education-and-Training-Requirements-for-Clinically-Qualified-Medical-Physicists. (Accessed January 23, 2015.)

International Atomic Energy Agency (IAEA). 2013e. SAFRON https://rpop.iaea.org/RPOP/RPoP/Modules/login/safron-register.htm. (Accessed January 23, 2015.)

International Atomic Energy Agency (IAEA). 2014a. Diagnostic Radiology Physics: A Handbook for Teachers and Students. http://www-pub.iaea.org/books/IAEABooks/8841/Diagnostic-Radiology-Physics-A-Handbook-for-Teachers-and-Students. (Accessed December 4, 2014.)

International Atomic Energy Agency (IAEA). 2014b. IAEA Resources in Dosimetry and Medical Radiation Physics. http://www-naweb.iaea.org/nahu/DMRP/documents/ IAEA_Resources_in_Dosimetry_and_Medical_Radiation_Physics.pdf. (Accessed January 23, 2015).

International Atomic Energy Agency (IAEA). 2014c. Radiation Protection and Safety of Radiation Sources: International Basic Safety Standards, General Safety Requirements Parts 3. IAEA Safety Standards Series GSR Part 3. http://www-pub.iaea.org/books/IAEABooks/8930/Radiation-Protection-and-Safety-of-Radiation-Sources-International-Basic-Safety-Standards. (Accessed January 23, 2015.)

International Bureau of Weights and Measures (BIPM). 1999. International Committee for Weights and Measures (CIPM) Mutual Recognition Arrangement (MRA). http://www.bipm.org/en/cipm-mra/cipm-mra-text/. (Accessed December 5, 2014.)

International Bureau of Weights and Measures (BIPM). 2002. Key Comparison Database of the International Bureau of Weights and Measures (BIPM). http://kcdb.bipm.org/appendixC/default.asp. (Accessed December 5, 2014.)

R. Alfosno et al. 2008. A new formalism for reference dosimetry of small and nonstandard fields. Medical Physics. 35, No. 11 (November): 5179–86. http://scitation.aip.org/content/aapm/journal/medphys/35/11/10.1118/1.3005481.

T. Tanaka et al. 2014. Key Comparison APMP.RI(I)-K2 of air kerma standards for the CCRI reference radiation qualities for low-energy X rays, including a supplementary comparison for the ISO 4037 narrow spectrum series. Metrologia. 51, Tech. Suppl. 06019: 1–56. http://iopscience.iop.org/0026-1394/51/1A/06019/.

United Nations Scientific Committee on the Effects of Atomic Radiation (UN-SCEAR). 2010a. Sources and Effects of Ionizing Radiation. Volume 1. UNSCEAR 2008 Report to the General Assembly, with Scientific Annexes. New York: United Nations.

United Nations Scientific Committee on the Effects of Atomic Radiation (UN-SCEAR). 2010b. Sources and Effects of Ionizing Radiation. Volume 2. UNSCEAR 2008 Report to the General Assembly, with Scbientific Annexes. New York: United Nations.

CHAPTER 15

The World Health Organization Global Initiative on Radiation Safety in Healthcare Settings

Maria del Rosario Pérez
World Health Organization, Department of Public Health, Environmental and Social Determinants of Health

Miriam Mikhail
World Health Organization, Department of Essential Medicines and Health Products, Medical Devices

CONTENTS

THE WORLD HEALTH ORGANIZATION (WHO) is the coordinating authority for health within the United Nations system. Its governing body is the World Health Assembly, made up of 194 Member States. It has a decentralized structure with 147 country offices, 6 regional offices, and headquarters in Geneva.

The WHO's core functions include:

• Articulating ethical and evidence-based policy positions

• Setting norms and standards, and promoting and monitoring their implementation

• Shaping the research agenda, and stimulating the generation, translation and dissemination of valuable knowledge

- Providing technical support, catalyzing change, and developing sustainable institutional capacity

- Monitoring the health situation and assessing health trends

- Providing leadership on matters critical to health and engaging in partnerships where joint action is needed

WHO's objective is the attainment by all peoples of the highest possible level of health, defined as "a state of complete physical, mental and social well-being and not merely the absence of disease or infirmity" (WHO 1948). Health is a human right that people rate as one of their highest priorities. Health promotion and protection is essential to human welfare and sustained economic and social development; it contributes to a better quality of life and also to global peace and security (WHO 1978).

15.1 THE GLOBAL INITIATIVE ON RADIATION SAFETY IN HEALTHCARE SETTINGS

Environmental risk factors are key determinants of human health. Approximately one-quarter of the global burden of disease, and more than one-third of the burden among children, are due to modifiable physical, chemical, and biological environmental factors. Although the burden of disease associated to radiation exposure, as such, has not been quantified globally, there is long-standing recognition of the health risks associated with radiation exposure.

Ionizing radiation is an essential tool for the diagnosis and treatment of human diseases. As the benefits for patients gain recognition, the use of ionizing radiation in medicine continues to increase (UNSCEAR 2010). Like all medical procedures, radiological medical procedures present both benefits and risks. On the benefit side, new technologies, applications, and equipment are constantly being developed to improve the safety and efficacy of procedures. At the same time, incorrect or inappropriate handling of these increasingly complex technologies can also introduce potential health hazards for patients and staff. This demands public health policies that both recognize the multiple health benefits that can be obtained, while addressing and minimizing health risks.

WHO's Radiation Program is aimed to protect patients, workers, and the public under planned, existing, and emergency exposure situations. As part of this program, in December 2008 WHO launched a Global Initiative on Radiation Safety in Health Care Settings (RSHCS), to mobilize the health sector towards safe and appropriate use of radiation in medicine. This initiative brings together health authorities, international organizations, professional bodies, scientific societies, and academic institutions in concerted action to support the implementation of radiation safety standards in healthcare settings.

The ultimate goal of this initiative is to strengthen radiation protection of patients and health workers by enhancing the safety and quality of radiological medical procedures. In order to maximize benefits with the least possible level of risk, the Global Initiative on RSHCS aspires to integrate radiation protection into the concepts of good medical practice and healthcare service quality. In the framework of this initiative, collaboration between relevant departments of WHO is in place to coordinate actions, promote synergies, and avoid duplication of efforts.

The Global Initiative on Radiation Safety in Health Care Setting is conducted by the WHO Department of Public Health, Environmental and Social Determinants of Health. Collaboration under this initiative is implemented with the WHO Department of Essential Medical Products and the WHO Department of Service Delivery and Safety.

15.2 RADIATION SAFETY STANDARDS

The development of norms and standards is fundamental to good governance in the field of radiation safety. The new International Basic Safety Standards (BSS) for Radiation Protection and Safety of Radiation Sources (BSS 2014) are co-sponsored by eight international agencies: the European Commission (EC), the Food and Agriculture Organization (FAO), the International Atomic Energy Agency (IAEA), the International Labour Organization (ILO), the Nuclear Energy Agency (NEA/OECD), the Pan American Health Organization (PAHO), the United Nations Environment Program (UNEP), and WHO. The BSS provide the leading international benchmark for safety standard-setting and policy making in medical, occupational, and public radiation exposures.

The section concerning safety requirements for medical exposures has been particularly expanded in the new international BSS. Safety requirements concerning medical exposures also represent a major component of the new European Commission Council Directive 2013/59/Euratom (EC 2013).

In line with its core functions, WHO was actively involved in the process of revision and update of the international BSS. WHO adopted the new BSS in May 2012 and it is currently cooperating with all cosponsors to support the implementation of the BSS in its Member States. A Task Group was established within the Inter-Agency Committee on Radiation Safety (IACRS) and a joint strategic plan is being implemented. This plan includes the development of safety guides, training packages, and information materials, as well as the joint organization of regional and national BSS workshops. Supporting the implementation of the new BSS in the medical sector is a major objective of the WHO Global Initiative on RSHCS.

15.3 RADIATION PROTECTION IN MEDICINE: CHALLENGES AND OPPORTUNITIES

The WHO Global Initiative on Radiation Safety in Health Care Settings includes activities in the field of radiation risk assessment, management, and communication (Table 15.1).

Table 15.1: WHO Global Initiative on Radiation Safety in Healthcare settings activities

Area of work	Activities
Radiation risk assessment	• Assessing population dose distribution due to the use of radiation in healthcare (in collaboration with UNSCEAR) • Shaping a global research agenda on radiation protection in medicine (priority on children)
Radiation risk management	• Reducing unnecessary radiation exposures (justification of radiological medical procedures and optimization of protection) • Promoting occupational health in healthcare settings • Addressing health workforce needs (scaling-up the role of medical physicists and radiological technologists) • Preventing accidental and unintended exposures, promoting reporting and learning systems • Fostering cooperation between health authorities and regulatory bodies
Radiation risk communication	Implementing a communication strategy (communication tools and information products for patients and health workers)

Promoting safe and appropriate use of radiation in medicine implies challenges and opportunities that are the focus of the activities and multiple collaborations in the framework of the WHO Global Initiative on RSHCS. Some of them are briefly described below alongside relevant ongoing collaborations. The text that follows provides examples but is not an exhaustive description of all the activities of the Global Initiative on Radiation Safety in Health Care Settings.

15.3.1 Research to Inform Policy and Actions

In the area of risk assessment, there is a recognized need for more focused research to evaluate health risks following medical exposures, including cancer and non-cancer effects. Most vulnerable populations such as children, young adults and pregnant women require particular consideration. To support such research, there is a parallel need to improve data collection on frequency of radiological medical procedures and population dose distribution.

The WHO Global Initiative on RSHCS includes a strategy for establishing a global research agenda on radiation protection in medicine. The European Commission (EC) and several European countries set up a High Level Expert Group (HLEG) to create a platform dedicated to low dose risk research. This platform, called MELODI (Multidisciplinary European Low Dose Initiative), seeks to facilitate a dialogue between stakeholders for the development and implementation of a long-term strategic research agenda on the effects of low doses of ionizing radiation. The European HLEG, as well as research institutions in the United States of America and Japan, in turn has provided technical support to WHO to foster a global research agenda. The collaboration is ongoing, and new achievements are being proposed to manage radiation protection research under the Open Project for European Radiation Research Area (OPERRA).

In the context of the Global Initiative on RSHCS, WHO collaborates with the United Nations Scientific Committee on the Effects of Atomic Radiation (UNSCEAR) to improve data collection on medical exposures, particularly in developing countries where this information is still scarce. A bilateral arrangement for cooperation to coordinate the periodic collection of data on global use of ionizing radiation in medicine was signed. The UNSCEAR-WHO collaboration includes development/update of the questionnaires, dissemination of information, capacity-building, and technical support to conduct national surveys. Joint regional workshops in Asia, Africa, Europe and the Americas have been organized in 2014 and 2015.

15.3.2 Appropriate Use of Radiation in Healthcare

Promoting appropriate use of radiation in medicine is one of the priorities of the WHO Global Initiative on RSHCS. While the concept of appropriateness applies to both diagnostic and therapeutic uses of radiation, it constitutes a matter of concern, particularly in medical imaging, because it has been reported that a substantial fraction of medical imaging procedures may be inappropriate and may thus result in unnecessary radiation exposures and preventable risks (Malone et al. 2012). A medical imaging examination is useful if its outcome, either positive or negative, influences the management of the patient or strengthens confidence in the diagnosis. When choosing a procedure utilizing ionizing radiation the benefit/risk balance must be carefully considered through the process of justification. When indicated and available, imaging modalities that do not use ionizing radiation, e.g., ultrasound or magnetic resonance imaging, are preferred, especially in children. The possibility of deferring imaging to a later time if/when the patient's condition may change also must be considered. The final decision may also be influenced by cost, expertise, availability of resources, and/or patient values.

The ultimate purpose of healthcare systems is to deliver the best care to every person any time and everywhere. Poor-quality care may come in the form of overuse (i.e., giving people care they do not need), underuse (i.e., failing to give people care they need), and misuse (i.e., making errors that can damage people). All three problems should be addressed to improve quality and safety of healthcare. Considerable disparities exist today between and within countries with respect to the use of radiation technologies in healthcare. While most developing countries still lack adequate capacity, access, and resources, and are facing the challenge of underuse, developed countries are increasingly facing the risk of overuse. An area of special concern is the unnecessary use of radiation imaging in cases where clinical evaluation or other imaging modalities could provide an accurate diagnosis. This is particularly critical in the context of pediatric healthcare, since children are especially vulnerable to environmental threats and have a longer life-span to develop long-term radiation-induced health effects like cancer.

Evidence-based referral guidelines for appropriate use of medical imaging can significantly improve the use of healthcare resources and reduce unnecessary radiation exposure. As decision-aiding tools, imaging referral guidelines provide a basis for good medical practice for referrers and medical imaging practitioners. They provide physicians with information on which procedure is most likely to yield the most informative results, and whether another modality is equally or more effective, and therefore more appropriate (Oakeshott et al. 1994, Hadley et al. 2006, Remedios et al. 2014).

Evidence-based imaging referral guidelines have been developed by professional bodies in several countries (ACR 2014, RCR 2013, DIP 2014, SFR/SFMN 2013). However, they are not available worldwide, particularly in developing countries. Even in those countries where guidelines exist, concerted efforts are still needed to integrate them into daily medical practice. An international collaboration involving 23 international, regional, and national agencies and professional societies was established within the WHO Global Initiative on RSHCS to make available evidence-based imaging referral guidelines, facilitate their implementation, monitor their use, and evaluate their impact in different clinical settings.

15.3.3 Optimization of Protection

Radiation protection in medicine is built on two principles: justification of procedures and optimization of protection (ICRP, 2007a and 2007b). Although these two pillars of radiological protection in healthcare are implicit in the notion of good medical practice, some health professionals are not familiar with them, and have a low awareness of radiation doses and risks. The WHO Global Initiative on Radiation Safety in Healthcare Settings (RSHCS) aims to enhance the implementation of the optimization principle in healthcare. Optimization of protection in medical exposures requires the management of the radiation dose to the patient to be commensurate with the medical purpose. This is applied at two levels: (i) the design and construction of equipment, software, and installations, and (ii) the working procedures and operational parameters.

In the case of optimization in medical imaging, methods for dose reduction should be applied and protocols should be tailored according to the patient size, and the level of acceptable noise according to the clinical indication. This is particularly important in pediatric imaging. Diagnostic Reference Levels (DRLs) are tools for optimization in radiodiagnosis, nuclear medicine, and interventional radiology. They help ensure that the doses for a given procedure do not deviate significantly from those achieved at peer departments. The establishment of DRLs and the implementation of the DRL concept are part of the BSS requirements and are therefore considered within the activities of the WHO Global Initiative on RSHCS.

Optimization in radiotherapy implies a compromise between the dose to the target volume and the dose to normal tissues, in order to maximize tumor control and minimize health risks such as radiation toxicity and second cancers. Biological models in radiotherapy allow predicting the tumor control probability (TCP) as well as the normal tissue complication probability (NTCP) already at the stage of treatment planning, to optimize the protection of each individual patient. Enhancing optimization of protection in therapeutic use of radiation is also considered in the Global Initiative on RSHCS.

15.3.4 Patient Safety in Medical Uses of Radiation

A review of radiation accidents occurring between 1945 and 2007 indicated that a large number of fatalities (46) and the highest number of cases of acute injuries (623 cases) were due to accidents that occurred during the use of radiation in healthcare (UNSCEAR 2008). Most of these accidents involved patients undergoing radiotherapy. Many other cases may occur that are not reported or even not recognized. Accidental and unintended exposures also occur in diagnostic imaging and nuclear medicine. Skin burns and other injuries are increasingly observed in patients undergoing fluoroscopic-guided interventional procedures. The development of new technologies has, meanwhile, introduced new challenges in terms of quality assurance, equipment safety, education, training, and staffing, which require a stronger culture of safety amongst healthcare providers.

Quality assurance and continuing education are two major measures that can help prevent incidents and accidents within the health care system. Even small breaks in the quality chain can compromise patient safety and treatment outcome if allowed to go undetected. A coincidence of several errors can lead to radiation incidents and accidents. In radiotherapy, this may result in large groups of patients being overexposed or receiving under-dosage, which denies them the chance of cure.

Minimization of adverse events in the medical use of radiation is addressed by the WHO Global Initiative on RSHCS. Primary prevention being essential, adverse event reporting systems are a cornerstone to improve safety culture, by translating reporting into learning and using this knowledge to improve the safety of frontline care. WHO has started an inter-cluster project based on the Minimum Information Model (MIM) for patient safety adverse event reporting and learning that can serve as the basis for harmonization of safety taxonomy across different medical disciplines. As part of MIM and in the framework of the Global Initiative on RSHCS, WHO promotes the use of Safety in Radiation Oncology (SAFRON), a web-based system for reporting incidents and near-misses in radiotherapy developed by the IAEA.

15.3.5 Radiation Safety Culture in Healthcare

Establishing a radiation-protection culture in the medical practice is crucial to ensure that patients benefit from the medical use of radiation, and contributes to a more cost-effectively allocation of health resources. Radiation protection education and training is a key measure to improve radiation safety culture in health professionals. The inclusion of radiation-protection contents in the curricula of medical and dental schools is advocated in the WHO Global Initiative on RSHCS. Guidance on radiation protection education and training of health professionals has been provided by the International Commission on Radiological Protection (ICRP 2009). The European Commission conducted the project MEDical RAdiation Protection Education and Training (MEDRAPET). The WHO contributed to this project, which concluded with the publication of guidelines on this topic (EC 2014). Through the Global Initiative on RSHCS, WHO facilitates the dissemination of those products at a global level.

15.4   THE BONN CALL FOR ACTION

The "International Conference on Radiation Protection in Medicine: Setting the Scene for the Next Decade" was held in Bonn, Germany, in December 2012. It was organized by the IAEA, cosponsored by the WHO, and hosted by the Government of Germany through the Federal Ministry for the Environment, Nature Conservation, and Nuclear Safety (BMU). This conference was attended by 536 participants and observers from 77 countries and 16 international organizations. The main outcome of this conference was the so-called "Bonn Call for Action" that identifies priority actions to enhance radiation protection in medicine for the next decade. The ten priority actions and related sub-actions identified in the Bonn Call for Action are summarized below. The WHO Global Initiative on RSHCS is currently focused on supporting the implementation of these ten priority actions in Member States.

15.4.0.1 Action 1: Enhance the Implementation of the Principle of Justification

a) Introduce and apply the three A's (awareness, appropriateness, and audit), which are seen as tools that are likely to facilitate and enhance justification in practice;

b) Develop harmonized evidence-based criteria to strengthen the appropriateness of clinical imaging, including diagnostic nuclear medicine and non-ionizing radiation procedures, and involve all stakeholders in this development;

c) Implement clinical imaging referral guidelines globally, keeping local and regional variations in mind, and ensure regular updating, sustainability, and availability of these guidelines;

d) Strengthen the application of clinical audit in relation to justification, ensuring that justification becomes an effective, transparent, and accountable part of normal radiological practice;

e) Introduce information technology solutions, such as decision support tools in clinical imaging, and ensure that these are available and freely accessible at the point of care;

f) Further develop criteria for justification of health-screening programs for asymptomatic populations (e.g., mammography screening) and for medical imaging of asymptomatic individuals who are not participating in approved health-screening programs (e.g., use of CT for individual health surveillance).

15.4.0.2 Action 2: Enhance the Implementation of the Principle of Optimization of Protection and Safety

a) Ensure establishment, use of, and regular update of diagnostic reference levels for radiological procedures, including interventional procedures, in particular for children;

b) Strengthen the establishment of quality assurance programs for medical exposures, as part of the application of comprehensive quality management systems;

c) Implement harmonized criteria for release of patients after radionuclide therapy, and develop further detailed guidance as necessary;

d) Develop and apply technological solutions for patient exposure records, harmonize the dose data formats provided by imaging equipment, and increase utilization of electronic health records.

15.4.0.3   Action 3: Strengthen Manufacturers' Role in Contributing to the Overall Safety Regime

a) Ensure improved safety of medical devices by enhancing the radiation protection features in the design of both physical equipment and software and to make these available as default features rather than optional extra features;

b) Support development of technical solutions for reduction of radiation exposure of patients while maintaining clinical outcome, as well as of health workers;

c) Enhance the provision of tools and support in order to give training for users that is specific to the particular medical devices, taking into account radiation protection and safety aspects;

d) Reinforce the conformance to applicable standards of equipment with regard to performance, safety, and dose parameters;

e) Address the special needs of healthcare settings with limited infrastructure, such as sustainability and performance of equipment, whether new or refurbished;

f) Strengthen cooperation and communication between manufacturers and other stakeholders, such as health professionals and professional societies;

g) Support usage of platforms for interaction between manufacturers and health and radiation regulatory authorities and their representative organizations.

15.4.0.4   Action 4: Strengthen Radiation Protection Education and Training of Health Professionals

a) Prioritize radiation protection education and training for health professionals globally, targeting professionals using radiation in all medical and dental areas;

b) Further develop the use of newer platforms such as specific training applications on the Internet for reaching larger groups for training purposes;

c) Integrate radiation protection into the curricula of medical and dental schools, ensuring the establishment of a core competency in these areas;

d) Strengthen collaboration in relation to education and training among education providers in healthcare settings with limited infrastructure as well as among these providers and international organizations and professional societies;

e) Pay particular attention to the training of health professionals in situations of implementing new technology.

15.4.0.5   Action 5: Shape and Promote a Strategic Research Agenda for Radiation Protection in Medicine

a) Explore the re-balancing of radiation research budgets in recognition of the fact that an overwhelming percentage of human exposure to man-made sources is medical;

b) Strengthen investigations into low-dose health effects and radiological risks from external and internal exposures, especially in children and pregnant women, with an aim to reduce uncertainties in risk estimates at low doses;

c) Study the occurrence of and mechanisms for individual differences in radiosensitivity and hyper-sensitivity to ionizing radiation, and their potential impact on the radiation protection system and practices;

d) Explore the possibilities of identifying biological markers specific to ionizing radiation;

e) Advance research in specialized areas of radiation effects, such as characterization of deterministic health effects, cardiovascular effects, and post-accident treatment of overexposed individuals;

f) Promote research to improve methods for organ dose assessment, including patient dosimetry when using unsealed radioactive sources, as well as external beam small-field dosimetry.

15.4.0.6 Action 6: Increase Availability of Improved Global Information on Medical Exposures and Occupational Exposures in Medicine

a) Improve collection of dose data and trends on medical exposures globally, and especially in low- and middle-income countries, by fostering international co-operation;

b) Improve data collection on occupational exposures in medicine globally, also focusing on corresponding radiation protection measures taken in practice;

c) Make the data available as a tool for quality management and for trend analysis, decision making, and resource allocation.

15.4.0.7 Action 7: Improve Prevention of Medical Radiation Incidents and Accidents

a) Implement and support voluntary educational safety reporting systems for the purpose of learning from the return of experience of safety-related events in medical uses of radiation;

b) Harmonize taxonomy in relation to medical radiation incidents and accidents, as well as related communication tools such as severity scales, and consider harmonization with safety taxonomy in other medical areas;

c) Work towards inclusion of all modalities of medical usage of ionizing radiation in voluntary safety reporting, with an emphasis on brachytherapy, interventional radiology, and therapeutic nuclear medicine, in addition to external beam radiotherapy;

d) Implement prospective risk analysis methods to enhance safety in clinical practice;

e) Ensure prioritization of independent verification of safety at critical steps, as an essential component of safety measures in medical uses of radiation.

15.4.0.8   Action 8: Strengthen Radiation Safety Culture in Healthcare

a) Establish patient safety as a strategic priority in medical uses of ionizing radiation, and recognize leadership as a critical element of strengthening radiation safety culture;

b) Foster closer co-operation between radiation regulatory authorities, health authorities, and professional societies;

c) Foster closer co-operation on radiation protection between different disciplines of medical radiation applications as well as between different areas of radiation protection overall, including professional societies and patient associations;

d) Learn about best practices for instilling a safety culture from other areas, such as the nuclear power industry and the aviation industry;

e) Support integration of radiation-protection aspects in health technology assessment;

f) Work towards recognition of medical physics as an independent profession in health care, with radiation-protection responsibilities;

g) Enhance information exchange among peers on radiation protection and safety-related issues, utilizing advances in information technology.

15.4.0.9   Action 9: Foster an Improved Radiation Benefit–Risk Dialogue

a) Increase awareness about radiation benefits and risks among health professionals, patients, and the public;

b) Support improvement of risk communication skills of healthcare providers and radiation protection professionals — involve both technical and communication experts, in collaboration with patient associations, in a concerted action to develop clear messages tailored to specific target groups;

c) Work towards an active informed decision-making process for patients.

15.4.0.10   Action 10: Strengthen the Implementation of Safety Requirements Globally

a) Develop practical guidance to provide for the implementation of the International Basic Safety Standards in healthcare globally;

b) Further the establishment of sufficient legislative and administrative frameworks for the protection of patients, workers, and the public at a national level, including enforcing requirements for radiation protection education and training of health professionals, and performing on-site inspections to identify deficits in the application of the requirements of this framework.

15.5   FINAL CONSIDERATIONS

Improving healthcare requires a multi-sectoral approach and partnerships with a range of stakeholders. While WHO can play a unique stewardship role in bringing together diverse stakeholders in the medical sector to promote the review and translation of evidence into global policies and standards, Member States are the essential partners, both as initiators and implementers of new policies.

Over the past decade, a range of important expert networks have been established at the regional and global levels to address specifics topics related to radiation protection in healthcare. Even though most of them were created in some specific regions (e.g., North America, Europe), many activities organized within the framework of these networks are inclusive of other countries outside those regions. Different stakeholders within the medical sector thus have the opportunity to discuss and to exchange information, and such integration serves as a powerful catalyst to harmonization and benchmarking.

Countries exhibit many diverse levels of development and socio-economic conditions, requiring flexibility in the adaptation of new policies and approaches. At the same time, the highest standards of technical excellence, patient safety, and healthcare remain as the benchmark for all countries. Experiences in one region can thus be relevant to policymaking in other parts of the world. Global partnership should be expanded and a stronger collaboration between radiation protection and healthcare communities should be fostered to improve the radiation-protection culture in the medical sector. The WHO Global Initiative on RSHCS provides a platform to support and strengthen such global partnership and collaboration.

15.6 REFERENCES

ACR. 2014. Appropriateness Criteria of the American College of Radiology. http://www.acr.org/Quality-Safety/Appropriateness-Criteria. (Accessed December 19, 2014.)

BSS. 2014. Radiation Protection and Safety of Radiation Sources: International Basic Safety Standards IAEA Safety Standards Series GSR Part 3. http://www-pub.iaea.org/MTCD/Publications/PDF/Pub1578_web-57265295.pdf. (Accessed July 24, 2014).

DIP. 2014. Government of Western Australia, Department of Health, Diagnostic Imaging Pathways (DIP), A Clinical Decision Support Tool and Educational Resource for Diagnostic Imaging endorsed by the Royal Australian and New Zealand College of Radiology (RANZCR). http://www.imagingpathways.health.wa.gov.au. (Accessed December 20, 2014.)

EC. 2013. Council Directive 2013/59/Euratom of 5 December 2013, laying down basic safety standards for protection against the dangers arising from exposure to ionising radiation, and repealing Directives 89/618/Euratom, 90/641/Euratom, 96/29/Euratom, 97/43/Euratom and 2003/122/Eu-

ratom.   http://eur-lex.europa.eu/LexUriServ/LexUriServ.do?uri=
OJ:L:2014:013:0001:0073:EN:PDF. (Accessed December 20, 2014.)

EC. 2014. Radiation Protection No. 175 Guidelines on Radiation Protection
Education and Training of Medical Professionals in the European Union.
http://ec.europa.eu/energy/nuclear/radiation_protection/doc/ publica-
tion/175.pdf. (Accessed December 20, 2014.)

Hadley JL, Agola J, and Wong P. 2006. Potential Impact of the American
College of Radiology Appropriateness Criteria on CT for Trauma. AJR
186:937–942.

ICRP. 2007a. The 2007 Recommendations of the International Commission on
Radiological Protection. ICRP Publication 103; Ann ICRP 37 (2–4).

ICRP. 2007b. Radiological Protection in Medicine. ICRP Publication 105.
Ann. ICRP 37 (6).

ICRP. 2009. Education and Training in Radiological Protection for Diagnostic
and Interventional Procedures. ICRP Publication 113. Ann. ICRP 39.

Malone J, Guleria R, Craven C, et al. 2012. Justification of diagnostic medical exposures: some practical issues. Report of an International Atomic
Energy Agency Consultation. Br J Radiol. 85(1013):523–38.

Oakeshott P, Kerry SM, Williams JE. 1994. Randomized controlled trial of the
effect of the Royal College of Radiologists' guidelines on general practitioners' referrals for radiographic examination. Br J Gen Pract;44: 427e8.

RCR. 2013. iRefer: Making the best use of clinical radiology, 7th Edition
- essential radiological investigation guidelines tool from The Royal
College of Radiologists freely accessible to NHS professionals in the
UK. http://www.rcr.ac.uk/content.aspx?PageID=995. (Accessed December 20, 2014.)

Remedios D, Drinkwater K. and Warwick R. 2014. National audit of appropriate imaging. Clinical Radiology Audit Committee (CRAC), The Royal
College of Radiologists, London. Clin Radiol. 69(10):1039–44.

SFR/SFMN. 2013. Guide du Bon Usage des examens d'imagerie médicale
de la Societé Française de Radiologie (SFMN) et la Societé Française
de Médecine Nucléaire (SFMN), 2me Edition. http://gbu.radiologie.fr/.
(Accessed December 13, 2014.)

UNSCEAR. 2008. United Nations Scientific Committee on the Effects of
Atomic Radiation (UNSCEAR) Report to the General Assembly Sources
and effects of ionizing radiation, Volume II Annex C - Radiation
exposures in accidents. http://www.unscear.org/docs/reports/2008/11-
80076_Report_2008_Annex_D.pdf. (Accessed December 20, 2014.)

UNSCEAR. 2010. United Nations Scientific Committee on the Effects of Atomic Radiation (UNSCEAR) Report to the General Assembly. Sources and effects of ionizing radiation Annexe A: Medical Radiation Exposures. http://www.unscear.org/docs/reports/2008/09-86753_Report_2008_Annex_A.pdf. (Accessed December 20, 2014.)

WHO. 1948. Constitution of the World Health Organization, Geneva 1948 Available from: http://www.opbw.org/int_inst/health_docs/WHO-CONSTITUTION.pdf. (Accessed December 20, 2014).

WHO. 1978. Declaration of Alma-Ata, International Conference on Primary Health Care, Alma-Ata, September 6–12, 1978. http://www.who.int/publications/almaata_declaration_en.pdf. (Accessed December 20, 2014.)

CHAPTER 16

The Medical Physicists and the International Organization for Medical Physics (IOMP)

Kin Yin Cheung

President of the International Organization for Medical Physics 2012–2015
Hong Kong Sanatorium and Hospital, Hong Kong

CONTENTS

CONTEMPORARY HEALTHCARE relies on rapid and accurate diagnostic information and powerful therapeutic tools in patient management. Medical imaging and radiation oncology are vital tools for rapid and accurate diagnostic evaluation and treatment of an illness. All medical exposures, irrespective of normal or abnormal irradiation, can in theory induce a certain degree of health risks to the patients. The rapid increase in the number of medical imaging exposures around the world deserves some attention. There is a need to minimize patient dose in imaging procedures. To be able to achieve this goal, a concerted effort among the healthcare professionals involved in performing the radiological procedures and the associated quality assurance work, the referring practitioners, the equipment manufacturers, and health and radiation safety regulators is essential. Radiation health and safety regulators are responsible for establishing, in their own countries, appropriate legislative control of all activities involving the manufacture, sale, transportation, storage, use, and disposal of radioactive substances and irradiating equipment. They provide a legislative framework to ensure radiation safety in a country. The legal standards and limits they established, which are normally based on national or international safety guidelines or recommendations, can have significant impact on what and how radiation is used in healthcare. Equipment manufacturers play their part in dose reduction in medical exposure by developing appropriate new technologies that can better meet clinical requirements with less medical and occupational radiation exposure.

Medical practitioners are responsible for justifying medical exposures to individual patients. Medical physicists are key members of the radiology and radiation oncology teams involved in performing the radiological procedures. They play a leading role in radiation safety and protection in the clinics. They ensure that all medical exposures are properly performed and that they are individually optimized such that the radiation dose and the risks to an individual patient are small as compared with the risks arising from the disease if the exposure is refused or delayed. They are responsible for implementation and management of a robust system of radiation protection in the clinics so as to ensure that the principles of radiation protection are applied in every exposure and that the patients can benefit from any medical exposure in the management of their illnesses.

National and international medical physics organizations also play an important role in enhancing radiation safety in healthcare. They provide guidelines and recommendations on such issues as qualification and standard of practice of medical physicists and safety procedures or code of practice in the clinics. The International Organization of for Medical Physics (IOMP) is a professional organization that represents all medical physicists in the world (IOMP). This organization plays an important role in promoting the global development of medical physics, including the professional status and standard of practice of the medical physicists. The roles and responsibilities of the medical physicists and those of the IOMP are briefly discussed in this chapter.

16.1 THE ROLE OF MEDICAL PHYSICISTS IN RADIATION PROTECTION IN MEDICINE

Medical physicists practicing in healthcare with clinical responsibilities are health professionals. They are qualified with education and specialist training in the concepts and techniques of applying physics in medicine, competent to practice independently in the specific specialties of medical physics. They are qualified with the particular knowledge and skills to develop and implement radiation safety and protection programs in medical institutions and perform or supervise others to perform specific aspects in applying radiation protection principles in the clinics. They work as a team together with the radiologists, nuclear medicine physicians, radiation oncologists, radiation therapists, imaging technologists, or radiographers and other allied health professionals, to provide a diagnostic or therapeutic radiology service that best meet institutional pledges. The basic requirements for education and specialist training of medical physicists working as health professionals and their roles and responsibilities in healthcare are set out in IOMP Policy Statements Nos. 1 and 2 (IOMP PS1 2010, IOMP PS2 2010) and IAEA Report No. 25 in the Human Health Series (IAEA 2013). As described in IOMP Policy Statement No. 1 (IOMP PS1 2010), medical physics consists of a number of sub-fields or specialties, including:

- Radiation oncology physics

- Medical imaging physics

- Nuclear medicine physics

- Medical health physics (radiation protection in medicine)

- Non-ionizing medical radiation physics

- Physiological measurement

Radiation protection in medicine is an important part of the duties of the medical physicists. Traditionally, radiation protection in medicine is concerned mainly with safe use of radiation in diagnostic radiology, nuclear medicine, and radiation oncology. With increasing use of imaging and interventional radiology in other clinical specialties and departments, ionizing and non-ionizing radiation has found its applications in a wider range of medical procedures, particularly in cardiology, neurology, urology, and orthopedic surgery. Thus, radiation safety and protection services provided by medical physicists cover a wide range of radiation user departments than ever before. The key roles and responsibilities of medical physicists in radiation safety in healthcare include the following:

- Radiation safety of the patients receiving medical exposures, the staff involved with radiation work, members of the public, and other persons in the vicinity of the radiological facilities.

- Radiological risk assessment, including impact on the environment of new radiological facilities, and they conduct radiation shielding design and calculation with respect to radiological installations.

- Assessment and optimization of patient dose in radiological procedures and monitoring of occupational dose received by members of staff under normal and emergency situations.

- Counseling of the patients and their relatives, especially when young patients are involved, and members of staff on radiation dose and safety issues.

- Supervising the safe use, management, custody, and maintenance of radiation sources and radiological equipment.

- Formulating radiation safety guidelines and procedures and conducting specialized assessment and measurements.

- Supporting the formulation of protocols for dose optimization in medical exposure, and minimizing radiation dose to patient, staff, and other persons.

- Research, development, and implementation of new patient dose reduction techniques and technologies in diagnostic or therapeutic procedures.

- Supporting the radiation monitoring and management of contaminated patients in the event of nuclear accident and emergency.

- Teaching and training of medical and allied health professionals on radiation safety and protection.

The safety measures in hospitals should be comprehensive and systematic. Essentially, a formal and structured radiation safety program should be implemented one that should include a system for identification of controlled areas where radiation is used, the radiation protection staff and their roles and responsibilities, and the type of radiation sources used and stored. It should include a set of case-, site-, and equipment-specific safe operation or working procedural guidelines for each of the controlled areas under normal and emergency situations. There should also be a system for protection of the patient, staff, and members of the public. Protection of patients should include a system for dose optimization based on ICRP principles of radiation protection and its guiding principles, which are as follows:

16.1.1   Justification

Every medical exposure should be justified. This should be based on, among other considerations, clinical needs, what other diagnostic or therapeutic options are available, and patient's choice and conditions. Justification is in general a case-specific clinical decision normally made by the referring practitioner or the radiologist or radiation oncologist in consultation with the patient and, in some cases such as pregnancy and young patients, medical physicist.

16.1.2   Optimization

In the case of diagnostic radiology procedures, the exposure dose to the patient should be optimized such that the required diagnostic information needed for making clinical diagnosis and supporting clinical decision-making is acquired with the minimum amount of radiation exposure. The optimized radiation dose is determined and can be affected by the functionality, quality, and conditions of the radiological equipment used; the machine exposure parameters used in acquiring the image data, and the skills of the practitioner and other staff involved in performing the radiological procedures. In therapeutic radiological procedure, the key objective of dose optimization is to deliver the prescribed therapeutic dose to the treatment target volume as accurately as possible with the minimum amount of dose to any normal tissues and organs. Optimization in radiation therapy is a more complex and multidisciplinary process involving the practitioner, medical physicist, and radiation therapist performing the radiological procedure. The medical physicist plays a very important role in this process by ensuring that the equipment is calibrated, and operating properly and safely according to specification; and that all the machine exposure parameters such as beam energy, intensity, exposure time, etc. used in the diagnostic or therapeutic procedure are optimal for a particular radiological exposure.

16.1.3   ALARA Principle

The radiation dose to the patient under a diagnostic radiological examination, the practitioner, and other operating staff performing the examination procedure, as well as other persons in the vicinity of the radiological facility, should be kept to as low as reasonably achievable (ALARA). Application of the ALARA principle in the clinics should take into account other site- and case-specific factors and conditions, including socioeconomic, service, environmental, religious, and cultural factors. The ALARA principle appears simple but is often misinterpreted and inappropriately applied in radiation protection. It is often interpreted "as low as possible" without taking into consideration the above-mentioned balancing factors. One may notice that radiation safety regulators in some countries appear to make the same misinterpretation. Taking radiation shielding requirements as an example, the regulators in some countries set a very stringent legislative requirement based on instantaneous dose rate without taking into consideration workload, occupancy, and use factors. This results in the demand for excessive shielding in their radiological facilities, even for infrequently used equipment. Such an approach usually cannot make any significant difference in the overall radiation safety of the facility if it has any benefit at all. On the contrary, excessive unnecessary shielding with materials such as lead, steel, or concrete is a waste of global resources. This in turn can cause other damages such as chemical harm and pollution to the environment, which have proven damages to human health. It is more meaningful to make use of the wasted resources for service improvement, such as reducing the charges to patients, investment in better medical technologies, and training of staff.

Appropriate interpretation and application of the ICRP radiation protection principles in practice are crucial for effective and optimal implementation of a radiation safety program for protection of the patients, staff, and other persons in the clinics. Effective implementation of these protection principles requires special knowledge and skills from the staff responsible for radiation safety and those who are directly or indirectly involved in performing the radiological procedures. Medical physicists who are qualified by IOMP definition (IOMP PS2 2010) are competent to perform or advise others to perform or implement radiation safety measures in the clinics by applying appropriately and effectively the principles of radiation protection. The ability and performance of the medical physicists has a direct impact on the quality and safety of the radiological services. They must be fully qualified with the academic knowledge, professional skills, and competency to perform their duties effectively and safely. Apart from acquiring the basic academic qualifications and going through the clinical training, medical physicists should also be subject to an appropriate continuing professional development program to maintain their professional competency, so as to face future professional and technological challenges in the clinics.

16.2 THE INTERNATIONAL ORGANIZATION FOR MEDICAL PHYSICS

The IOMP was founded in 1963, initially with four national medical physics organizations from Canada, Sweden, UK, and USA. Currently IOMP has 84 National Member Organizations (NMO) with more than 18,000 individual members practicing in 84 countries or regions around the world. The organization has, together with the respective NMOs, formed six Regional Organizations (ROs) in different regions of the world. These are:

- European Federation of Organizations for Medical Physics (EFOMP)

- Asian-Oceania Federation of Organizations for Medical Physics (AFOMP)

- Latin American Medical Physics Association (ALFIM)

- Southeast Asian Federation for Medical Physics (SEAFOMP)

- Federation of African Medical Physics Organizations (FAMPO)

- Middle East Federation of Organizations for Medical Physics (MEFOMP)

The mission of IOMP is to advance medical physics practice worldwide by disseminating scientific and technical information, fostering the educational and professional development of medical physics, and promoting the highest quality medical services for patients. Radiation safety and protection is an important part of medical physics service in healthcare. IOMP is actively engaged in promoting the global development of medical physics in support of the development of radiation medicine in different parts of the world, particularly in developing countries. Improving medical physics and radiation safety services in developing countries has always been a high priority in the IOMP agenda.

Raising the standard of practice, professional status, and visibility of the medical physicists are some of the key agenda items for IOMP. IOMP issues guidance documents on professional, scientific, and educational matters. IOMP provides guidance and motivates NMOs to improve their standards of practice and raise their visibility in the healthcare system. In collaboration with statutory international organizations such as IAEA, IOMP sensitizes state healthcare officials on the important role of medical physicists in radiation medicine. The organization has issued a number of policy documents or position statements covering recommendation or guidelines on such issues as roles and responsibilities of medical physicists (IOMP PS1 2010), standards and requirements for education, training, and professional accreditation of medical physicists (IOMP PS2 2010), cancer risks to patients due to exposure to ionizing radiation during medical imaging procedures (IOMP PS3 2013), and the role of medical physicists in case of a nuclear or radiological emergency (IOMP 2011). The professional status and standard of practice of the medical physicists in a particular country is often related to the level of recognition and awareness of the healthcare administration and service planners on the important role they play in radiation medicine. As an initiative to raising the visibility and awareness of medical physicists in the community, in 2013 IOMP launched 2013 the International Day of Medical Physics (IDMP). November 7, the birthday of Marie Sklodowska-Curie has been selected to mark this day. This is an annual event in which a series of scientific, educational, media, and publicity activities are organized around the world by the ROs and NMOs to celebrate the event.

IOMP motivates the interaction of medical physicists for exchange of scientific information and professional experience. It provides a platform to facilitate networking and interaction between medical physicists from different parts of the world. The organization, in collaboration with NMOs and ROs, organizes international scientific conferences and educational programs on medical physics and related topics, including radiation safety and protection. It publishes its own journal Medical Physics International Journal (MPIJ) and newsletter Medical Physics World (MPW), which are freely accessible through the Internet. It has a special interest in the development of medical physics in the developing countries and is aware of the limitations the medical physicists in these countries are facing as to resources for such needs as opportunity for training and professional development, access to library books and journals, and equipment and laboratory facilities. To address these issues, IOMP supports medical physicists from developing countries so they may attend scientific conferences and educational and training programs. In collaboration with publisher Taylor & Francis and the American Association of Medical Physicists (AAPM), IOMP donates books and journals to libraries in developing countries through the AAPM–IOMP Library Program. IOMP also collaborates with AAPM on donation of equipment to medical and educational institutions in developing countries through the IOMP Equipment Donation Program (IOMP Donation). Apart from donation of books to AAPM-IOMP

Libraries, IOMP also has an agreement with Taylor & Francis on discounts for IOMP members on purchases of books they publish in the Medical Physics and Biomedical Engineering Series.

IOMP, in collaboration with international organizations such as IAEA, the World Health Organization (WHO), and the International Radiation Protection Association (IRPA) has in the past few years organized a number of scientific and professional sessions or workshops on improving medical physics and radiation safety in developing countries, at major international conferences. Similar workshops are being planned for upcoming meetings such as the World Congress on Medical Physics and Biomedical Engineering, the IRPA regional conference, and the IOMP International Conference on Medical Physics (ICMP), to be held in the coming years. IOMP has collaborated with IAEA on a number of their projects, including projects on strengthening medical physics in radiation medicine, education and training of medical physicists, and implementation of the IAEA Basic Safety Standards (IAEA 2014) in medicine. IOMP has signed an MOU with WHO and a statement of collaboration with IRPA on improving radiation safety in healthcare. The goals of these collaborations are to promote the global development of medical physics, particularly in developing countries, and to foster and enhance radiation protection culture in health care.

16.3 CONCLUSION

All medical exposures can potentially induce a certain degree of health risks to the patients. The rapid increase in the number of medical imaging exposures around the world, particularly in the use of CT scans, deserves some attention. There is a need to minimize patient dose in imaging procedures. To achieve this goal, a concerted effort among healthcare professionals involved in performing the radiological procedures and the associated quality assurance work, the referring practitioners, the equipment manufacturers, and health and radiation safety regulators is essential.

Medical physicists are key members of the radiology, nuclear medicine, and radiation oncology teams. They play a leading role in radiation safety and protection in the clinics. They are also responsible for implementing a radiation safety framework to ensure safe and effective use of ionizing radiation in these clinics. In order to perform their duties effectively and be competent to face any clinical challenges, medical physicists should be qualified with an advanced university degree, such as an M.Sc. or Ph.D., followed by specialized clinical training as recommended by IOMP and IAEA. They are competent to provide the required scientific support to all clinical departments in patient management involving the use of radiation. Apart from contributing to routine clinical services, medical physicists play a key role in research, development, and implementation of improved or new imaging and radiation therapy modalities, techniques, and procedures. Radiation safety in the clinics is an important part of their duties. They are familiar with the installation requirements and operation and exposure characteristics of radiological equipment and the specific operations, workflows, and clinical needs of a medical institution. This allows them to perform, reliably and appropriately, radiological risk assessment and radiation shielding design of radiological facilities in the hospital. They are responsible for licensing of radiological facilities, safe management of radioactive sources and wastes, implementation of safety guidelines, and training of staff on radiation safety. They are familiar with the principle of operation, functionality, performance, and operational and dosimetric characteristics of radiological equipment and radioactive sources. Such expertise is essential for effective patient dose optimization in both diagnostic and therapeutic procedures without compromising the quality of the radiological procedures. They are responsible for assessment and monitoring of patient and staff dose. They also play a leading role in patient dose optimization, including the development of improved or new dose reduction radiological techniques and procedures for individual facilities. They are responsible for implementation of an appropriate QA program to ensure service quality and avoid the occurrence of abnormal exposure or radiation accidents. Medical physicists help ensure the success of radiation therapy and diagnostic imaging procedures and minimize the amount of radiation dose to staff and patients.

The IOMP represents all medical physicists in the world. It plays a leading role in promoting the global advancement of medical physics and radiation safety in medicine through a number of globally coordinated actions. It provides guidance on educational and professional training requirements, and on accreditation of the professional qualifications of medical physicists. It provides a platform to facilitate the networking and interaction among medical physicists for exchange of expertise and scientific information in medical physics. It collaborates with NMOs, ROs, and other international and professional organizations in hosting international scientific conferences, training workshops, and education programs in different parts of the world. It also provides guidance on standards of practice and roles and responsibilities of medical physicists in healthcare. IOMP collaborates with NMOs, ROs, and international organizations such as IAEA, WHO, and IRPA in strengthening medical physics and radiation safety cultures in healthcare. It advocates the use of sound professional judgment when applying ALARA, dose optimization, justification, and other radiation protection principles in solving specific radiation protection problems. IOMP emphasizes the importance of correct interpretation and application of the radiation protection principles to be exercised by both regulators and radiation users in the clinics, so as to optimize patient benefits in all medical exposures. To improve this aspect of radiation protection, a good dialogue and collaboration between medical physicists, health physicists, and regulators on relevant issues would be helpful.

16.4 REFERENCES

International Atomic Energy Agency. 2013. Human Health Series No. 25, Roles and Responsibilities, and Education and Training Requirements for Clinically Qualified Medical Physicists, International Atomic Energy Agency.

International Atomic Energy Agency. 2014. Radiation Protection and Safety of Radiation Sources: International Basic Safety Standards, General Safety Requirements Part 3, IAEA Safety Standards No. GSR Part 3, IAEA, Vienna.

IOMP-AAPM Library Program, http://www.aapm.org/international/ LibraryProgram.asp. (Accessed January 25, 2015).

IOMP Equipment Donations Programme, http://www.iomp.org/?q=node/46. (Accessed January 25, 2015).

International Organization for Medical Physics. www.iomp.org. (Accessed January 25, 2015).

International Organization for Medical Physics. 2010. Policy Statement No. 1, The Medical Physicist: Role and Responsibilities, International Organization for Medical Physics.

International Organization for Medical Physics. 2010. Policy Statement No. 2, Basic Requirements for Education and Training of Medical Physicists, International Organization for Medical Physics.

International Organization for Medical Physics. 2011. Do medical physicists have a role in case of a nuclear or radiological emergency? International Organization for Medical Physicist. http://www.iomp.org/sites/default/files/ med_phys_and_nuclear_emergency_2011-03-29.pdf (Accessed January 25, 2015).

International Organization for Medical Physics. 2013. Policy Statement No. 3, Cancer risks to patients due to ionizing radiation during medical imaging procedures, International Organization for Medical Physicist.

Medical Physics International, International Organization for Medical Physics, www.mpijournal.org. (Accessed January 25, 2015).

Medical Physics World, International Organization for Medical Physics, www.iomp.org. (Accessed January 25, 2015).

CHAPTER 17

The International Union for Physical and Engineering Sciences in Medicine (IUPESM)

Fridtjof Nüsslin

Technical University Munich, Germany

CONTENTS

S INCE THE 18TH CENTURY, and still today, progress in healthcare is mainly based on achievements in science and engineering. In particular, with the discovery of radioactivity and X-rays, the application of ionizing radiation in medicine initiated numerous physical and technical developments to significantly improve diagnosis of nearly all diseases, and to provide effective and cost-efficient methods in cancer treatment. With the technical revolution in medicine, the role of physicists and engineers became ever more prominent and paved the way towards new scientific fields in medicine such as radiology, radiotherapy and nuclear medicine, and, most importantly, medical physics and biomedical engineering.

17.1   INTRODUCTION

Traditionally, medical physicists have played a major role in all physical and technical aspects of the clinical application of ionizing and non-ionizing radiation, the development of new methods and instrumentation, and particularly, in radiation protection across the entire clinical environment. A good definition and description of typical task for a medical physicist can be found at the ISCO-08 plan issued by the International Labor Organization (ILO) (Smith, Nüsslin, 2013). On the other hand, the domain of biomedical engineering is the application of engineering principles in the development and design of instrumentation in healthcare. In practice, there is a large overlap in both disciplines, and medical physicists and biomedical engineers complement each other in their wide-ranging professional fields. Hence it is no wonder that their international professional organizations, the IOMP (International Organization for Medical Physics) and the IFMBE (International Federation for Medical and Biological Engineering), both founded in the second half of the last century, explored opportunities for closer relationships by organizing a back-to-back congress, in 1976 in Ottawa and a joint one in 1979 in Jerusalem (Nagel, 2007). The success of this collaboration led to forming an umbrella organization, the International Union for Physical and Engineering Sciences in Medicine (IUPESM), in 1980. The first "World Congress for Medical Physics and Biomedical Engineering" was held in 1982 in Hamburg and initiated the series of triennial IUPESM congresses as the key gatherings of all professionals engaged with such healthcare technologies.

17.2   THE ROLE OF IUPESM

Currently, the IUPESM, via its constituents IOMP and IFMBE, comprises nearly 140,000 professionals working in clinical environment, in patient services, in research and development and in industry, administration, and governmental positions.

Besides organizing the series of World Congresses, the main objectives of the IUPESM are:

- Representation of healthcare professionals in the International Council for Science (ICSU), a non-governmental organization with a global membership of national scientific bodies and International Scientific Unions, recognized by the United Nations as the voice of international science in policy formulation. Since biomedical sciences and engineering are central in the realization of a healthy world, IUPESM is specifically involved in the new ICSU initiatives "Future Earth" and "Health and Well-Being in the Urban Environment."

- Collaboration with other international scientific, professional, and medical organizations with common interests.

- Publishing scientific journals, newsletters, books, and electronic documents to enhance progress. Their book series in Medical Physics and Biomedical Engineering was launched in 1991. The most recent publication and official organ of IUPESM is "Health and Technology," a cross-disciplinary journal for all medical and scientific research professionals involved in health technologies.

- Disseminating, promoting, and/or developing standards of practice in the fields of medical physics and biomedical engineering, to enhance the quality of healthcare worldwide. Since the 1980's, IUPESM has specifically collaborated with the World Health Organization, with a major focus on healthcare technologies in developing countries. To underpin this priority, at the World Congress 2006, IUPESM founded the Health Technology Task Group (HTTG). This group was tasked "to assist countries in defining their health technology needs, and identifying and rectifying health system constraints for adequate management and utilization of health technology, particularly through training, capacity building and the development and application of appropriate technology" (HTTG Mission).

In summary, the role of IUPESM in radiation protection is less directly visible by its own activities, except via the programs of the HTTG, than by the many initiatives of one of its two constituents, the IOMP.

17.3   REFERENCES

Nagel JS. 2007. IUPESM: the international umbrella organization for biomedical engineering and medical physics. Biomed Imaging Interv J 2007; 3(3):e56.

Smith PHS, Nüsslin F. 2013. Benefits to Medical Physics from the Recent Inclusion of Medical Physicists in the International Classification of Standard Occupations (ISCO-08). Medical Physics International Journal 2013, 1(1):10–14.

CHAPTER 18

Education and Training in Radiation Protection

Eliseo Vaño

Radiology Department, Medical School, Complutense University, Madrid, Spain

CONTENTS

T HE NUMBER OF MEDICAL PROCEDURES USING IONIZING RADIATION is
rising, and imaging procedures resulting in higher patient and staff doses
are being performed more frequently. The need for education and training of
medical staff and other healthcare professionals in the principles of radiation
protection (RP) is becoming even more necessary than in the past (ICRP
2009b). Medical physicists should have the highest levels of education and
competence in RP, and need to be closely involved in the education and training programs on RP for other health professionals.

18.1 THE IMPORTANCE OF EDUCATION AND TRAINING IN RADIATION PROTECTION

Training in RP is widely recognized as one of the basic components of optimization programs for medical exposures. This training plays a key role in
preventing incidents and accidents in radiotherapy (ICRP 2000b, ICRP 2005,
ICRP 2009a) and in the medical management of radiological or nuclear accidents (Vaño et al. 2011). The International Commission on Radiological
Protection (ICRP) and most of the international bodies acknowledge the importance of education and training in reducing patient and staff doses while
maintaining the desired level of quality in medical exposures (ICRP 2007b,
2007c, 2009a, 2009b).

It is important that medical and other healthcare professionals understand
the hazards of radiation in order to avoid unnecessary risks to the population
as a whole. Lack of awareness of these crucial issues may result in requesting more ionizing radiation imaging tests than necessary when lower-dose
imaging tests or even non-radiation tests could be performed. This is particularly important in the case of computed tomography scans and interventional
radiology that involve relatively high doses to patients and staff, when dealing with interventional procedures (ICRP 2000a, 2000c, 2007a, 2010, 2013).
To start with, professionals must be aware that a proper education in RP
is already necessary for selecting the appropriate imaging and radiotherapy
equipment. Imaging equipment is often offered with several additional options
that can impact on radiation protection, such as radiation protection tools,
and software for patient dose reports and quality control (QC). These options
need to be evaluated and, if mentioned, included in the purchase specification.
Implementation of the optimization principle of radiation protection requires
that a systematic approach be taken to establish protocols on how to use a
given imaging device during a given procedure. The medical physicist is an
important contributor to this process, but medical doctors (radiologists, cardiologists, etc.) should also acquire enough knowledge regarding these issues
(Meghzifene et al. 2010).

The medical physicist has the expertise to provide specialist radiation protection training for staff, including equipment and room-specific practical radiation protection training. There is a strong inter-relationship between patient

dose and personal dose in image-guided interventional procedures, and the medical physicist is best placed to provide training to ensure both parties are afforded optimized protection. This includes training new staff and also the provision of refresher training as part of continuing professional development (Meghzifene et al. 2010).

In recent years, several scientific and professional societies have produced guidelines on radiation safety, including aspects on patient dosimetry, occupational protection, and protection during pregnancy. Some of these guidelines have been adopted simultaneously by the American and European societies of interventional radiology (Dauer et al. 2012, Miller et al. 2010, 2012). Other recommendations have been produced by expert groups and later endorsed by the professional societies (Law et al. 2014, Duran et al. 2013). The role of the European Commission and the International Atomic Energy Agency (IAEA) in the publication of guidelines, reports, and books is particularly important in the optimization of digital and interventional procedures.

18.2 EDUCATION AND TRAINING IN RP FOR INTERVENTIONAL PROCEDURES AND DIGITAL RADIOLOGY

During interventional radiology and cardiology procedures, occupational and patient radiation risks can be quite high. Radiation injuries have been extensively described in the scientific literature (ICRP 2000a, 2010, 2013, Shope 1996, Koenig et al. 2001, Vaño et al. 1998). International concern on this issue is reflected in the literature, and the ICRP has made specific recommendations to improve radiation safety. Several European actions promoted specific training courses for interventionists following the ICRP recommendations, with a wide and positive acceptance among the medical professionals (Vaño et al. 2001). The International Basic Safety Standards (IAEA 2014), the European Council Directive 97/43/EURATOM on medical exposures (EC 1997), and the new European Directive 2013/59/EURATOM (EC 2014a) consider interventional radiology as a "special practice" that involves high doses to the patient, and require practitioners and other involved staff to have adequate theoretical and practical training as well as relevant competence in RP. The directives also state that continuing education and training after qualification should be provided, with a provision of appropriate training for practitioners conducting special practices.

Digital radiology may also be seen as a type of special practice with various specific requirements in relation to knowledge and skill for all professionals involved (radiologists, medical physicists, and radiographers) (ICRP 2004). These requirements are quite different from daily practice in conventional radiology. With the continuing integration of digital technology into radiology departments and outpatient practices, the need for specialized training and continuing education in topics relevant for digital radiology is increasing (Peer et al. 2005). The ICRP has published specific recommendations on RP for Digital Radiology and CT (ICRP 2004, 2000c, 2007a).

The European DIMOND research project (Dose and Image Quality in Digital Imaging and Interventional Radiology), involving 13 European centers of excellence from 11 member states of the European Union, promoted various aspects relevant to the advancement of digital imaging, with the main focus on quality improvement and RP. A full work package addressed the "training needs for professionals (radiology technicians, physicists and radiologists) in digital radiology." Organized post-graduate training programs on RP for these professional groups were found to be missing, and it was advised that RP training should be included in the scientific programs of congresses and meetings or within the industry activities introducing digital technology in medical imaging (Peer et al. 2005).

18.3   INTERNATIONAL RECOMMENDATIONS

The ICRP made recommendations for education and training in different publications (ICRP 2000a, 2004, 2007c, 2009b). Education on RP should be planned for several categories of medical practitioners and other healthcare professionals who perform or provide support for therapeutic, diagnostic, and interventional procedures utilizing ionizing radiation.

Education and training in RP should be provided to:

- Cognizant regulators, health authorities, medical institutions, and professional bodies with responsibility for radiological protection in medicine;

- Industry that produces and markets the equipment used in these procedures; and

- Universities and other academic institutions responsible for the education of professionals involved in the use of ionizing radiation in healthcare.

The "Bonn Call for Action" (WHO 2014) issued by the IAEA and the WHO after the International Conference on Radiation Protection in Medicine held in Bonn, in December 2012, included one specific action to:

"Strengthen radiation protection education and training of health professionals," with 5 items:

- Prioritize radiation protection education and training for health professionals globally, targeting professionals using radiation in all medical and dental areas;

- Further develop the use of newer platforms such as specific training applications on the Internet for reaching larger groups for training purposes;

- Integrate radiation protection into the curricula of medical and dental schools, ensuring the establishment of a core competency in these areas;

- Strengthen collaboration in relation to education and training among education providers in healthcare settings with limited infrastructure as well as among these providers and international organizations and professional societies;

- Pay particular attention to the training of health professionals in situations of implementing new technology.

18.4 TERMINOLOGY: EDUCATION, TRAINING, ACCREDITATION, AND CERTIFICATION

ICRP uses the term "education" in referring to imparting knowledge and understanding on the topics of radiation health effects, radiation quantities and units, principles of radiological protection, radiological protection legislation, and the factors in practice that affect patient and staff doses. The term "training" refers to providing instruction with regard to radiological protection for the justified application of the specific ionizing radiation modalities (e.g., computed tomography, fluoroscopy) that a medical practitioner or other healthcare or support professional will utilize in that individual's role during medical practice (ICRP 2009b).

ICRP also provides advice on the accreditation and certification of the recommended education and training. For ICRP, the term "accreditation" means that an organization has been approved by an authorized body to provide education or training on the radiological protection aspects of the use of radiation procedures in medicine. The accredited organization is required to meet standards that have been set by the authorized body. The term "certification" means that an individual medical or clinical professional has successfully completed the education or training provided by an accredited organization. The individual must demonstrate competence in the subject matter in a manner required by the accredited body (ICRP 2009b).

18.5 THE EFFECT OF TRAINING ON RADIATION DOSE

It is recognized that education and training in RP is one of the main aspects of any quality assurance program in clinical services using ionizing radiation. Education and training programs have a relevant impact in the justification of medical imaging and interventional procedures (e.g., proper use of referral guidelines and clinical decision support tools), but they are also a valuable help for selecting and optimizing the appropriate protocols, for using the new imaging technology in the best way, and for following the patient and staff dose values. The registry of patient dose values and their periodic comparison with diagnostic reference levels (DRLs) have demonstrated a capability of contributing effectively to the reduction of patient and staff dose values (ICRP 2001b, Picano et al. 2007).

The retrospective evaluation of occupational doses carried out at a large university hospital in Madrid over a period of 15 years (Vaño et al. 2006) shows how training impacted interventional cardiologists. During the last 5 years of the study, after the implementation of radiation protection actions and of a program of patient-dose optimization, the occupational doses recorded under the lead apron were 14% of those recorded at the beginning of the study and those recorded over the apron were 14-fold less than those recorded at the beginning of the study. The most effective actions involved in reducing the radiation risk proved to be training in radiation protection and a program of patient-dose reduction.

18.6 SUGGESTED LEVEL OF TRAINING FOR SPECIALTIES AND RECOMMENDATIONS OF THE ICRP

There should be RP training requirements for physicians, dentists, and other health professionals who request, conduct, or assist in medical or dental procedures that utilize ionizing radiation in diagnostic and interventional procedures, nuclear medicine, and radiation therapy. These professionals should be aware of the risks and benefits of the procedures involved, as the final responsibility for the radiation exposure lies with the physician or with a regulated healthcare professional capable of providing proper justification for the exposure being carried out.

Education in RP also needs to be given to referrers of imaging techniques using ionizing radiation, and to medical and dental students. Referrers need to be familiar with referral criteria appropriate for the range of examinations that they are likely to request. The ICRP recommends that a stronger emphasis should be placed on transfer of knowledge of RP and its application to referrers (ICRP 2009b). This recommendation applies particularly to practitioners and medical specialists outside radiological specializations. Since all medical professionals are likely to refer for medical exposures, the ICRP recommends that basic education in RP for physicians should be given as part of the medical degree (ICRP 2001a).

Professionals involved more directly in the use of ionizing radiation should receive education and training in RP at the start of their career, and the education process should continue throughout their professional life as the collective knowledge of the subject develops. It should include specific training on related RP aspects as new equipment or techniques are introduced into a center. These staff should be registered into a continuing professional development schedule.

Interventional procedures can involve high doses of radiation, and the special radiological risk needs to be taken into account if deterministic effects on the skin are to be avoided. In ICRP Publication 85 (ICRP 2000a), it is proposed that a second level of RP training be provided for interventional radiologists and cardiologists, in addition to the training recommended for other physicians who use X-rays. This should also be applied to other medical doctors conducting interventional fluoroscopy-guided procedures (ICRP 2009b, 2010).

Training in RP given to interventional cardiologists and other medical doctors conducting interventional fluoroscopy-guided procedures (e.g., vascular surgeons) in most countries is limited. The ICRP considers that provision of more RP training for these groups should be a priority.

Medical physicists working in RP, nuclear medicine, and diagnostic radiology should have the highest level of training in RP, as they have additional responsibilities as trainers in RP for most clinicians. Medical physicists should have proven knowledge and professional competency by way of professional certification or state registration before they are allowed to practice independently and to teach other medical professionals. They should also enter into a continuing professional development program.

Nurses and other healthcare professionals assisting in fluoroscopic procedures require knowledge of the risks and precautions to minimize their exposure and that of others.

Maintenance engineers and applications specialists currently receive some training in RP, but this may be primarily focused on RP of staff. Training on RP of patients needs to be expanded, particularly in relation to digital radiology and new equipment.

The ICRP recommends training related to their training, related to their practical work, in the hospital for radionuclide laboratory staff. This may be of rather longer duration, as staff members may work with radionuclides on a full-time basis (ICRP 2009b).

It is essential that courses on RP for medical professionals are perceived as relevant and necessary, and only require a limited time commitment so that individuals can be persuaded of the advantages of attending. Training for healthcare professionals in RP should be related to their specific jobs and roles.

A key component in the success of any training program is to convince the engaged personnel about the importance of the principle of optimization in RP so that they implement it in their routine practice. In order to achieve this, the training material must be relevant and presented in a manner that the clinicians can relate to their own situation.

Priority topics to be included in the training will depend on the involvement of the different professionals in medical exposures. A useful orientation on some of the topics to be included in the education program on RP for medical students could be ICRP Supporting Guidance 2, "Radiation and Your Patient: a Guide for Medical Practitioners" (ICRP, 2001b).

Training programs need to be devised for a variety of different categories of medical and clinical staff based on the greater or lesser involvement with medical exposures.

A training program in RP for healthcare professionals has to be oriented towards the type of training to which the target audience is accustomed.

Practical training should be in a similar environment to that in which the participants will be practicing.

RP training should be updated when there is a significant change in radiology technique or radiation risk, and at intervals not exceeding 36 months.

18.7 EVALUATION OF THE RP KNOWLEDGE ACQUIRED

ICRP recommends (ICRP 2009b) that training activities in RP should be followed by an evaluation of the knowledge acquired from the training program. This will allow the accreditation of the training for the attendants. Basic details should be given in the diplomas or certificates awarded to those attending a training program in RP. Education and training in RP should be complemented by formal examination systems to test competency before the person is awarded certification.

If certification in RP is required for some practices (e.g., interventional cardiology), the certificate should be obtained before a professional is involved in practicing the specialty at a specific center (Vaño 2010). If the requirement is introduced in a country once the professionals are already working in the specialty, the different healthcare providers will need to make the resources available to train their own professionals in RP.

Part of the follow-up to maintain the accreditation of organizations providing training should be analyses of results from surveys of participant responses at the end of training courses or training activities.

Training programs should include initial training for all incoming staff, regular updating and retraining, and accreditation of the training.

18.8 RESOURCES FOR EDUCATION AND TRAINING

The need to provide adequate resources for education and training in RP for future professional and technical staff that request or partake of radiological practices in medicine must be recognized.

The minimum requirements for accreditation of a training program should take account of sufficient administrative support, guarantees for the archiving of files, diplomas, etc., for a minimum number of years, sufficient didactic support, teachers qualified in the topics to be taught and with experience in hospital medical physics, instrumentation for practical exercises, and availability of clinical installations for practical sessions.

18.9 LECTURERS AND TRAINERS

The primary trainer in RP should be a person who is an expert in RP in the practice with which he or she is dealing. This means a person who, in addition to having a detailed understanding of radiological protection, has knowledge about the clinical practice in the use of radiation.

Lecturers in training courses should be competent in RP; this is best demonstrated by professional certification, state registration, or an equivalent professional recognition system. They must also have experience in RP in medical installations and in practical work in a clinical environment (e.g., medical physicists, radiographers, etc.).

Training of those using radiation-imaging equipment should be provided by a team involving radiological professionals, each of whom bring their specific knowledge.

Trainers participating in these activities should meet the local requirements and demonstrate sufficient knowledge in the RP aspects of the procedures performed by the medical specialists involved in the training activity.

Lectures and training programs organized by professional bodies, universities, and other medical institutions will play a key role in enabling continuing professional development.

18.10 ONLINE EDUCATION SYSTEMS AND COMPUTER-BASED TOOLS

It may be worthwhile for organizations to develop online evaluation systems because of the magnitude of the requirements for RP training. Such online methods are currently available mainly from organizations that deal with examinations carried out on a large scale. The development of self-assessment examination systems should also be encouraged.

With many medical schools using computer-based tools for their curricula as well as continuing education, it seems reasonable that the same approach could be employed for continuing education on radiation biology and radiation exposures in medicine (ICRP 2009b).

18.11 EUROPEAN APPROACH TO RP TRAINING BASED ON KNOWLEDGE, SKILLS, AND COMPETENCES

In 2000 the European Commission published "Radiation Protection 116: Guidelines on education and training in radiation protection for medical exposures" (EC 2000). This document was updated in 2014 (EC 2014b) to take into account the scientific, technological, and regulatory developments of the past decade. The new guidelines bring several additional improvements: a) the document follows the modern format and terminology of the European Qualifications Framework for Lifelong Learning; b) detailed requirements for initial and continuing training are specified for each of the included professions; and, perhaps most importantly, c) the document was developed and endorsed by the major European professional societies in the area.

The European regulations require that practitioners and individuals involved in the practical aspects of medical radiological procedures have adequate education, information, and theoretical and practical training for the purposes of medical radiological practices, as well as relevant competence in RP. For this purpose, Member States of the European Union (EU) shall ensure that appropriate curricula are established and shall recognize the corresponding diplomas, certificates, or formal qualifications. Member States shall also ensure that continuing education and training after qualification is provided, and, in the special case of the clinical use of new techniques, training is provided on these techniques and the relevant RP requirements. In addition, Member States shall encourage the introduction of a course on RP in the basic curriculum of medical and dental schools.

According to the EU recommendations on the establishment of the European Qualifications Framework (EQF) for Lifelong Learning, professional qualifications have been classified into eight levels (European Parliament 2008). Each of the eight levels is defined by a set of descriptors indicating the learning outcomes relevant to the qualifications at that level in terms of Knowledge, Skills, and Competences (KSC). For the area of RP, the level of learning outcomes depends very much on the level of involvement of a particular health profession with ionizing radiation.

For example, while entry into the profession as a medical doctor requires at least KSC level 7 for the medical subject areas, radiation protection KSC level 5 may be sufficient if the particular medical doctor acts as referrer for the use of ionizing radiation. Education and training guidelines and KSC tables should be updated regularly to reflect technological and other advances in the field of medical RP. The European guidelines have been divided into sections according to the healthcare profession in question, and each section includes KSC and continuous professional development (CPD) at the required level.

The new European Guidelines on Radiation Protection education and training of medical professionals in the European Union contain the core learning outcomes for RP, and specific learning outcomes (entry requirements and continuous professional development in RP) for:

- Referrers.

- Diagnostic radiologists.

- Interventional radiologists.

- Non-radiologist specialists employing ionizing radiation in interventional techniques.

- Nuclear medicine specialists.

- Radiation oncologists.

- Dentists and dental surgeons.

- Radiographers.

- Medical physicists and medical physics experts.

- Nurses and other healthcare workers.

- Maintenance engineers and maintenance technicians.

18.12 MEDICAL AND HEALTHCARE PROFESSIONALS REQUIRING RP TRAINING

ICRP has identified a total of 17 professional categories of healthcare professionals requiring RP training (ICRP 2009b). These categories will be identified below in Tables 18.1 and 18.2 to indicate the suggested level of knowledge (low, medium, or high) to be included in the education and training programs for the different topical areas.

Category 1 — radiologists: physicians who are going to take up a career in which the major component involves the use of ionizing radiation in radiology. This includes those performing interventional radiology procedures.

Category 2 — nuclear medicine specialists: physicians who are going to take up a career in which the major component involves the use of radiopharmaceuticals in nuclear medicine for diagnosis and treatment including PET or PET/CT.

Category 3 — cardiologists and interventionalists from other specialties: physicians whose occupation involves a fairly high level of ionizing radiation use, although it is not the major part of their work, such as interventional cardiologists. The specialties involved vary around the globe, but may include vascular surgeons and neurosurgeons.

Category 4 — other medical specialists using X-rays: physicians whose occupation involves the use of X-ray fluoroscopy in urology, gastroenterology, orthopaedic surgery, neurosurgery, or other specialties.

Category 5 — other medical specialties using nuclear medicine: physicians whose occupation involves prescription and use of a narrow range of nuclear medicine tests.

Category 6 — other physicians who assist with radiation procedures: physicians such as anaesthetists who have involvement in fluoroscopy procedures directed by others, and occupational health physicians who review records of radiation workers.

Category 7 — dentists: dentists who take and interpret dental X-ray images routinely.

Category 8 — medical referrers: physicians who request examinations and procedures involving ionizing radiations, and medical students who may refer for examinations in the future.

Category 9 — medical physicists: medical physicists specialising in RP, nuclear medicine, or diagnostic radiology.

Category 10 — radiographers, nuclear medicine technologists, and X-ray technologists: individuals who are going to take up a career in which a major component is involved with operating and/or testing X-ray units, including those carrying out some tests on a range of X-ray units in different hospitals and operating radionuclide imaging equipment.

Category 11 — maintenance engineers and clinical applications specialists: individuals with responsibilities for maintaining the X-ray and imaging systems (including nuclear medicine), or advising on the clinical application of such systems.

Category 12 — other healthcare professionals: other professionals such as podiatrists, physiotherapists, and speech therapists who may be involved in the use of radiology techniques to assess patients.

Category 13 — nurses: nursing staff and other healthcare professionals assisting in diagnostic and interventional X-ray fluoroscopy procedures, radiopharmaceutical administration, or the care of nuclear medicine patients.

Category 14 — dental care professionals: dental hygienists, dental nurses, and dental care assistants who take dental radiographs and process images.

Category 15 — chiropractors: chiropractors and other healthcare professionals who may refer for, justify, and take radiographic exposures.

Category 16 — radiopharmacists and radionuclide laboratory staff: radiopharmacists and individuals who use radionuclides for diagnostic purposes such as radioimmunoassay.

Category 17 — regulators: individuals with responsibility for enforcing ionizing radiation legislation.

Table 18.1: Recommended RP training requirements for different categories of physicians and for dentists

Training Area	1DR	2NM	3CDI MDI	4MDX	5MDN	6MDA	7DT	8MD
Atomic structure, X-ray production and interaction of radiation	m	h	l	l	l	l	l	-
Nuclear structure and radioactivity	m	h	l	l	m	l	-	-
Radiological quantities and units	m	h	m	m	m	l	l	l
Physical characteristics of the X-ray machines	m	l	m	m	l	l	m	-
Fundamentals of radiation detection	m	h	l	l	m	-	l	-
Principle and process of justification	h	h	h	h	h	h	h	m
Fundamentals of radiobiology, biological effects of radiation	h	h	m	m	m	l	l	l
Risks of cancer and hereditary disease	h	h	m	m	m	l	m	m
Risk of deterministic effects	h	h	h	m	l	l	m	l
General principles of RP including optimization	h	h	h	m	m	m	m	l
Operational RP	h	h	h	m	h	m	m	l
Particular patient RP aspects	h	h	h	h	h	m	h	l
Particular staff RP aspects	h	h	h	h	h	m	h	l
Typical doses from diagnostic procedures	h	h	m	m	m	m	m	m
Risks from fetal exposure	h	h	l	m	m	l	l	l
Quality control and quality assurance	m	h	m	l	l	-	l	-
National regulations and international standards	m	m	m	m	m	l	m	l
Suggested number of training hours	30-50	30-50	20-30	15-20	15-20	8-12	10-15	5-10

DR — Diagnostic Radiology Specialists

NM — Nuclear Medicine Specialists

CDI — Interventional Cardiologists

MDI — Interventionalists from other specialties

MDX — Other Medical Specialists using X-ray systems

MDN — Other Medical Specialists using nuclear medicine

MDA — Other Medical Doctors assisting with fluoroscopy procedures such as anaesthetists and occupational health physicians

DT — Dentists

MD — Medical Doctors referring for medical exposures as well as medical students who may refer in the future

Level of knowledge

l - Low level of knowledge indicating a general awareness and understanding of principles.

m - Medium level of knowledge indicating a basic understanding of the topic, sufficient to influence practices undertaken.

h - High level of detailed knowledge and understanding, sufficient to be able to educate others.

Acronyms in Table 18.1: DR, diagnostic radiology specialists; NM, nuclear medicine specialists; CDI, interventional cardiologists; MDI, interventionalists from other specialties; MDX, other medical specialists using X-ray systems; MDN, other medical specialists using nuclear medicine; MDA, other medical doctors assisting with fluoroscopy procedures such as anaesthetists and occupational health physicians; DT, dentists; MD, medical doctors referring for medical exposures and medical students; l, low level of knowledge indicating a general awareness and understanding of principles; m, medium level of knowledge indicating a basic understanding of the topic, sufficient to influence practices undertaken; h, high level of detailed knowledge and understanding, sufficient to be able to educate others.

Table 18.2: Recommended RP training requirements for different categories of healthcare professionals other than physicians or dentists

Training Area	9MP	10RDNM	11ME	12HCP	13NU	14DCP	15CH	16RL	17REG
Atomic structure, X-ray production, and interaction of radiation	h	m	m	l	l	m	l	m	l
Nuclear Structure and radioactivity	h	m	m	-	-	-	-	m	l
Radiological quantities and units	h	m	m	l	l	l	m	m	m
Physical characteristics of the X-ray machines	h	h	h	m	-	l	m	m	l
Fundamentals of radiation detection	h	h	h	l	l	l	l	m	l
Principle and process of justification	h	h	-	l	l	l	h	-	m
Fundamentals of radiobiology, biological effects of radiation	h	m	l	m	l	l	m	m	l
Risks of cancer and hereditary disease	h	h	l	m	l	m	m	m	m
Risks of deterministic effects	h	h	-	l	l	l	m	l	m
General principles of RP including optimization	h	h	m	m	m	m	m	m	m
Operational RP	h	h	m	m	m	m	m	h	m
Particular-patient RP aspects	h	h	m	h	m	m	h	-	m
Particular-staff RP aspects	h	h	m	h	m	m	h	h	m

Continued on next page

Table 18.2 — *Continued from previous page*

Training Area	9MP	10RDNM	11ME	12HCP	13NU	14DCP	15CH	16RL	17REG
Typical doses from diagnostic procedures	h	h		l	-	l	m	-	l
Risks from fetal exposure	h	h	l	m	l	l	m	m	l
Quality control and quality assurance	h	h	h	l	-	m	m	l	m
National regulations and international standards	h	m	h	m	l	l	m	m	h
Suggested number of training hours	150–200	100–140	30–40	15–20	8–12	10–15	10–30	20–40	15–20

MP — Medical physicists specializing in RP, nuclear medicine, and diagnostic radiology

RDNM — Radiographers, nuclear medicine technologists, and X-ray technologists

HCP — Healthcare professionals directly involved in X-ray procedures

NU — Nurses assisting in X-ray or nuclear medicine procedures

DCP — Dental care professionals including hygenists, dental nurses, and dental-care assistants

ME — Maintenance engineers and applications specialists

CH — Chiropractors and other healthcare professionals referring for, justifying, and delivering radiography procedures. Amount of training depends on range of tasks performed.

RL — Radiopharmacists and radionuclide laboratory staff

REG — Regulators

Acronyms in Table 18.2: MP, medical physicists specialising in RP, nuclear medicine, and diagnostic radiology; RDNM, radiographers, nuclear medicine technologists, and X-ray technologists; HCP, healthcare professionals directly involved in X-ray procedures; NU, nurses assisting in X-ray or nuclear medicine procedures; DCP, dental-care professionals including hygienists, dental nurses, and dental-care assistants; ME, maintenance engineers and applications specialists; CH, chiropractors and other healthcare professionals referring for, justifying, and delivering radiography procedures (amount of training depends on range of tasks performed); RL, radiopharmacists and radionuclide laboratory staff; REG, regulators; l, low level of knowledge indicating a general awareness and understanding of principles; m, medium level of knowledge indicating a basic understanding of the topic, sufficient to influence practices undertaken; h, high level of detailed knowledge and understanding, sufficient to be able to educate others.

18.13 EDUCATION AND TRAINING PROVIDERS AND ROLE OF THE COMPETENT AUTHORITIES

RP education and training for medical staff should be promoted by the regulatory and health authorities, and by professional bodies and scientific societies. RP education programs should be implemented by healthcare providers and universities and coordinated at local and national levels to provide courses based on agreed syllabuses and similar standards. Various sources of education and training materials for radiation protection in medicine are presented in Table 18.3.

Education and training should be given at medical schools during the medical studies and later, appropriate to the role of each category of physician, during the residency, and in focused specific courses. There should be an evaluation of the training, and appropriate recognition that the individual has completed the training successfully. In addition, there should be corresponding RP training requirements for other clinical personnel that participate in the conduct of procedures utilizing ionizing radiation, or in the care of patients undergoing diagnoses or treatments with ionizing radiation.

Regulatory and health authorities have the capability to enforce some levels of RP training and certification for those involved in medical exposures, and to decide if a periodic update could be necessary for some groups of specialists. They also have the capacity to direct resources for these training programs, to promote and co-ordinate the preparation of training material, and, in some cases, to maintain a register of the certified professionals.

The critical issues that have to be taken into account by the regulatory bodies and health authorities when requiring certification in RP for medical professionals are the available infrastructure for organization of the training programs and the financial requirements.

Staff from the regulatory authority will need to receive a limited amount of RP training. This should include aspects of optimization and practical RP.

Scientific and professional societies should contribute to the development of syllabuses to ensure a consistent approach, as well as to the promotion and support of education and training. Scientific congresses should include refresher courses on RP, attendance at which could be a requirement for continuing professional development for professionals using ionizing radiation.

The ICRP has suggested that professional medical and RP societies work together to develop continuing education in collaboration with healthcare providers. Professional bodies are encouraged to promote lectures on RP relevant to their specialty in medical congresses to facilitate continuing professional development.

The radiology equipment manufacturers have an important role to play in RP training for new technologies. The radiology industry should produce training material in parallel with the introduction of new X-ray or imaging systems to promote the advances in RP of patients. The equipment manufacturers should alert operators about the impact of their technologies on patient doses if the equipment is not used properly. Equipment manufacturers have a responsibility to develop and make available appropriate tools that are built into radiological equipment to facilitate easy and convenient determination and recording of exposure with reasonable accuracy.

Equipment manufacturers should ensure that maintenance engineers with responsibilities for imaging systems and clinical applications specialists have training in RP of patients. It is important that they understand how the settings of the X-ray systems and adjustments that they may make influence the radiation doses to patients.

Table 18.3: Sources of educational material for radiation protection in medicine

Source	URL
American Association of Physicists in Medicine	http://www.aapm.org/ http://www.aapm.org/meetings/virtual_library/
American College of Radiology	http://www.acr.org/Quality-Safety
Australian Department of Health of Western Australia	http://www.imagingpathways.health.wa.gov.au/
EFOMP (European Federation of Organizations For Medical Physics)	http://www.efomp.org/
EMERALD portal	http://www.emerald2.eu
European Commission	http://ec.europa.eu/energy/nuclear/radiation_protection/publications_en.htm http://bookshop.europa.eu/is-bin/INTERSHOP.enfinity/WFS/EU-Bookshop-Site
European Society of Radiology	http://www.eurosafeimaging.org/about
European Society for Therapeutic Radiology and Oncology	http://www.estro.org/school/about-the-school/educational-material
EUTEMPE (European Training and Education for Medical Physics Experts in Radiology)	http://www.estro.org/school/articles/e-learning/index http://www.eutempe-rx.eu/
Image Gently	http://www.imagegently.org/Education.aspx
Image Wisely	http://www.imagewisely.org/

Continued on next page

Table 18.3 — *Continued from previous page*

International Atomic Energy Agency	https://rpop.iaea.org/RPoP/RPoP/Content/index.htm http://www.iaea.org/Publications/ http://www-pub.iaea.org/MTCD/Publications/PDF/Pub1564web-8272546.pdf
International Commission on Radiological Protection	http://www.icrp.org/publications.asp http://www.icrp.org/page.asp?id=111
International Radiation Protection Association	http://www.irpa.net/page.asp?id=37
International Organization for Medical Physics (official Medical Physics References)	http://www.iomp.org/?q=node/34
Medical Physics Encyclopaedia and Multilingual Dictionary — Online version	http://www.iomp.org/?q=content/encyclopaedia-and-multilingual-dictionary-0 http://www.emitel2.eu/
MEDRAPET — Medical Radiation Protection Education and Training	http://www.eurosafeimaging.org/medrapet
National Council on Radiation Protection & Measurements	http://www.ncrponline.org/
Perry Sprawls	http://www.sprawls.org/resources/#radiation
Office of Radiation Protection, Washington State Department of Health, USA	http://www.doh.wa.gov/CommunityandEnvironment/Radiation
The Medical Physics Encyclopaedia — Paper version	http://www.crcpress.com/product/isbn/9781439846520
University of Washington	http://www.ehs.washington.edu/rsotrain/ http://courses.washington.edu/radxphys/PhysicsCourse.html

Web addresses (Accessed December 23, 2014) of organizations with training material for radiation protection in medicine.

18.14   REFERENCES

Dauer LT, Thornton RH, Miller DL et al. 2012. Radiation management for interventions using fluoroscopic or computed tomographic guidance during pregnancy: a joint guideline of the Society of Interventional Radiology and the Cardiovascular and Interventional Radiological Society of Europe with Endorsement by the Canadian Interventional Radiology Association. J Vasc Intervent Radiol 23(1):19–32.

Duran A, Hian SK, Miller DL et al. 2013. A summary of recommendations for occupational radiation protection in interventional cardiology. Catheter Cardiovasc Interv 81(3):562–567.

European Commission. 1997. Council Directive 97/43/Euratom on health protection of individuals against the dangers of ionizing radiation in relation to medical exposure. Off. J. Eur. Commun. L180, 22–27.

European Commission. 2000. Guidelines on education and training in radiation protection for medical exposures. RP Publication 109. Directorate General for Energy. http://ec.europa.eu/energy/nuclear/radiation_protection/doc/publication/116.pdf (Accessed December 23, 2014).

European Commission. 2014a. European Directive 2013/59/Euratom on basic safety standards for protection against the dangers arising from exposure to ionizing radiation and repealing Directives 89/618/Euratom, 90/641/Euratom, 96/29/Euratom, 97/43/Euratom and 2003/122/Euratom. Off J Eur Commun. L13; 57: 1–73 (2014).

European Commission. 2014b. Guidelines on Radiation Protection Education and Training of Medical Professionals in the European Union. Directorate General for Energy. Radiation Protection 175. Available at: http://ec.europa.eu/energy/nuclear/radiation_protection/doc/ publication/175.pdf (Accessed December 23, 2014).

European Parliament and Council. 2008. Recommendation of 23 April 2008 on the establishment of the European Qualifications Framework for Lifelong Learning. Official Journal of the European Union 6.5.2008-C111:1-7. http://eur-lex.europa.eu/legal-content/EN/TXT/PDF/?uri=CELEX:32008H0506%2801%29&from=EN (Accessed December 23, 2014).

IAEA. 2014. International Basic Safety Standards (BSS). IAEA Safety Standards Series GSR part 3. Vienna.

ICRP. 2000a. Avoidance of radiation injuries from medical interventional procedures. ICRP Publication 85. Ann. ICRP 30(2).

ICRP. 2000b. Prevention of accidents to patients undergoing radiation therapy. ICRP Publication 86. Ann. ICRP 30(3).

ICRP. 2000c. Managing patient dose in computed tomography. ICRP Publication 87. Ann. ICRP 30(4).

ICRP. 2001a. Radiation and your patient - a guide for medical practitioners. ICRP Supporting Guidance 2. Ann. ICRP 31(4):1–31.

ICRP. 2001b. Reference levels in medical imaging: review and additional advice. ICRP Supporting Guidance 2. Ann ICRP 31(4): 33–52.

ICRP. 2004. Managing patient dose in digital radiology. ICRP Publication 93. Ann. ICRP 34(1).

ICRP. 2005. Prevention of high-dose-rate brachytherapy accidents. ICRP Publication 97. Ann. ICRP 35(2).

ICRP. 2007a. Managing patient dose in multi-detector computed tomography (MDCT). ICRP Publication 102. Ann. ICRP 37(1).

ICRP. 2007b. The 2007 Recommendations of the International Commission on Radiological Protection. ICRP Publication 103. Ann. ICRP 37(2–4).

ICRP. 2007c. Radiological protection in medicine. ICRP Publication 105. Ann. ICRP 37(6).

ICRP. 2009a. Preventing accidental exposures from new external beam radiation therapy technologies. ICRP Publication 112. Ann. ICRP 39(4).

ICRP. 2009b. Education and training in radiological protection for diagnostic and interventional procedures. ICRP Publication 113. Ann. ICRP 39(5).

ICRP. 2010. Radiological protection in fluoroscopically guided procedures performed outside the imaging department. Publication 117. Ann ICRP. 40(6):1–102.

ICRP. 2013. International Commission on Radiological Protection. ICRP Publication 120. Radiological protection in cardiology. Ann ICRP. 2013;42(1):1–125.

Koenig TR, Mettler FA, Wagner LK. 2001. Skin injuries from fluoroscopically guided procedures: part 2, review of 73 cases and recommendations for minimizing dose delivered to patient. AJR Am J Roentgenol. 177(1):13–20.

Law L, Ng KH, Editors, 2014. Radiological Safety and Quality. Paradigms in Leadership and Innovation. Dordrecht: Heilderberg, Springer.

Meghzifene A, Vaño E, Le Heron J et al. 2010. Roles and responsibilities of medical physicists in radiation protection. Eur J Radiol. 76(1):24–7.

Miller DL, Vaño E, Bartal G et al., on behalf of the Cardiovascular and Interventional Radiology Society of Europe and the Society of Interventional Radiology. 2010. Occupational radiation protection in interventional radiology: a joint guideline of the Cardiovascular and Interventional Radiology Society of Europe and the Society of Interventional Radiology. Cardiovasc Intervent Radiol 33(2):230–239.

Miller DL, Balter S, Dixon RG et al. 2012. Quality improvement guidelines for recording patient radiation dose in the medical record for fluoroscopically guided procedures. J Vasc Intervent Radiol 23(1):11–18.

Peer S, Faulkner K, Torbica P et al. 2005. Relevant training issues for introduction of digital radiology: results of a survey. Radiat Prot Dosimetry. 117(1–3):154–61.

Picano E, Vaño E, Semelka R et al. 2007. The American College of Radiology white paper on radiation dose in medicine: deep impact on the practice of cardiovascular imaging. Cardiovascular Ultrasound. 5:37. doi:10.1186/1476-7120-5-37.

Shope TB. 1996. Radiation-induced skin injuries from fluoroscopy. Radiographics. 16(5):1195–9.

Vaño E, Arranz L, Sastre JM et al. 1998. Dosimetric and radiation protection considerations based on some cases of patient skin injuries in interventional cardiology. Br J Radiol. 71(845):510–6.

Vaño E, Gonzalez L, Faulkner K et al. 2001. Training and accreditation in radiation protection for interventional radiology. Radiat Prot Dosimetry. 94(1-2):137–42.

Vaño E, Gonzalez L, Fernandez JM et al. 2006. Occupational radiation doses in interventional cardiology: a 15-year follow-up. Br J Radiol. 79(941):383–8.

Vaño E, Ohno K, Cousins C et al. 2011. Radiation risks and radiation protection training for healthcare professionals: ICRP and the Fukushima experience. J Radiol Prot. 31(3):285–7.

Vaño E. 2010. Mandatory radiation safety training for interventionalists: the European perspective. Tech Vasc Interv Radiol. 13(3):200–3.

WHO. 2014. Bonn Call for Action. http://www.who.int/ionizing_radiation/ medical_exposure/ Bonn_call_action.pdf (Accessed December 23, 2014).

CHAPTER 19

Medical Exposures: Adverse Consequences and Unintended Exposures

Fred A. Mettler, Jr.

Department of Radiology, University of New Mexico, School of Medicine, NM, USA

CONTENTS

M EDICAL RADIATION USE accounts for more than 99.9% of the per capita dose from man-made sources (NCRP 2009). Worldwide, medical procedures using radiation grew from an estimated 1.7 billion in 1980 to a staggering almost 4 billion in 2007 (UNSCEAR 2006). Overall, the benefit from medical exposure vastly outweighs the risks (Holmberg 2010). Adverse consequences and unintended radiation exposures are rare in medicine but they do occur and the consequences can range from often negligible to rarely fatal. At low doses, there is the question of cancer induction and at higher doses there can be various tissue effects. Although it is largely unappreciated, high dose accidents in medical exposure have resulted in more acute radiation deaths than from any other accidental or occupational source including Chernobyl (Gusev 2001).

The major categories of exposure in medicine are diagnostic and interventional radiology, nuclear medicine, and radiation therapy. In the past these were relatively separate disciplines, but more recently there has been evolution of hybrid technology, e.g., CT scanning combined with particle accelerators or CT scanning combined with positron emission tomography (PET) scans.

19.1   TYPES OF RADIATION HEALTH EFFECTS

19.1.1   Stochastic Effects

Stochastic effects are probabilistic, they occur without obvious dose threshold, and the severity of the effect is unrelated to the magnitude of the radiation exposure. The most commonly discussed stochastic effect is cancer induction and, less commonly, possible hereditary effects.

19.1.1.1   Cancer

It is well known that radiation can cause cancer; however, radiation is a relatively poor carcinogen. The largest epidemiological study from which radiation cancer risks are estimated is the follow-up of atomic bomb survivors in Hiroshima and Nagasaki. In the latest report on the approximately 86,000 survivors followed for 60 years, only 853 (4.9%) of 17,448 incident solid cancers are attributed to radiation from the bombs (Cullings 2014). Radiation does not cause cancer in various tissues to the same degree. Tissues such as breast and lung are relativity high in radiation-induced cancer sensitivity, while others are less so (kidney and bladder) and for some tumors (pancreas, prostate cervix, lymphoma), radiation does not appear to result in a discernable increase. Radiation-induced cancers are also characterized by a latent period between exposure and appearance of the cancer. The minimal latent period is somewhat variable (e.g., a few years for non-CLL leukemias, about 5 years for sarcomas and thyroid cancer, and 10 or more years for most other tissues. The risk of an incident cancer after a whole body dose is about 10% per Gy and the risk of fatal cancer is about half that. Radiation-induced excess cancer becomes statistically significant at doses at or above about 150 mSv. At lower doses the risk is difficult to determine, in part, because of the high incidence of cancer from other causes. Based on cellular and molecular insights, a number of scientists hypothesize that the risk at low doses does exist but is too small to detect from epidemiological studies. It is possible we may never know the answer unless a radiation signature can someday be identified in the cancer tissue (UNSCEAR 2006).

19.1.1.2   Hereditary

Radiation exposure was of major concern after early experiments with fruit flies showed that X-rays are a mutagen. Subsequent numerous and large human studies involving the offspring of the atomic bomb survivors and children of childhood cancer survivors(treated with radiation therapy) have not shown an increase in a variety of studied heritable effects. In addition, human studies of pre-conception exposure have not shown an increase in cancer or other abnormalities in subsequent offspring (UNSCEAR 2013).

19.1.2   Deterministic Effects

Deterministic effects are sometimes called tissue reactions (ICRP 2012). These effects are predominantly due to cell killing. Well-known deterministic effects include radiation skin burns, ulceration, and necrosis. These effects are characterized by a clinical threshold dose (below which no effect is apparent). In addition, the severity of the effect increases with increasing dose.

Deterministic effects have a variable time course depending on the tissue and the dose. There may be effects apparent within minutes, hours, days, or even a few weeks. This is often the result of chemical changes and killing of rapidly dividing cells. Subacute changes occur over a few months and chronic changes can occur over months to years. The chronic changes often are due to fibrosis and small blood vessel narrowing or occlusion, causing subsequent tissue necrosis (Mettler and Upton 2007).

Some of the most common deterministic complications from diagnostic and interventional medical X-ray procedures include skin reddening (erythema), hair loss (epilation), and necrosis with ulceration (ICRP 2000a). These changes are variable with dose, fractionation, and time since exposure, and also depend upon the area or volume of tissue irradiated. For most patients, significant skin reactions are not seen at acute doses of < 5 Gy although there can be some transient prompt erythema and early epilation. At 5–10 Gy there is prompt erythema, early erythema, and early and midterm epilation and possible mid- and long-term epilation or dermal atrophy. At 10–15 Gy there is prompt transient erythema, early erythema, and epilation associated with dry or moist desquamation. Long-term there can be telangiectasia and dermal atrophy or induration. At acute skin doses > 15 Gy there is prompt erythema, edema, and ulceration, which may heal and then develop necrosis in the mid- and long-term. For these effects the word prompt refers to < 2 weeks, early 2–8 weeks, midterm 6–52 weeks, and long-term > 40 weeks (Balter et al. 2010).

Complications from radiation therapy are usually quite different from adverse effects of diagnostic or interventional X-ray procedures, because in radiation therapy skin-sparing protocols are used, the highest doses are internal, and the effects are often difficult to clinically appreciate. Tissues are also very variable in radiation sensitivity, with rapidly dividing systems (e.g., mucosa and intestine) being quite sensitive, and slowly dividing cellular systems being more resistant (e.g., brain and bone). Small blood vessels are intermediate in sensitivity and as such, can result in decreased blood supply, causing late necrosis in otherwise resistant tissues.

19.1.2.1 Cataracts

Vision-impairing cataracts occurring after high doses to the lens of the eye have been known for many decades. Acute doses of fractionated doses over weeks that exceed several Gy can cause cataracts within a few years. These radiation cataracts sometimes can be differentiated from senile cataracts by first appearing in the posterior aspect of the lens. More recently, with more sophisticated technology, it has become apparent that lens opacities can be seen after doses as low as 0.5 Gy. These may not impair vision and whether there is a true dose threshold for such findings and whether they are progressive remains a matter of debate (Ainsbury et al. 2009).

19.1.2.2 Cardiovascular Disease

Increased risk of cardiovascular disease (including myocardial infarction, coronary artery disease, and stroke) are well-documented effects after high radiation doses to the heart or neck that may occur with radiation therapy. Recently there is evidence of increased risk of cardiovascular disease in atomic bomb survivors at lower doses (down to about 0.5Gy). The issue is complicated because cardiovascular disease is not a single entity, and there are many potential confounding factors including diet, tobacco use, and genetics. The nature and magnitude of the risk (if any) at doses less than 0.5 Gy is unresolved (Darby et al. 2010, Takahashi et al. 2013).

19.1.2.3 Pregnancy

There are radiation risks throughout pregnancy that are related to fetal absorbed dose and stage of pregnancy. Radiation risks are highest during organogenesis and the early fetal period, less in the second trimester, and least in the third trimester. Doses in the range of 1 Gy to the central nervous system at 8–15 weeks gestation result in a high probability of mental retardation. The risk is somewhat lower at 16–25 weeks and is largely gone after that. Malformations have not been seen in humans or animals as a result of fetal doses of less than 100–200 mGy. The risk of leukemia and childhood cancer after in-utero exposure is felt to be generally the same as for children (UNSCEAR 2013). The relative risk is 1.4 per 10 mGy or lower. Even with a relatively high RR of 1.4, the total individual risk after 10 mGy is low (about 0.3–0.4%) since the background incidence of childhood cancer is so low (about 0.2–0.3%) (ICRP 2000b, NCRP 2013).

19.2 DIAGNOSTIC RADIOLOGY

Diagnostic radiology includes simple radiographic procedures, fluoroscopy, and fluoroscopically or CT-guided interventional procedures. There is no detectable risk or adverse consequence associated with properly conventional diagnostic X-ray procedures.

19.2.1 Computed Tomography

CT scans commonly produce absorbed doses in the range of 5–30 mGy. There have been reported rare instances in which CT perfusion protocols have been adjusted and inadvertently resulted in doses high enough to cause epilation, and accidental exposures that have resulted in skin erythema (Figure 19.1). None of these appears to have resulted in a long-term deterministic effect. There is little question that multiple CT scans can result in tissue doses that are associated with a detectable increase in cancer in the atomic bomb survivors. There are a few preliminary studies of CT and cancer risk in children, but the results have been questioned and additional epidemiological studies are underway. Prevention of adverse effects from CT involves use of standardized and low-dose protocols, periodic review, and a quality assurance program. It is especially important to have specific pediatric protocols and not use adult protocols on children, as it will result in an unnecessarily high dose (ICRP 2000c, ICRP 2007).

19.2.2 Fluoroscopy

Fluoroscopes are used in a wide variety of hospital departments and clinics including radiology, pain clinics, operating rooms, cardiology, gastroenterology, and orthopedic surgery. Unfortunately, often the operating physicians have limited appreciation of what the radiation dose actually is or what the

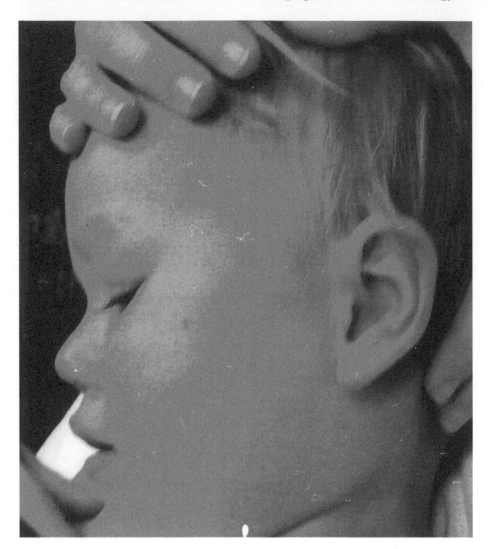

Figure 19.1 Skin erythema

consequences might be (ICRP 2000a). Fluoroscopes are capable of producing very high doses to limited parts of the patient's skin as well as high doses to the unwary operator. Most adverse consequences have occurred as a result of applications in interventional radiology, cardiology, and vascular surgery (Koenig et al. 2001a, Koenig et al. 2001b). The most common causes of these adverse events are placing the X-ray tube too close to the skin, not collimating the field, using magnification mode when not necessary, not using pulsed fluoroscopic mode, using too much fluoroscopy time in complicated procedures, using "boost or turbo" mode when not needed, keeping the tube and entrance beam on one spot, using too many cine frames, and not using last image hold mode. The patients most at risk are those who have had prior procedures with irradiation the same spot, as well as large and diabetic patients. The primary way to integrate all this information is to check the dose indices as the procedure progresses. These are displayed on most machines and include fluoroscopy time (a poor indicator), or the much preferable quantities of dose-area product (DAP) expressed in cGy-cm^2, or better yet, total air kerma (TAK) expressed in mGy. Skin dose can be approximated by multiplying TAK by 1.4. Thus, a TAK of 1400 mGy is equivalent to a skin dose of about 2 Gy, which is the threshold for skin effects, and a TAK of about 10,500 mGy may have resulted in a skin dose of 15Gy, with a high likelihood of necrosis and ulceration.

The common high-dose fluoroscopic skin injuries are square, round, or rarely elliptical lesions, and most often occur on the back of the patient (since the tube is usually under the patient when the patient is supine on the table). The majority of these injuries go unrecognized since the main effects of desquamation and necrosis do not occur for days or weeks after exposure (Figure 19.2). As a result, it is important to identify those patients at high risk as indicated by the dose metrics and have them return for examination of the skin in 3–4 weeks. If there is persistent moist desquamation, then there is a significant likelihood of a long-term ulcer that will require surgical intervention (usually a full thickness graft).

19.3 NUCLEAR MEDICINE

Nuclear medicine involves the administration of unsealed radiopharmaceuticals for either diagnosis or therapy. Adverse consequences are very rare as a result of diagnostic nuclear medicine examinations, but can occur with therapeutic procedures.

19.3.1 Administered Activity

There are typical or suggested guidelines for the administered activity for almost all diagnostic examinations. These are generally a function of the needed photon flux and length of the examination. The prescribed amounts may be adjusted by the attending physician for body size, disease state, or other factors. Events in which the wrong amount of administered activity is given, is administered by the wrong route, or to the wrong patient, are often called misadministrations, or, less commonly, medical events. Depending upon the circumstances and national or local regulations, these events may be required to be reported to regulatory authorities, the patient's physician, and the patient.

There is particular need to adjust administered activity when procedures are being performed on infants and children. The smaller body size, closer proximity of organs, and other factors result in higher absorbed doses per unit

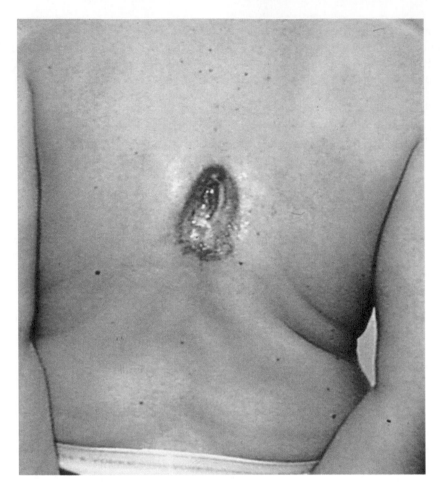

Figure 19.2 Necrosis

administered activity than in adults. There are suggested pediatric guidelines published by both European and North American Consensus groups (Lassman et al. 2007, Mettler and Guiberteau 2013).

19.3.2 Marrow Depression

There are a number of radionuclides used to provide palliative treatment of metastatic cancer, particularly bone metastases. These include ^{89}Sr, ^{186}Re, ^{153}Sm, and ^{223}Ra. Because these cause large absorbed doses to the bone marrow, it is important to calculate the organ dosimetry and check the blood count/marrow status before and for 8 weeks after such therapy. Similar considerations apply to therapy done with other agents such as monoclonal antibodies, and repeated treatments with ^{131}I.

19.3.3   Pulmonary Fibrosis

Thyroid cancer commonly metastasizes to the lungs and repeated treatments with very high activities of ^{131}I raise concerns about development of radiation-induced pulmonary fibrosis. The situation is often complicated by reduced pulmonary function as a result of the numerous metastatic deposits and arteriovenous shunting that occurs in the tumor tissue.

19.3.4   Pregnancy

Pregnancy is an important issue and has probably been involved in the highest number of adverse consequences in nuclear medicine. A serum pregnancy test should be done before any therapeutic nuclear medicine procedure in a potentially pregnant female, and some authors suggest this is a good idea before a diagnostic examination. The major source of clinically significant problems is administration of large administered activities of ^{131}I for treatment of hyperthyroidism or thyroid cancer to women who were not recognized to be pregnant. Before 10 weeks of pregnancy, the fetal thyroid does not accumulate significant amounts of iodine; however, the fetus will still be exposed to gamma rays from placental transfer and from the mother's tissues and bladder. Such fetal doses may potentially cause a small increase in the probability of cancer. At later times of pregnancy, dose to the fetal thyroid is very high and will typically result in thyroid ablation or significant hypothyroidism. If the patient is not treated with thyroid hormone replacement, the child will be born with significant mental impairment (cretinism).

19.3.5   Breastfeeding

A number of radiopharmaceuticals are excreted in breast milk and can be transferred to infants if breastfeeding is not temporarily or permanently discontinued. The issue of breastfeeding should be inquired about for all females in the child-bearing age group before administration of radiopharmaceuticals. For ^{99m}Tc-based agents that are excreted in breast milk, discontinuing breastfeeding for a number of hours or a few days is sufficient. For longer-lived radionuclides (particularly ^{131}I), breast feeding can result in very high doses to the infant thyroid, raising the probability of thyroid cancer or possibly hypothyroidism (ICRP 2008). If ^{131}I is involved and the situation is recognized, early administration of stable potassium iodide to the infant can be helpful.

19.3.6   Release of Patients

Patients who have had nuclear medicine procedures and are released can potentially expose members of their families and the public. For diagnostic examinations (including positron-emission tomography) the doses to other persons are typically negligible. However, after administration of large quantities of radioiodine (as described above), it is important to restrict contact with other persons (especially infants and children) for several days as well as to control saliva, sweat, and urine. Local and national regulations vary significantly in this regard (ICRP 2000c, NCRP 2006). The risks to others are predominantly related to thyroid uptake and as yet an undetectable potential risk of cancer.

19.4   RADIATION THERAPY

Radiation therapy involves either the use of external radiation or internal placement of sealed radioactive sources to treat cancer (and occasionally benign conditions). It is estimated that about 50% of cancer patients would benefit from radiotherapy in the treatment or progression of their disease. The worldwide annual frequency of radiotherapy is about 0.7/1000 population but varies widely by country, and the total annual number of patients is estimated to be about 5 million. Since radiotherapy intentionally involves administration of lethal doses to tumor cells, and there is not much difference between tolerance of normal and cancer cells to radiation, adverse consequences are common, and in fact, are often an expected and accepted part of the treatment. Differences in prescribed dose of more than 10% can result is an unacceptable number of complications and an underdosage of 10% can result in not curing the tumor. With such little tolerance, it is not surprising that there have been a number of severe radiotherapy accidents. Evaluation of adverse consequences from radiotherapy is complicated by the fact that most patients also receive chemotherapy or surgery as part of their treatment.

19.4.1   Second Malignancies

Since radiation is known to increase the risk of certain types of cancers and leukemias, it is an obvious concern as to whether and to what degree second malignancies may occur in patients who have had cancer therapy. The first issue is that there is a normal risk (unrelated to any specific therapy) of a second cancer just because the patient is living longer. Depending upon the age of the patient and residual lifespan, this risk is about 10% but can range from 0 to 25%. There are also some cancer types that seem to predispose to a higher risk than normal for development of a second cancer. These include Hodgkin's disease, and breast, ovarian, and testicular cancer. Chemotherapy with a number of drugs including alkylating agents, cisplatin, topoisomerase II inhibitors, and anthracyclines is also known to increase the risk of second malignancies. Overall a relatively small proportion (about 8%) of second cancers occurring later in radiotherapy patients are due to the radiotherapy. The absolute risk is about 5 excess cancers per 1000 patients (0.5%) treated with radiotherapy by 15 years after diagnosis (Berrington de Gonzalez et al. 2011).

19.4.2 Consequences in Surrounding Normal Tissues

High doses will necessarily occur to normal tissues around any tumor that is being treated with radiotherapy. The effects on normal tissue vary greatly depending upon the treatment volume, location, and specifics of the treatment protocol. Normal tissues vary greatly in their response both to acute, subacute, and long-term effects. Radiation oncologists have used the concept of tolerance dose (TD) for adverse effects. This is usually expressed in Gy. A tolerance dose with 5% severe complications within 5 years is referred to as TD5/5 and a tolerance dose with 50% severe complication in 5 years is referred to as TD50/5. These values will vary significantly on the fractionation scheme and the volume or length of the organ irradiated. Complication rates of up to 10% occur even in the best radiotherapy practices. In fact, if there were no complications in normal tissues it would be unlikely that tumors would be adequately treated. There is a vast literature on potential complications in normal tissues, which is beyond the scope of this chapter (Mettler and Upton 2007). One well-documented example, however, is the risk of cardiac disease after radiotherapy for breast cancer. Historically, the risk of cardiac death was increased by almost 30% and increased by about 3% per Gy cardiac. Fortunately, more recent radiotherapy protocols have attempted to reduce the cardiac dose and myocardial volume.

19.4.3 Accidents

While side effects and expected complications are accepted in normal radiotherapy practice, accidents in radiotherapy are rare. When they occur the outcomes can be devastating, long-lasting, and even lethal (Figure 19.3). There have been more than 100 published radiotherapy accidents involving teletherapy, intensity modulated radiotherapy, and brachytherapy. There are undoubtedly many more accidents that have occurred and have not been recognized or reported. The accidents have occurred due to a variety of problems with radiation measurement systems, commission and calibration, treatment planning, patient setup and treatment, decommissioning, and equipment malfunction. The majority of recognized accidents involve overdosage (ICRP 2009). A few examples will be given below.

In 1996 there was an accident in Costa Rica when a new cobalt-60 source was installed and miscalibrated, resulting in overdosage of patients by 60% per fraction. There were 115 patients involved and at least 7 died as a result of the overexposure, with many others suffering severe complications (IAEA 1998). In 2000–2001 there was an accident in Panama due to improper use of a treatment planning system for shielding blocks on 28 patients. At least 5 patients died as a direct result of overexposure (IAEA 2001). A similar accident resulting from inadequate training in the treatment planning system occurred in Toulouse, France in 2006 involving 23 patients and resulting in 16 patients with acute complications and 1 death due to overexposure. A machine

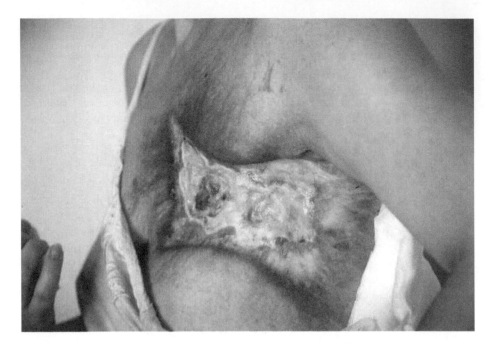

Figure 19.3 Radiotherapy accident

malfunction due to electrical outage occurred in Poland in 2001, resulting in severe overexposure of 5 patients being treated for breast cancer. All 5 patients developed significant chest-wall injuries and complications (IAEA 2004b). In 1990 there was an accident in Zaragoza, Spain due to faulty maintenance procedures. Eleven deaths were attributed to this accident. Accidents that involve underexposure are difficult to identify since the only finding would be less-than-expected tumor cure or control. One notable such accident occurred in the United Kingdom as a result of improper understanding of a computerized treatment planning system. The accident continued unrecognized for almost a decade from 1982 to 1990 and resulted in underdosage of 1045 patients. The effect was dependent upon tumor type, but for bladder and prostate cancer patients, a 20% reduction in dose accounted for a 50% reduction in 5-year disease-free survival (Ash and Bates 1994).

19.5   SUMMARY

Worldwide, there are about 4 billion medical radiation procedures done annually. The incidence of adverse effects is extremely low and overall the benefits of medical radiation clearly outweigh the potential or actual adverse effects. For specific (and rare) patients, there are well-documented adverse effects that can range from trivial to fatal. It is incumbent for the medical, physics, and radiation protection communities to be aware of the potential for such effects and to try to minimize these without unduly restricting medical benefits.

19.6 REFERENCES

Ainsbury EA, SD Bouffler, W Dorr, et al., 2009, Radiation cataractogenesis: A review of recent studies, Rad Res 172(1):1–9.

Ash D and T Bates, 1994, Report on the clinical effects of inadvertent radiation underdosage in 1045 patients. Clin. Oncol. 6:214–225.

Balter S, JW Hopewell, D Miller, et al., 2010, Fluoroscopically Guided Interventional Procedures: A review of radiation effects on skin and hair, Radiology 254(2) 326–341.

Berrington de Gonzalez A, RE Curtis, SF Kry, et al., Proportion of second cancers attributable to radiotherapy treatment in adults: A cohort study in the US SEER cancer registries, Lancet Oncol. 12(4):353–60.

Cullings HM, 2014, Impact of the Japanese atomic bomb survivors of radiation received from the bombs, Health Phys 106(2):281–293.

Darby S, DJ Cutter, M Boerma, et al., 2010, Cardiovascular disease and radiation exposure: the Beebe Symposium. Int. J. Radiat. Oncol. and Phys. 76(3):656–665.

Gusev I, AK Guskova, FA Mettler, 2001, Medical management of Radiation Accidents (2nd ed) CRC Press, Orlando FL.

Holmberg O, R Czarwinski, FA Mettler, 2010, The importance and unique aspects of radiation protection in medicine. Eur. J. Radiol. 76:6–10.

International Atomic Energy Agency (IAEA), 1998, Accidental overexposure of radiotherapy patients in San Jose, Costa Rica, IAEA, Vienna, Austria.

International Atomic Energy Agency (IAEA), 2001, Investigation of an accidental exposure of radiotherapy patients in Panama, IAEA, Vienna, Austria.

International Commission on Radiological Protection (ICRP), 2000a, Avoidance of radiation injuries from medical interventional procedures, Publication 85, Annals of the ICRP 30(2) Pergamon Press. Oxford.

International Commission on Radiological Protection (ICRP), 2000b, Pregnancy and medical radiation, Publication 84, Annals of the ICRP 30(1), Pergamon Press. Oxford.

International Commission on Radiological Protection (ICRP), 2000c, Managing patient dose in computed tomography, Publication 87, Annals of the ICRP 30(4) Pergamon Press. Oxford.

International Commission on Radiological Protection (ICRP), 2004a, Release of patients after therapy with unsealed radionuclides, Publication 94, Annals of the ICRP, Pergamon Press. Oxford.

International Atomic Energy Agency (IAEA), 2004b, Accidental overexposure of radiotherapy patients in Bialystok, IAEA, Vienna, Austria.

International Commission on Radiological Protection (ICRP), 2007, Managing patient dose in multidetector computed tomography, Publication 102, Annals of the ICRP 37(1), Elsevier Press. Orlando FLA.

International Commission on Radiological Protection (ICRP), 2008, Radiation dose to patients from radiopharmaceuticals, Publication 106, Annals of the ICRP, 30(1–2), Elsevier Press. Orlando FLA.

International Commission on Radiological Protection (ICRP), 2009a, Education and training in radiological protection for diagnostic and interventional procedures, Publication 113, Annals of the ICRP 39(5), Elsevier Press. Orlando FLA.

International Commission on Radiological Protection (ICRP), 2009b, Preventing accidental exposures from new external beam radiation therapy technologies, Publication 112, Annals of the ICRP, Elsevier Press. Orlando FLA.

International Commission on Radiological Protection (ICRP), 2012, Statement on tissue reactions and early and late effects of radiation in normal organs and tissues, Publication 118, Annals of the ICRP 41(1-2), Elsevier Press. Orlando FLA.

Koenig TK, LK Wagner, FA Mettler, 2001, Radiation injury of the skin following diagnostic and interventional fluoroscopic procedures: Part 1: Characteristics of Radiation Injury, AJR 177:3–11.

Koenig TK, LK Wagner, FA Mettler, 2001, Radiation Injury from Fluoroscopically Guided Procedures: Part 2, Review of 79 Cases, AJR 177: 13–20.

Lassman M, L Biassoni, M Monsieur, et al., 2007, The new EANM pediatric dosage card, Eur. J Nuc Med Mol Imaging 2007, 34(5):796–798.

Mettler FA and AC Upton, 2007, Medical Effects of Ionizing Radiation (3rd ed) Elsevier/Saunders Phila. PA.

Mettler FA and MJ Guiberteau, 2013, Essentials of Nuclear Medicine (6th ed) Elsevier/Saunders Phila. PA.

National Council on Radiation Protection and Measurements (NCRP), 2006, Management of radionuclide therapy patients, Report No. 155, NCRP, Bethesda, Maryland USA.

National Council on Radiation Protection and Measurements (NCRP), 2009, Ionizing exposure of the population of the United States, Report No.160, Bethesda, Maryland USA.

National Council on Radiation Protection and Measurements (NCRP), 2013, Preconception and prenatal radiation exposure: Health effects and protective guidance, Report No.174, NCRP, Bethesda, Maryland USA.

Takahashi I, W Ohishi, FA Mettler et al., 2013, Radiation and Cardiovascular Disease: Meeting Report, J. Radiol Prot., 33(4): 869–880.

United Nations Scientific Committee on the Effects of Atomic Radiation (UNSCEAR), 2006, Effects of ionizing radiation, 2006 Report, United Nations, Vienna.

United Nations Scientific Committee on the Effects of Atomic Radiation (UNSCEAR), 2013, Effects of radiation exposure of children, 2013 Report, Volume 2, United Nations, New York.

IX

Ethical Use of Human Subjects in Medical Research

IX

CHAPTER 20

Informed Consent in Radiation Medicine Practice and Research

Kevin Nelson

Mayo Clinic, Arizona, USA

Kelly Classic

Mayo Clinic, Arizona, USA

CONTENTS

THE CONCEPTS of "inherent dignity" and "equal and unalienable rights of all members of the human family" resulted from the atrocities that occurred in World War II, and were formalized as part of the Universal Declaration of Human Rights in 1948. These concepts were the building blocks of what is now known as "informed consent."

The term "informed consent" was first coined by Paul G. Gebhard in 1957 (Merck, 2011) and is defined as the autonomous act of a patient or research subject to expressly permit a person to perform a medical action on a patient or to include a person in a research project (Terry, 2007). Informed consent is recognized in the majority of industrialized countries.

In North America and in Europe at least four models of patient or informed consent have been used. These include the paternalistic model, the informative model, the interpretive model, and the deliberate model (WHO, 2014). The paternalistic model, where the physician made the determination for the patient, has in large part been replaced with models that support what a reasonable person might want to know, and has sometimes been termed the prudent-patient standard. Court decisions in last decade have also supported the prudent-patient concept.

Because patient rights will vary from country to county due to prevailing societal and cultural norms, physician responsibilities to a patient or research subject will also vary. In 2002, a consortium of concerned professional societies and organizations, including the American Board of Internal Medicine (ABIM) Foundation, ACP-ASIM Foundation, and the European Federation of Internal Medicine (ABIM, 2002) included in their set of professional responsibilities the commitment of professional competence and ensuring patients are "completely and honestly" informed prior to the start of a procedure or treatment.

Due to recent changes in the healthcare delivery system across the world and changing business pressures in medicine, the Charter on Medical Professionalism, created by a consortium of concerned professional medical organizations including the ABIM, American College of Physicians–American Society of Internal Medicine Foundation, and the European Federation of Internal Medicine stressed three fundamental principles (ABIM, 2002):

- The needs of the patient come first;

- The medical profession must promote justice in the health care system; and,

- Clinicians must be honest with their patients and empower them to make informed decisions about their treatment.

Many countries have passed common laws or guidance to be followed by practitioners related to informed consent. In the United Kingdom, the Human Rights Act of 1998 required public authorities to act in accordance with certain patient rights, including proper informed consent (UKDOH, 2009). Article 3 of the Charter of Fundamental Rights of the European Union (European Commission, 2000) requires, in the fields of medicine and biology, "free and informed consent of the person concerned." In the United States, no national standards exist regarding which procedures or decisions require informed consent.

Clearly, the patient's right to determine their own course of action for any given medical treatment or research protocol is fundamental.

The amount of information provided to a patient in the informed consent process has been hotly debated over the years. The proposed procedure complexity as well as associated risks and the patient's wishes will determine the course of the discussion. The General Medical Council (GMC) and Department of Health in the United Kingdom have provided guidance on this subject (Barnett et al., 2004). True informed consent, at a minimum, should address the following:

- Disclosing relevant medical facts and alternative options;

- Ensuring patients understand the medical information provided to them;

- Ensuring patient capacity to understand the ramifications of their decision;

- Discussion occurs in absence of coercion or manipulation; and

- Ability to consent

For many procedures, there is no threshold for which informed consent is required. Bioethicists have suggested that risks greater than those encountered in everyday life should require informed consent.

A discussion of informed consent also requires a review of human research consent. Historically, various philosophies influenced how human research was performed: informed consent was optional; if research was performed during war time, consent wasn't needed; individuals who were expected to soon die could be used without regard to what the drug, device, or withholding of treatment would do; if the results were good results, the experiment was ethical. It was these seeming indiscretions that led to ethical codes of conduct and laws designed to protect human subjects (Brandt, 1978; Lock, 1995; ACHRE, 1996).

20.1 DIAGNOSTIC PROCEDURES

20.1.1 Background

The average amount of ionizing radiation received by the US population undergoing medical procedures has approximately doubled in the last 30 years. Similar increases have been seen across the world (NCRP, 2009). Approximately 1 in 4 persons have had a recent computed tomography (CT) or nuclear medicine procedure. The use of CT has grown annually at a rate of about 10% (Gerber, 2009) and now accounts for 50% of the collective dose from all imaging procedures, while nuclear medicine accounts for about 16% of the collective dose.

Half of the diagnostic CT scans conducted in adults are of the body, with about 1/3 of the total being head CT's. Approximately 75% of CT's conducted in the United States are obtained in the hospital setting and 25% in a single-specialty clinic setting (Brenner, 2007). The largest increase in CT use has been with pediatric procedures and adult screening programs such as virtual colonoscopy, lung screening of former and current smokers, and cardiac calcium screening.

Cardiac imaging using ^{99m}Tc or ^{201}Tl and PET CT exams for oncologic staging account for the greatest increase in nuclear medicine collective dose.

Because of the speed and ease of conducting CT scans, many exams are ordered without commensurate benefit to patients. CT exams ordered for management of blunt trauma, seizures, chronic headaches, and acute appendicitis in children are often questioned in the medical literature (Brenner, 2007).

The increased use of CT and other imaging modalities involving ionizing radiation for diagnostic and screening purposes has triggered concern from patients and regulators and an increased interest in informed consent for these diagnostic procedures. Scientific organizations such as ICRP, NCRP, and BEIR have held that for any level of ionizing radiation, some level of risk, however small, exists. In the United States, the Department of Health and Human Services has listed X-rays as a carcinogen (USDHHS, 2006). The American College of Radiology White Paper on Radiation Dose in Medicine noted: "The rapid growth of CT and certain nuclear medicine studies may result in an increased incidence of radiation-induced cancer in the not-too-distant future" (ACR, 2007). As the medical use of ionizing radiation has increased, so has the postulated associated radiation-induced cancer risk, and this has triggered a public health concern (Brenner, 2002; 2004).

Informed consent for diagnostic imaging studies has not been examined in great detail in the medical literature. Although there is widespread agreement in the radiology community for consenting interventional procedures when inserting devices or performing therapeutic procedures (Beditti, 2007; ACR, 2014), such agreement does not exist for diagnostic exams. The increased population risk from these exams mandates that informed consent be considered for ethical and legal reasons.

Leonard Berlin, M.D., in a series of articles published in the American Journal of Roentgenology and Journal of the American College of Radiology, has done an excellent job in summarizing the legal and ethical dilemmas facing radiologists regarding malpractice and informed consent, particularly in the United States (Berlin, 2000; 2001; 2003a; 2003b; 2011; 2014).

Requirement 36 of the IAEA General Safety Standards (IAEA, 2014) requires that member states ensure (1) an appropriate referral has been made, and (2) the patient "has been informed of the expected diagnostic and therapeutic benefits of the radiological procedure as well as the radiation risks."

In the European Union, the use of radiation for diagnostic purposes is subject to legislation that mandates the use of non-ionizing imaging whenever possible. In addition, CT dose information must accompany electronically stored image data for patients (Euratom, 2013).

In the United States, no national standards exist for informing patients about their radiation exposure to diagnostic CT scans. It has been estimated by Lee et al., that 92% to 95% of patients are not informed of any radiation risks prior to their CT scan in the United States (Lee, 2004). Except for mammography, only a few states currently require the tracking of patient dose, although accreditation organizations, such as the American College of Radiology and The Joint Commission, may require action on this topic in the future.

Strong opinions exist regarding informed consent for imaging studies. Most dissenting opinions originate from the United States. Alternatively, in their review on patient safety in June 2009 (European Union, 2009), the European ministers recommended, "When ionizing radiation is used for medical diagnosis or treatment, patients should receive adequate information on the benefits and limitations of the procedure and the potential radiation exposure effects to enable them to take informed decisions." Additional education of the public and practitioners on radiation effects and risk was also part of their recommendation.

The discussion on what constitutes informed consent and procedures requiring informed consent should continually be evaluated, as failure to obtain proper informed consent could result in legal action by the patient against the practitioner and/or the employer, and jeopardize the practitioner's professional licensure.

20.1.2 Justification and Optimization of Radiological Procedures

To better understand the interplay between radiation risk of medical procedures and the informed consent process, one must first understand the radiation protection principles of justification and optimization. These principles were first introduced by the ICRP in 1977 (ICRP, 1977) and have been incorporated into Euratom Directive 2013/59 (Euratom, 2013).

In Article 55 of Euratom Directive 2013/59, justification is defined as: "Medical exposure shall show a sufficient net benefit weighing the potential diagnostic or therapeutic benefits it produces, including the direct benefits to health to an individual and the benefits to society, against the individual detriment that the exposure might cause, taking into account the efficacy, benefits and risks of available alternative techniques having the same objective but involving no or less exposure to ionizing radiation." Further, Euratom Directive 2013/59 specifies that the referrer and practitioner should be involved with the justification process with the referrer responsible for providing enough information to assist the practitioner in justifying the appropriateness of the proposed exam.

Requirement 37 of the IAEA General Safety Standards (IAEA, 2014) requires that medical exposures be justified and that national or international referral guidelines be taken into consideration in justifying the radiological procedure.

The ICRP has identified three levels at which justification operates (ICRP, 2007):

- Level 1: Justification of the use of radiation in medicine

- Level 2: Justification of a specific defined radiological procedure

- Level 3: Justification of the application of the procedure to an individual patient

In practice, the second and third levels of justification are encountered most typically in diagnostic radiology.

Appropriateness criteria or referral guidelines are the terms commonly used to describe justification by radiologists. Professional societies such as the Royal College of Radiologists (RCR), the American College of Radiology (ACR), and the European Commission in the European Union have issued appropriateness criteria guidelines. In Europe, compliance with the Medical Exposures Directive (MED) requires that member states produce referral criteria for medical exposures and include radiation doses (Malone, 2012).

20.1.3   Why Is Justification Important in the Imaging Process?

The European Commission estimated that at least 1/5 of the 4 billion X-ray exams conducted annually around the world are inappropriate (European Commission, 2014). Others estimate that at least 30% of all ionizing exams remain inappropriate in clinical practice, in spite of the existing European law and European Commission recommendations (Semelka, 2007 and Gibbons, 2007). In 2009 IAEA suggested that 50% of examinations may not be necessary. In the United States, Brenner et al (Brenner, 2010), suggested that several studies indicate that 20 to 40% of CT scans could be avoided if clinical decision guidelines were followed.

Reasons identified in the IAEA Radiation Protection of Patients document (IAEA, 2014, European Commission, 2014) for the non-justified use of radiation include:

- The exam was previously conducted.

- The exam isn't required and won't change the patient management.

- The exam could be delayed allowing for further assessment of symptoms and previous treatments.

- The exam isn't ideal.

- The exam has not been justified by the referring physician.

- The exam is one of too many investigations being performed.

Reasons for over-investigation suggested by IAEA (IAEA, 2014) include:

- Patient wishes

- Financial considerations

- Defensive medicine

- Role of media

- Role of industry

- Convenience

Once a procedure has been justified, it must be optimized. Optimization is defined by the ICRP (ICRP, 2007) as the source-related process to keep the number of exposures and magnitude of exposure as low as reasonable achievable taking economic and societal factors into account. Constraints or diagnostic reference levels provide an upper bound for the optimization process.

The European Union through Euratom Directive 2013/59 (Euratom, 2013) requires member states to establish diagnostic reference levels for radiological examinations with the primary objective of improving clinical practice. IAEA

General Safety Requirements Part 3 (IAEA, 2014) mentions that the national government "shall ensure that relevant parties are authorized to assume their roles and responsibilities, and that diagnostic reference levels, dose constraints, and criteria and guidelines for the release of patients are established." In the United States, no federal requirement exists for the formal establishment of diagnostic reference levels or constraints.

Optimized protection is the result of an evaluation, which balances the risk of the exposure with the resources available for the protection of individuals. As noted in ICRP Report No. 103, the total collective effective dose is not a useful tool for making decisions over large populations, large geographic areas, or long time periods.

Radiation doses in medical imaging can be optimized by (Brenner, 2010):

- Improved quality control and quality assurance. Radiation doses can vary significantly from facility to facility. Brenner et al., suggests that since the MSQA requirement has greatly reduced mammography doses, perhaps similar national legislative requirements for other modalities in the United States would have a similar effect.

- Additional radiation safety and risk education for practitioners.

- Incorporating decision-making criteria into computerized ordering systems.

20.1.4   Radiation Risk

The risk posed by radiation has been studied extensively for over 100 years. Many cohorts have been studied, some exposed for medical purposes. The largest cohort of exposed individuals is the atomic bomb survivors, and they have been followed since 1950 as part of the Life Span Study (LSS). There is no consensus on whether the effects of high levels of whole-body acute exposures can be extrapolated to radiation workers or patients receiving much lower levels of radiation, or to different ethnic origin.

A linear dose response model fits must solid cancers in the Japanese survivors as well as other radiation-exposed cohorts, with the notable exception of leukemia, which follows a linear quadratic relationship. There is a latency period ranging from years to decades between the initiation of a solid cancer caused by radiation and the appearance of clinical disease (Balter, et al, 2011). Leukemia has the earliest latency period, with increased risks being noted two to five years following radiation exposure (Semelka, 2007). Individuals exposed early in life have a greater likelihood of expressing a biologic effect.

Radiation is considered a weak carcinogen when compared with other physical and chemical agents. Many believe that even small amounts of radiation can cause detriment. The European Commission in 2007 (European Commission, 2007) stated that "a small fraction of the genetic mutations and malignant diseases occurring in the population can be attributed to natural background radiation," and to assist referring clinicians, have classified the effective dose received from common imaging procedures in bands (Table 20.1).

Exams resulting in similar effective dose to background radiation include an intravenous ureterogram, lumbar spine X-ray, and head and neck CT.

Table 20.1: Band classification of the typical effective dose from common imaging procedures

Band	Typical Effective Dose, mSv	Examples
0	0	US, MRI
I	<1	CXR, XR limb, XR pelvis
II*	1-5	IVU, XR lumbar spine, CT head and neck
III	5-10	CT chest and abdomen, NM (e.g., cardiac)
IV	>10	NM (e.g., PET)

*The average annual background dose in most parts of Europe fall in band II.

20.1.4.1   Stochastic and Deterministic Effects

Radiation effects are divided into effects that have a threshold and those that do not.

Stochastic effects do not have a threshold; the probability of an effect is proportional to the dose. With the current linear, non-threshold theory of radiation response, receiving even small doses of radiation incurs some level of risk. Radiation-induced cancer is the stochastic effect of greatest concern. Depending on the type of cancer and population studied, cancer mortality occurs at approximately half of the incidence rate and cancer incidence increases with age.

Deterministic effects have a threshold and the severity of the effect will increase after the threshold has been reached. Examples of deterministic effects include skin erythema and cataract formation.

Examples of medical procedures that could exceed the skin erythema threshold of 5 Gy (cumulative air kerma) include (ACR, 2014):

- Transjugular intrahepatic portosystemic shunt creation (TIPS)

- Percutaneous transluminal coronary angioplasty (PTCA)

- Embolization (any location, any lesion)

- Stent-graft placement

- Cardiac stent placement

- Radiofrequency cardiac ablation

- Percutaneous coronary intervention

The Society of Interventional Radiology (Stecker, 2010) recommends placing directions in the patient's discharge instructions when the threshold for skin erythema is exceeded.

20.1.4.2 Effective vs. Organ Dose

Effective dose, E, as introduced by the ICRP in 1977 (ICRP, 1977), reflects the risk of a nonuniform dose distribution compared to a uniform whole body exposure. It provides a way of quantifying non-uniform exposures that involve only a portion of a person's body, in terms of whole-body exposures received by the Japanese atomic bomb survivors. Effective dose represents a weighted summation of organ and tissue doses for a defined set of radiosensitive organs and tissues averaged over all ages and sexes (Table 20.2). It cannot be measured directly, only derived by computation.

Effective dose was intended to be used in the field of radiation protection for procedural optimization, e.g., CT protocol review, and for broad comparisons of the relative risks from different procedures that use ionizing radiation against each other or background radiation (Gerber, 2009). The uncertainty in the relative value for effective dose in a reference patient has been estimated to be $\pm$ 40% (Martin, 2007). Because it uses a standardized model and can't be measured directly or quantified precisely, effective dose should not be used to determine the risk for an individual. It often has been used incorrectly in this manner in medical imaging literature.

Table 20.2: Tissue weighting factors, w_T (ICRP Report No. 103)

Organ	w_T
Bone marrow	0.12
Colon	0.12
Lung	0.12
Stomach	0.12
Remainder tissues*	0.12
Gonads	0.08
Esophagus	0.04
Liver	0.04
Thyroid	0.04
Bone surface	0.01
Brain	0.01
Salivary glands	0.01

Continued on next page

Table 20.2 – *Continued from previous page*

Organ	w_T
Skin	0.01

* Remainder tissues — arithmetic mean of the dose to the following organs: adrenals, extrathoracic (ET) region, gall bladder, heart, kidneys, lymph nodes, muscle, oral mucosa, pancreas, prostate (for males), small intestine, spleen, thyroid, uterus/cervix (for females).

Table 20.3: Higher effective dose adult radiologic exams and cancer incidence risk

Examination	Average Effective Dose, mSv*	Range, mSv*	Risk of Cancer Incidence**
Upper GI series	6	1.5–12	4.4×10^{-4}
Barium enema	8	2.0–18.0	5.9×10^{-4}
Chest CT	7	4.0–18.0	5.2×10^{-4}
Chest CT for pulmonary embolism	15	13–40	11.1×10^{-4}
Abdomen CT	8	3.5–25	5.9×10^{-4}
Pelvis CT	6	3.3–10	4.4×10^{-4}
Three-phase CT liver study	15		11.1×10^{-4}
Spine CT	6	1.5–10	4.4×10^{-4}
Coronary angiography CT	16	5.0–32	11.7×10^{-4}
Virtual colonoscopy CT	10	4.0–13.2	7.3×10^{-4}
Coronary percutaneous transluminal angioplasty, stent placement, or Rf ablation	15	6.9–57	11.1×10^{-4}
Abdominal angiography or aortography	12	4.0–48.0	8.8×10^{-4}
TIPS placement	70	20–180	51.0×10^{-4}
Pelvic vein embolization	60	44–78	44.0×10^{-4}

*Me, 2008. **ICRP Report No. 103 - Table A.4.4 using detriment, adjusted nominal risk coefficient for whole population of 7.3×10^{-2} Sv^{-1} for cancer and adjusted for a DDREF of 1.5 (ICRP, 2007).

Examples of radiologic exams that generate higher effective doses in an adult population and the associated cancer incidence risk can be found in Table 20.3.

The risk to the standardized population using the average effective dose in Table 20.3 ranges from approximately 4 in 10,000 to 50 in 10,000 for certain radiographic exams.

Organ doses, stratified by age and gender, are the preferred method of risk assessment for individual patients undergoing medical diagnosis and treatment (ICRP, 2007; Gerber, 2009; Brenner, 2007).Organ specific absolute risk data based on age and gender can be found in the BEIR VII Report on the Health Risks from Exposure to Low Levels of Ionizing Radiation (BEIR, 2006). BEIR VII evaluated the lifetime attributable risk of cancer incidence and mortality for various sites based on gender and age (either at 5- or 10- year intervals) resulting from a single dose of 0.1 Gy. The lifetime attributable risk for females exposed to a single dose of 0.1 Gy for specific organs can be found in Figure 20.1. For males, this information can be found in Figure 20.2. As can be seen from the graphs, the largest risk for any solid tumor site occurs when individuals are exposed at an early age, and then decreases steadily until approximately the age of 30.

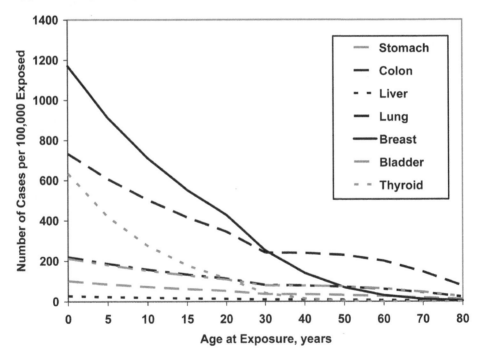

Figure 20.1 Lifetime attributable risk of cancer incidence for females exposed to a single dose of 0.1 Gy

The rate of decline is less after the age of 30. Rates are very low as an individual approaches the age of 80 because they are not likely to live long enough to experience an attributable effect.

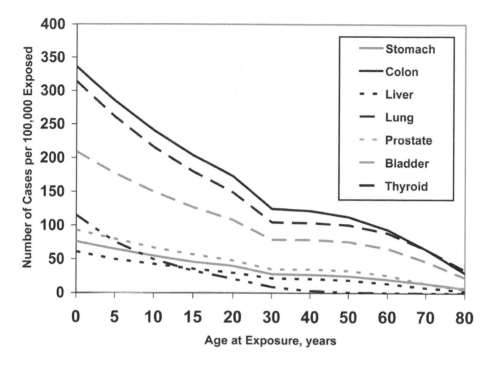

Figure 20.2 Lifetime attributable risk of cancer incidence for males exposed to a single dose of 0.1 Gy

Typical organ doses and associated cancer incidence risk for several common radiographic procedures can be found in Table 20.4.

In theory, for any given age and gender, one could estimate risk based on the exposure to the most radiosensitive organ(s) (see Table 20.2) or those receiving the highest exposure. As expected, the lifetime attributable risk to any exposed adult organ is much less that the risk calculated using effective dose.

The drawback to using this approach is that medical imaging exams will partially irradiate many organs and tissues, and it is the total patient risk that is of interest (Semelka, 2007). Also, age and gender-specific organ risks are based on pooled cohort study results. The exposure to organs in an individual patient may not reflect the pooled results.

Table 20.4: Typical organ doses and risks from common radiographic procedures

Procedure	Critical organ	Organ Dose, mGy*	BEIR VII Risk of Cancer Incidence per 100,000**			
			30-year-old male	60-year-old male	30-year-old female	60-year-old female
Chest X-ray (PA)	Lung	0.01	0.01	0.01	0.02	0.02
Mammography	Breast	3	-	-	7.6	0.9
Abdominal CT	Stomach	10	2.8	2.0	3.6	2.7
Barium enema	Colon	15	18.8	14.1	12.3	9.3

*Brenner 2007
**Assumes a DDREF of 1.5

As mentioned by Balter et al., (2011), "Any practicable published risk estimate is likely to be no better than within an order of magnitude of the actual radiogenic risk to which an individual patient would be exposed...." Possible reasons for this variance include:

• Experimental uncertainties in calculating or measuring dose using Monte Carlo calculations or anthropomorphic phantoms and radiation monitoring devices (McCollough, 2010).

• Dose delivered to the patient will vary from patient to patient, from procedure to procedure, and from facility to facility.

• Risk to any specific organ will vary based on gender and age of exposure.

• Radiosensitivity of individual patients.

Instead of calculating a numerical estimate, risk can also be described in broad categories. A "low" dose of radiation has been defined by the BEIR VII committee as ranging from 0 to 100 mSv effective dose (Semelka, 2007). Risk, at "low" levels of radiation and due to the associated uncertainty, could be described in broad categories (Martin, 2007):

• < 0.1 mSv Negligible

• 0.1-1 mSv Minimal

- 1-10 mSv Very low

- 10-100 mSv Low

Most scientists believe that for effective doses less than 100 mSv, it is extremely difficult to statistically detect a radiation-induced cancer above the high natural incidence of cancer (BEIR 2006; Cardis, 2007, Tubiana, 2009).

It has been suggested that at effective doses less than 1 mSv, implied consent should be sufficient. At effective doses greater than 10 mSv, the mandatory use of consent forms is suggested (Malone, 2012).

Others have argued that since there is no scientific evidence of a carcinogenic effect for acute irradiation at doses less than 100 mSv and for protracted irradiation at doses less than 500 mSv, the carcinogenic risk, if it exists, is too small to be clinically significant, and it is implied that informed consent would not be necessary (Tubiana, 2009).

20.1.5 Physician Awareness of Radiation

Although radiation effects may be well understood by radiologists and perhaps other clinicians, the terminology used to describe radiation and radiation risk is less understood. A clearer understanding of these terms and topics will help facilitate better communication between clinicians as well as between patients and their doctors. Increasing radiological awareness of healthcare providers will help facilitate a better informed consent process.

This section will discuss recent studies that have identified the lack of knowledge of radiation risk as a concern. Qualifications for practitioners will also be briefly discussed.

The results of the surveys recently performed on British physicians (Shiralkar, 2003), Israeli orthopedists (Finestone, 2003), Italian cardiologists (Correia, 2005), Canadian pediatricians (Thomas 2006), and US radiologists in large urban academic centers (Lee, 2004), show that the majority of doctors grossly underestimated the radiation doses (usually by up to 500 times) and corresponding cancer risks for most commonly requested investigations.

Shiralkar et al. (2003) conducted a study of 130 physicians from two different hospitals in South Wales and Oxford to better understand the level of knowledge physicians have concerning radiation doses received by patients undergoing commonly requested exams. Results indicate that none of the physicians knew the approximate dose of radiation received by a patient undergoing a chest X-ray exam. In all 97% of respondents underestimated the actual dose; 5% and 11% of respondents, respectively, were not aware that ionizing radiation was not used in either ultrasound or MRI.

At an Israeli Orthopedic Society Meeting, Finestone et al. (2003) found that physicians grossly underestimated the potential radiation risk from a bone scan, with only approximately 5% responding correctly in a multiple choice survey.

Using a one-page multiple choice questionnaire, Correia et al. (2005) found that out of 100 responding physicians working in a tertiary-care cardiology referral center, 89% wrongly estimated the contribution of nuclear and radiological tests in overall radiation exposure, 95% incorrectly estimated the risk of fatal cancer associated with a stress myocardial perfusion study, and 71% did not correctly estimate the dose exposure of a myocardial stress perfusion study. The average level of radiological awareness was not correlated to the number of exams performed or prescribed per year or to patient type (adult or pediatric). Few doctors have knowledge about the level of radiation that their patients are exposed to during radiological investigations.

Thomas et al. (2006) investigated the level of awareness among pediatricians regarding radiation risks in children. Out of 220 respondents, only 6% were correct in their estimate of the quoted lifetime excess cancer risk associated with radiation doses equivalent to pediatric CT. Only 15% were familiar with the ALARA principle.

Adult patients seen in the emergency department (ED) of a US academic medical center during a two-week period with mild to moderate abdominopelvic or flank pain and who underwent CT were surveyed after acquisition of the CT scan. Physicians involved in the management of the patient were also surveyed regarding radiation risk by Lee et al. ED physicians and radiologists believing there was an increased cancer risk from the CT scan accounted for only 9% and 47% of respondents, respectively. Seventy-eight percent of ED physicians stated they did not outline the risks and benefits of the CT scan with patients and 93% of patients reported that they did not receive such information. All patients and most ED physicians and radiologists were unable to accurately estimate the effective dose from one CT scan compared with that for one chest radiograph (Lee, 2004).

20.1.5.1 Qualifications for Practitioners

The European Commission, through Euratom Directive 2013/59, has mandated radiation protection education and training for medical specialists since the 1980s. This also includes referring clinicians requesting radiology examinations. The requirements have been periodically updated to keep up with new technologies, the education and training requirements of the Euratom Basic Safety Standards Directive, and the European qualifications framework. Basic curriculum includes biological effects of radiation, justification of exposures, risk–benefit analysis, typical doses for each type of examination, and knowledge of the advantages and disadvantages of the use of ionizing radiation in medicine. Radiation protection learning outcomes for diagnostic radiologists are given as an example in Appendix A. The contents have been endorsed by the major European professional societies impacted by the legislation (European Commission, 2014). In North America, members of the Canadian Association of Radiologists have communicated with the Federal Ministry of Health, and to the Royal College of Physicians and Surgeons of Canada, concerns regarding radiological procedures performed by physicians with inadequate training (Doris, 2006).

The American Council of Graduate Medical Education (ACGME) has specified program requirements for each program receiving its accreditation. Pertinent radiation protection/risk requirements for programs most likely to use ionizing radiation for diagnostic purposes are outlined in Table 20.5.

Table 20.5: ACGME program requirements related to radiation safety

Program	Effective Date	Radiation Safety Program Requirements	Required No. of hours
Interventional Cardiology	July 1, 2007	"The program must provide formal instruction for the fellows to acquire knowledge of the following content areas: ... Radiation physics, biology and safety related to the use of X-ray imaging equipment."	None stated
Vascular Surgery	July 1, 2007	"Residents should have education in the entire vascular system. Instruction in each area should be associated with relevant patient exposure."	None stated
Vascular and Interventional Radiology	January 2005	"The fundamentals of radiation physics, radiation biology and radiation protection should all be reviewed during the vascular and interventional training experience."	None stated
Diagnostic Radiology	July 1, 2008	"There must be didactic components that address the following subjects: diagnostic radiologic physics and radiation biology; patient and medical personnel safety (i.e., radiation protection, MRI safety)."	None stated except for nuclear medicine NRC training and education requirements

The learning requirements for European-trained medical specialists are much more radiation-safety-and-risk-focused than their American counterparts. This apparent deficit in radiation protection training and education could, in part, help explain the reluctance of U.S. trained clinicians in providing informed consent for radiologic procedures. Of course, actual radiation safety and risk knowledge gained through medical training programs may not be reflective of the stated program requirements.

20.2 HUMAN SUBJECT RESEARCH AND INFORMED CONSENT

20.2.1 Background

Human subject research has been conducted for over a century. The search for vaccines in the 1800s (CPP, 2014) provided some of the earliest documented research efforts. Children were most valuable for studying vaccines because it was less likely they had disease exposure (ACHRE, 1995b). Some so-called research studies were simply the testing of "cures" on diseased individuals to keep them from dying. Discussions of informed consent in research were documented as early as the 1920s, when the U.S. Army regulated the use of volunteers for medical research, and in 1932 when the U.S. Navy required that subjects for proposed experiments be "informed volunteers" (ACHRE, 1996). In the 1950s, the Committee on Medical Research (subsequently the National Institutes of Health) set forth the precedent "where risks are involved, volunteers only should be utilized as subjects and these only after signed statements have been obtained" (Lock, 1995). These references to informed consent are for normal volunteers as opposed to patient-subjects.

In the 1930s to 1950s, human subject research was being conducted with the first use of radionuclides. Plutonium, uranium, and polonium research was being performed to establish basic biokinetic information that could be used to estimate exposure if a worker (primarily those working on the atomic bomb) were accidentally exposed. In children, radioiodine studies were performed to estimate possible exposure from fallout and radioactive minerals were being administered to determine absorption rates. There is little evidence that informed consent methods were being used at the time, judging by this statement: "In 1941, human experiments without consent were permissible provided that the risk of death was remote" (Weindling, 1996). The plutonium studies were kept secret; some believe for national security and others believe it was due to public relations and liability concerns. The children, because they were institutionalized or developmentally disabled, were considered "available" for studies that could have societal benefit (ACHRE, 1995a; ACHRE,1995b).

As early as 1900, the United States Senate had proposals before them to stop human experiments on "children, insane persons and pregnant women" suggesting that, if these studies were made public, they would be viewed as unethical (ACHRE, 1995b). Little progress was made on the protection of human subjects, however until publication of the Nuremburg Code in 1947 (Kaufman, 1997). This code, published in the Journal of the American Medical Association, set forth ten tenets on research using human subjects (Appendix B). This was followed by the Declaration of Geneva, with an international code of ethics (1949), a US National Institutes of Health Clinical Center policy on ethical responsibility for medical experiments (1953), additional US regulations requiring consent (1962), the Declaration of Helsinki (1964), more United States regulations throughout the 1970s (USDHHS 2009), and the Belmont Report in 1979 (Sparks, 2014). From 1980 through today, additional laws, policies, and practice guidance have been developed along with updates of the Declaration of Helsinki, the latest being in 2013 (WMA, 2013).

In the mid-1990s, the United States established two groups to look at alleged unethical practices in government-sponsored human use research. The first group was the Advisory Committee on Radiation Experiments (ACHRE 1996). Their investigation included the highly publicized Tuskegee Study where, it was reported, 400 primarily African-American men with syphilis were left untreated to follow the course of the disease. Reports of research being performed to determine the best and quickest method to kill large numbers of people, studies using children, the mentally handicapped, and even U.S. troops were also reviewed in the investigation (Lock, 1995; Corbie-Smith, 1999). Their findings led to the establishment of a National Bioethics Committee.

Also, in the United States, Title 45, Part 46 of the Code of Federal Regulations (http://www.hhs.gov/ohrp/humansubjects/guidance/45cfr46.html) contains laws regarding conduct of research involving human subjects. These requirements establish minimal ethical and legal obligations for persons and institutions conducting or supporting research involving humans. It requires each institution conducting federally funded research to adhere to the principles of the Belmont report (Appendix C), and sets forth ethical principles, policies, and procedures for protecting the rights and welfare of humans involved in research. Subsections contain information regarding institutional review board (IRB) membership, functions and operations, protocol review procedures, informed consent elements, and requirements for certain subject types (e.g., vulnerable populations). Research conducted under 45 CFR 46 requires each institution engaged in this research to provide a written assurance of compliance; this is referred to as a project assurance. Each assurance must contain an institutional commitment to employ the ethical principles of the Belmont Report and to comply with regulations (USDHHS, 1979; USDHHS, 2009).

Similarly, the Council for International Organizations of Medical Sciences (CIOMS) and the World Health Organization (WHO) recommend that research studies involving human subjects be reviewed for scientific merit and ethical acceptability by one or more review boards qualified to do such reviews. These review groups can be national or international depending on the research being conducted. The committees' membership must:

- be independent from the research being reviewed;

- be qualified to determine scientific and ethical merit;

- have familiarity with the cultures and customs of the subject population; and

- include individuals from different communities or countries when applicable and possible.

Appendix D contains CIOMS informed consent elements (CIOMS, 2002).

20.2.2   Consent Process for Radiological Procedures

20.2.2.1   Explicit and Implicit Consent

As mentioned by the World Medical Association (WMA, 2009), "A mentally competent adult patient has the right to give or withhold consent to any diagnostic procedure or therapy. The patient has the right to the information necessary to make his/her decisions. The patient should understand clearly what would be the purpose of any test or treatment, what the results would imply, and what would be the implications of withholding consent."

At what level should risks be disclosed? Ethically, most agree that if the risk associated with a medical procedure is greater than what would be expected while performing normal daily activities, some form of consent should be obtained (Terry, 2007). However, risk acceptance will vary dramatically from individual to individual. Some people will willingly accept large risks such as smoking or risks where they think they are in control, such as driving a car. Others tend to overestimate small risks such as being hit by a meteor or incurring a malignancy from a diagnostic imaging exam. A risk of death of 1 in a million is generally ignored (Picano, 2004b).

Evidence of consent can be implicit (implied) or explicit. Implicit consent occurs when a patient, based on their behavior and actions, indicates a willingness to proceed. An example of implicit consent would be a patient rolling up their sleeve to have their blood pressure taken (GMC, 2008). Explicit consent is given orally or in writing (WMA, 2009). The type of consent to be used depends on the radiographic procedure to be performed. For low-dose and low-risk non-invasive procedures, oral or implicit consent may be sufficient as long as a patient with capacity or their guardian understands the purpose and risks associated with the procedure. For cases involving higher risk or more than mild discomfort, explicit written consent may be required. Depending on the laws and codes of medical practice within a jurisdiction, mandatory written consent may be required for certain procedures, e.g., fertility treatment (GMC, 2008) or high-dose interventional procedures.

The validity of consent does not depend on the form in which it is given (UKDOH, 2009). Written consent is not valid if the patient has not been properly informed or has not voluntarily consented to proceed with the procedure.

The General Medical Council of the United Kingdom has indicated that written consent should be obtained if any of the following conditions apply (GMC, 2008):

- Procedure is complex and involves significant risk

- Adverse consequences of the procedure may impact the patient's employment or personal life

- The procedure does not provide a clinical benefit to the patient

- Patient is participating in a research protocol or "innovative treatment"

Currently in the United States the most common reason for obtaining written consent in radiology is the insertion of devices into a patient (Semelka, 2012), and rarely has anything to do with accumulated radiation exposure or risk. Consenting for adverse effects that may occur more than one month after a non-invasive procedure has been performed in radiology have not been addressed by state or federal legislative mandates.

20.2.2.2 Communication

Individuals involved in communicating risk must be flexible as patients may require more or less information or involvement in the decision-making process. No single approach will satisfy all patients (GMC, 2008).

Factors that influence decisions made by individuals regarding medical programs and procedures include (Alaszewski, 2003):

- Social factors

- Extent to which the source of information is trusted

- Relevance of risk to everyday life and decision making

- Relationship to other perceived risks

- Fit with existing patient knowledge

- Acknowledgement by the consent giver of the difficulty in arriving at a decision

Communication of radiological risk is difficult among physicians. Possible reasons include lack of knowledge of the risk factors, which may account for a portion of inappropriate exams (Picano, 2004b). Clinicians who are unable to provide a clear understanding of the risks should not require patients to give signed "informed" consent (Baerlocher, 2011).

According to Picano (Picano 2004b), the three possible ways radiological risk can be communicated include no mention of risk, underestimating the risk, and providing specific details of the risk. Reasons given for not mentioning risk include the perception that radiologists are too busy to obtain informed consent and are too wise to undertake inappropriate exams. In addition, the practitioner may be concerned that mentioning risks may cause the patient not to choose to have the exam. However, this last concern is not supported in the literature (Larson, 2007).

Underestimating known risk factors can occur in situations where well-intentioned clinical staff want to reassure patients and avoid patient concern over unavoidable risks. Finally, providing specific, accurate details of all significant risks to the patient should always be the goal. This consistently occurs in research protocols.

Patient comprehension of medical risk communication can be impacted by confusion, panic, shock, fatigue, pain, or medication (UKDOH, 2009). Medications that impact the mental awareness of the patient should be kept at a minimum and not be given to the patient less than 4 hours prior to the patient's consent; however, no patient should be deprived of adequate pain management for the purpose of obtaining consent (ACR, 2014). Mental acuity is impacted less with chronic pain medication use.

20.2.2.3 Who Gives Consent for Radiological Procedures?

It is preferred that the clinician providing the treatment or performing the procedure be responsible for ensuring that the person has given valid consent before treatment begins. In actual practice this may be conducted by the radiological technologists or radiologic physician assistants, with radiologists serving as backup (Semelka, 2012; Lee, 2006). The GMC guidance states that the task of seeking consent may be delegated to another person as long as they are suitably trained and qualified (UKDOH, 2009).

For interventional radiology procedures, most authorities agree that the radiologists should obtain informed consent (Berlin, 1997).

20.2.2.4 Information Provided in the Consent

Various references have provided information on the suggested elements of a valid consent. The General Medical Council published the following elements in 2008 (GMC, 2008):

- The diagnosis and prognosis

- Any uncertainties about the diagnosis or prognosis, including options for further investigations

- Options for treating or managing the condition, including the option not to treat

- The purpose of any proposed investigation or treatment and what it will involve

- The potential benefits, risks and burdens, and the likelihood of success, for each option; this should include information, if available, about whether the benefits or risks are affected by which organization or doctor is chosen to provide care

- Whether a proposed investigation or treatment is part of a research program or is an innovative treatment designed specifically for their benefit

- The people who will be mainly responsible for and involved in their care, what their roles are, and to what extent students may be involved

- Their right to refuse to take part in teaching or research

- Their right to seek a second opinion

- Any bills they will have to pay

- Any conflicts of interest that you, or your organization, may have

- Any treatments that you believe have greater potential benefit for the patient than those you or your organization can offer

In addition, it was mentioned that consent must be given voluntarily without any coercion and patient views must be respected.

If consent is done properly, it may also reduce the unfounded concern patients and families often have with the term "radiation," as well as the pressure clinicians feel to order exams that are not required but are contemplated principally due to patient pressure.

To improve radiologic risk communication with patients or research subjects, the following suggestions should be followed:

- Give clear simple messages with no more than three key points. Mention key points at the beginning and end of the discussion, to help ensure patient understanding (Dauer, 2011).

- Information should be provided in advance of the exam.

- Emphasize the benefits in comparison of the small risk to be received from the radiation exam or procedure. The radiologic exam or procedure should not be conducted unless it can be justified.

- To the extent possible, use clear, non-technical terminology to reduce the amount of cognitive effort required to evaluate radiation risk comparisons (Walters, 2006)

- Express risk probability information as a graphical display and illustrations (Walters, 2006; Dauer, 2011; Picano, 2004b, Malone, 2012).

- Express risk information should be listed as a percentage rather than frequency and should be in a suggested format of 1 in X (Walters, 2006; Dauer, 2011; Cardinal, 2011, Malone, 2012).

- Use risks associated with ordinary life activities as comparisons for proposed exams/procedures (Picano, 2004b; Malone, 2012).

For higher levels of radiation, written consent is preferred. All items identified above should be included in the consent. Several have suggested that the written consent should also include the type of exam, expected effective dose, dose equivalent in number of chest X-rays, the loss of life expectancy, and risks of fatal and non-fatal cancer (Picano, 2004a; Beditti, 2007), although exact information to be covered should be left to the clinician.

According to Cardinal (2011), the final justification in the effort to improve risk communication with patients receiving radiation for medical purposes is "the steady progression of the expectations of patients and the legal system toward "full disclosure".

20.2.2.5 Acceptable Risk Level for Consent

The risk level most commonly referenced for obtaining informed consent for medical exams or procedures involving ionizing radiation is 1 in 10,000. This corresponds roughly to a 1 mSv effective dose. A blood transfusion, which routinely requires informed consent, has a risk level of 1 in 5000 to 1 in 10,000 (Merck, 2010).

The Veterans Health Administration in the United States as well as the IAEA and select European imaging specialists have independently agreed that consent should be obtained for medical exposures expected to exceed 1 mSv or roughly equivalent to a risk of 1 in 10,000 (Malone, 2012; Semelka, 2012).

The majority of pediatricians practicing in a wide variety of hospital and clinical settings in Toronto, Canada, believed that a risk of 1 in 10,000 or more should be discussed with the patient (Thomas, 2006).

Picano et al., (2004b) has suggested that it be made mandatory to obtain informed consent for radiological exams where the risk exceeds 1 in 10,000.

20.2.3 Consent Process for Research

20.2.3.1 Research Consent

As a result of the Nuremberg Proceedings (trials of war criminals for human rights atrocities before the Nuremberg Military Tribunals in the late 1940s), the Nuremberg Code first introduced the principle of voluntary consent of a human subject prior to participation in research (U.S. Holocaust Museum, 2002). In 1953, the U.S. Department of Defense applied the voluntary consent principle and the U.S. National Institutes of Health adopted a policy requiring voluntary agreement based on informed understanding from subjects if the research was deemed particularly hazardous (ACHRE, 1996). In 1964, the World Medical Association prepared recommendations as a guide to physicians conducting biomedical human use research, which formed the basis for the Declaration of Helsinki. The Declaration included the principle of consent and expanded that principle to issues of coercion and study of individuals who are legally incompetent (WMA, 2013). In 1974, the National Research Act was signed into law creating the National Commission for the Protection of Human Subjects of Biomedical and Behavioral Research (NCPHS), which had as one of its charges to identify the primary ethical principles that would become the basis for the conduct of biomedical and behavioral research (Appendix D). These ethical principles, known as the Belmont Report, contain the formal definition of consent and were published in the Federal Register in 1979 (NCPHSBBR 1978). In 2002, these principles became the General Ethical Principles used by CIOMS and WHO (CIOMS, 2002). In 1981 the concept of consent was extended to the fetus (e.g., if the subject is or may become pregnant) for Health and Human Services and FDA-related investigations (Curran, 1981).

Informed consent for protocols involving radiation must appropriately address radiation risk(s). When deterministic effects are anticipated, the consent should identify specific risks such as low blood-cell counts, skin erythema, risk of bleeding, clotting disorders, infection, etc., in language the general public can understand. Castronovo (1993) reviewed consent form radiation risk statements from fourteen large medical research institutions and concluded that comparisons to annual natural background radiation or occupational exposure limits received the highest approval rankings. Other information addressing the risks of radiation in consent has been addressed in earlier sections of this chapter.

20.2.3.2   Protection of Vulnerable Groups

Children

Special circumstances surround the determination of radiation risk when the research proposes to study children defined as under the age to give legal consent (CIOMS 2002). Freeman (1994) makes a strong case as to why healthy children should not be allowed to participate in research protocols involving radiation, showing that, according to the author's interpretation, current regulations simply do not allow it. Several rebuttals to this article demonstrate otherwise by stating that healthy children will normally encounter risk in their daily lives from X-ray procedures, blood draws, shots, etc. (Keens, 1994). Hence, some protocols involving radiation will produce a risk no greater than the child will encounter in their daily life. Protocols involving greater than minimal risk, however, cannot include healthy children (Sugarman, 1998; US-DHHS 2009).

Parental or guardian consent must be given recognizing that the child cannot give permission. Although parents or guardians are thought to have the best interests of the child as their highest priority, the question remains whether consenting for a child to participate in nontherapeutic research is acceptable given some possible level of harm by being in the study.

In the United States, this question was answered by the National Commission for the Protection of Human Subjects of Biomedical and Behavioral Research, with ten recommendations governing research followed by Federal regulations in 1983 (NCPHSBBR, 1977; USDHHS, 2009). These state, in part, that a local review committee (like an institutional review board) reviewing federally funded research that poses greater than minimal risk can approve children as participants if they find that the benefit outweighs the risk, that the benefit-to-risk relationship is equal to or greater than alternatives, and that permission by the parents/guardian and assent of the child are obtained.

CIOMS suggests that children are necessary for research studies where results in adults may not apply to children and where medical knowledge being sought is relevant to children's health issues. Research can be performed if consent is given by the parents or guardian and when the child, if able, agrees. They do have a caveat that, if a child refuses to participate or continue participation, the child's wishes should be respected (CIOMS 2002).

Pregnant Women/Fetus

Women who are pregnant should not be allowed to participate in protocols involving radiation unless it is the pregnancy that is under investigation, the purpose is to meet the health needs of the mother, and the risk to the fetus is minimal (ICRP, 1991; Sugarman, 1998).

The NCPHS recommendations allow the nontherapeutic research study of the fetus and/or the pregnant woman where:

- there is minimal or no risk to the fetus,

- the pregnant woman has been informed of the fetal impact,

- the pregnant woman consented and the father did not object,

- appropriate review of the study is performed (i.e., IRB). and, in the case of fetal study,

- the results of the study could lead to biomedical knowledge unobtainable otherwise and animal/adult human studies have been performed where applicable (NCPHSBBR, 1975).

CIOMS suggests that a pregnant woman should be allowed to participate in research studies as long as she is informed of potential risks to herself, her pregnancy, and her fetus and its future fertility and offspring. The research being performed should be relevant to the pregnant woman or her fetus (CIOMS, 2002).

Vulnerable Persons

Vulnerable persons include prisoners and those persons unable to give or decline consent. For vulnerable persons, CIOMS suggests that the research be ethically and scientifically justifiable and the risk no greater than routine medical examination(s) these individuals might experience (CIOMS, 2002).

The investigator must show that the research could not be done on a less vulnerable population, that the knowledge gained will benefit the population being studied, that the risks are not greater than those normally encountered, and that, if the drug/device/intervention is effective, the participant will be given reasonable access.

20.2.3.3   Developing Countries Research

"One of the greatest challenges in medical research is to conduct clinical trials in developing countries that will lead to therapies that benefit the citizens of these countries" was stated by Varmus in a paper published in 1997. Within that "greatest challenge" is the difficulty of obtaining informed consent. There are also questions surrounding the ethics of including placebo groups and making the treatment, if successful, available to the population after the study is concluded.

Prior to starting a research study in developing countries, many have recommended that the sponsoring group(s) obtain permission from country or community authorities to perform the research and involve the host country in the design and conduct of the trial (CIOMS, 2002; Shapiro, 2001; Benatar, 2000). This process could aid in obtaining meaningful consent from participants.

Consent elements should not deviate from those used in research in developed countries and should follow the recommendations of the Belmont Report and Declaration of Helsinki as adopted by the World Medical Association (WMA, 2013)

20.3   CONCLUSION

Disagreement will continue on whether consent is required for diagnostic imaging procedures. Some will argue that because stochastic effect such as cancer cannot be statistically proven below an effective dose of 100 mSv, consent is not required. Others will stress that due to the rapid increase in imaging exams in the past few decades, especially in exams exceeding an effective dose of 1 mSv, a proportional increase in cancer risk will occur in the population.

It is because of this latter concern that Euratom Directive 2013/59 requires justification and optimization. In addition, European Member States are required to provide patients with adequate information relating to the benefits and risks associated with the radiation dose from medical exposure.

Since malpractice risks arising from radiology examinations such as CT are yet unknown, and, thus, unlimited, it would be prudent for the radiology community to disclose possible radiation risks from a legal and ethical perspective.

As stated by Semelka et al., "The danger to the field of radiology in not regulating itself and requiring informed consent for medical procedures that use ionizing radiation is that we stand the very real chance of having regulations imposed on us by government as is already the case in Europe" (Semelka et al., 2012).

Recommendations to reduce unnecessary radiation risk and provide proper informed consent, when needed, are as follows:

- Healthcare providers should discuss the risks and benefits of planned imaging procedures with patients whenever practical and appropriate.

- To the extent possible, eliminate deterministic effects such as skin erythema. Provide consent for high-dose procedures deemed to exceed the skin erythema threshold of approximately 5 Gy (cumulative air kerma).

- Reduce the likelihood of stochastic effects such as skin cancer. The threshold for obtaining consent is debatable, but considers consent for radiographic procedures where the radiation risk of either the effective dose or age- and gender-specific organ dose, exceeds a risk of 1 in 10,000. Depending on a number of variables, including equipment limitations and practitioner experience, most CT examinations, some pelvic/abdominal radiographs, and certain nuclear medicine studies would exceed this suggested consent threshold.

- Because of the uncertainty involved, consider describing radiation risk in broad categories up to an effective dose of 100 mSv.

- Physician education and training related to radiation protection and risk should be stressed.

- Clinicians should be aware of the appropriateness criteria or referral guidelines used in their area of expertise as well as jurisdiction as part of the justification process. Screening and surveillance programs should require particular scrutiny.

- Continue to review imaging protocols to ensure they are age appropriate and provide the best images at the lowest dose.

- Use electronic ordering systems to ensure referrers are ordering appropriate exams.

- Continue to develop age-appropriate modeling to better approximate organ dose and therefore individual risk estimates.

- Continue to develop educational materials to educate patients on radiation risks. Use graphs and express risk information in percentages rather than frequencies. Compare with risks associated with ordinary life activities.

- Imaging experts and manufacturers should continue to standardize radiation dose output metrics and develop and incorporate organ dose estimates.

For proper protection of research subjects, informed consent must be obtained. Informed consent must adhere to the Belmont Report ethical principles and contain the Declaration of Helsinki consent items. There must be full disclosure of the potential study risks in a manner easily understood by the potential participant.

In addition, the researchers and those who approve research studies (ethics committees, etc.) should be educated on research ethics ensuring that those conducting research, or having oversight, understand their ethical obligation. Without these conditions, we cannot guarantee sound scientific research.

20.4 REFERENCES

ABIM Foundation, ACP-ASIM Foundation, and European Federation of Internal Medicine. 2002. Medical professionalism in the new millennium: a physician charter. Ann Intern Med 136:243–246.

ACR. 2007. American College of Radiology White Paper on Radiation Dose in Medicine. J Am Coll Radiol 4:272–284.

ACR. 2014. American College of Radiology, ACR-SIR Practice Parameter on Informed Consent for Image-Guided Procedures. http://www.acr.org//media/1A03224CA4894854800C516012B6DB5A.pdf (Accessed December 1, 2014).

Advisory Committee on Human Radiation Experiments. 1995a. Chapter 5: Experiments with plutonium, uranium, and polonium. https://archive.org/details/advisorycommitte00unit (Accessed October 27, 2014).

Advisory Committee on Human Radiation Experiments. 1995b. Chapter 7: Nontherapeutic Research in Children. https://archive.org/details/advisorycommitte00unit (Accessed October 27, 2014).

Advisory Committee on Human Radiation Experiments. 1996. Research ethics and the medical profession. JAMA 276:403–409.

Alaszewski, A. and Horlick-Jones, T. 2003. How can doctors communicate information about risk more effectively? BMJ 327:728–731.

Baerlocher, MO and Detsky, AS. 2011. Letters to the Editor: Informed consent for radiologic procedures. JAMA 305(9):888–890.

Baerlocher, M.O., Rajan, D.K., and Mok, P.S. 2012. Is it time to include radiation risks in the consent discussion for interventional radiology procedures? JVIR 23(8):1106–1107.

Balter, S., Zanzonico, P., Reiss, G.R., et al. 2011. Radiation is not the only risk. AJR 196:762–767.

Barnett, G.C., Charman, S.C., Sizer, B., et al. 2004. Information given to patients about adverse effects of radiotherapy: a survey of patients' views. Clinical Oncology 16:479–484.

Beditti, G. and Lore, C. 2007. Radiological informed consent in cardiovascular imaging: towards the medico-legal perfect storm. Cardiovascular Ultrasound 5:35.

BEIR, 2006. National Research Council. 2006. Health risks from exposure to low levels of ionizing radiation: BEIR VII Phase 2. Washington, D.C.: The National Academies Press.

Benatar, S.R. and Singer, P.A. 2000. A new look at international research ethics. BMJ 321:824–26.

Berlin, L. 1997. Malpractice issues in radiology: informed consent. AJR 169:15–18.

Berlin, L. 2000. Malpractice issues in radiology: iodine-131 and the pregnant patient. AJR 176:869–871.

Berlin, L. 2001. Malpractice issues in radiology: radiation-induced skin injuries and fluoroscopy. AJR 177:21–25.

Berlin, L. 2003a. Malpractice issues in radiology: radiologic malpractice litigation; a view of the past, a gaze at the present: a glimpse of the future. AJR 181:1481–1486.

Berlin, L. 2003b. Malpractice issues in radiology: potential legal ramifications of whole-body CT screening: taking a peek into Pandora's Box. AJR 180:317–322.

Berlin, L. 2011. Informing patients about the risks and benefits of radiology examinations utilizing ionizing radiation: a legal and moral dilemma. J Am Coll Radiol 8(11):742–743.

Berlin, L. 2014. Shared decision-making: is it time to obtain informed consent before radiologic examinations utilizing ionizing radiation? Legal and ethical implications. J Am Coll Radiol 11:246–251.

Brandt, A.M. 1978. Polio politics publicity and duplicity: ethical aspects in the development of the Salk vaccine. Int J Health Service 8:257–270.

Brenner, D.J., Sawant, S.G., Hande, M.P., et al. 2002. Routine screening mammography: how important is the radiation-risk side of the benefit-risk equation? Int J Radiat Biol 78(12):1065–1067.

Brenner, D.J. 2004. Radiation risks potentially associated with low-dose CT screening of adult smokers for lung cancer. Radiology 231:440–445.

Brenner, D.J. and Hall, E.J. 2007. Computed tomography—an increasing source of radiation exposure. N. Engl J Med. 357:2277–2284.

Brenner, D.J. and Hricak, H. 2010. Radiation exposure from medical imaging: time to regulate? JAMA 304(2):208–209.

Cardinal, J.S., Gunderman, R.B. and Tarver, R.D. 2011. Informing patients about risks and benefits of radiology examinations: a review article. J Am Coll Radiol 8:402–408.

Castronovo, F.P. 1993. An attempt to standardize the radiodiagnostic risk statement in an institutional review board consent form. Investigative Radiology 28:533–538.

Cardis, E. 2007. Current status and epidemiological research needs for achieving a better understanding of the consequences of the Chernobyl accident. Health Physics 93(5):542–546.

College of Physicians of Philadelphia. 2014. The History of Vaccines. http://www.historyofvaccines.org/content/timelines/all (Accessed October 30, 2014).

Corbie-Smith, G. 1999. The continuing legacy of the Tuskegee Syphilis Study: considerations for clinical investigation. Am. J. Med. Sci. 317:5–8.

Correia, M.J., Helllies, A, Andreassi, M.G., et al. 2005. Lack of radiological awareness among physicians working in a tertiary-care cardiological centre. Int J Cardio. 103:307–311.

Council for International Organizations of Medical Sciences. 2002. International ethical guidelines for biomedical research involving human subjects. CIOMS, Geneva.

Curran, W.J. 1981. New ethical review policy for clinical medical research. NEJM 304–952.

Dauer, L.T., Thornton, R.H., Hay, J.L. et al. 2011. Fears, feelings and facts: interactively communicating benefits and risks of medical radiation with patients. AJR 196:756–761.

Doris, I.C. 2006. What could the nematodes teach us about radiation protection? Plenty. JACR 57:22–24.

Euratom. 2013. Council Directive 2013/59/Euratom of 5 December 2013 laying down basic safety standards for the protection against the dangers arising from exposure to ionizing radiation, and repealing Directives 89/618/Euratom, 90/641/Euratom, 96/29/Euratom, 97/43/Euratom and 2003/122/Euratom. 2014. Official Journal of the European Union 57: January 17, 2014.

European Commission, 2000. Charter of Fundamental Rights of the European Union. http://eur-lex.europa.eu/LexUriServ/LexUriServ.do?uri=OJ:C:2010:083:0389:0403:en:PDF (Accessed October 1, 2014).

European Commission. 2007. Radiation Protection 118; Guidelines for healthcare professionals who prescribe imaging investigations involving ionizing radiation. http://ec.europa.eu/energy/nuclear/radioprotection/publication/doc/118_update_en.pdf (Accessed October 1, 2014).

European Commission. 2014. Radiation Protection 175; Guidelines on radiation protection education and training in medical professionals in the European Union. http://ec.europa.eu/energy/nuclear/radiation_protection/doc/publication/175.pdf (Accessed October 1, 2014).

European Union, 2009 Council Recommendation of 9 June 2009 on patient safety, including the prevention and control of healthcare associated infections. 2009. Official Journal of the European Union 52: July 3, 2009.

Finestone, A., Schlesinger, T., Amir, H., et al. 2003. Do physicians correctly estimate radiation risks from medical imaging? Arch Env Health 58(1):59–61.

Freeman, W.L. 1994. Research with radiation and healthy children: greater than minimal risk. IRB: a review of human subjects research 16:1–5.

Gerber, T.C., Carr, J.J., Arai, A.E., et al. 2009. Ionizing radiation in cardiac imaging: a science advisory from the American Heart Association Committee on cardiac imaging of the Council on Clinical Cardiology and Committee on Cardiovascular Imaging and Intervention of the Council on Cardiovascular Radiology and Intervention. Circulation, 119:1056–1065.

Gibbons, R.J. 2007. Leading the elephant out of the corner: the future of health care: presidential address at the American Heart Association 2006 Scientific Session. Circulation 115:2221–2230.

GMC. 2008. General Medical Council 2008, Consent: patients and doctors making decisions together. http://www.gmc-uk.org/static/documents/content/Consent_-_English_0914.pdf (Accessed July 17, 2014).

IAEA. 2014. International Atomic Energy Agency 2014, Radiation protection and safety of radiation sources: international basic safety standards, General Safety Requirements Part 3. International Atomic Energy Agency, Vienna.

IAEA. 2014. International Atomic Energy Agency 2014, Radiation Protection of Patients. https://rpop.iaea.org/RPOP/RPoP/Content/Information For/HealthProfessionals/6_OtherClinicalSpecialities/ referring-medical-practitioners/index.htm (accessed July 10, 2014).

ICRP. 1977. International Commission on Radiological Protection. 1977. Recommendations of the ICRP. Oxford: Pergamon Press; ICRP Publication 26. Annals of the ICRP 1 (3).

ICRP. 1991. International Commission on Radiological Protection. 1991. Radiological protection in biomedical research. Oxford: Pergamon Press; ICRP Publication 62; Ann. ICRP 22(3):1–18.

ICRP. 2007. International Commission on Radiological Protection. 2007. The 2007 Recommendations of the International Commission on Radiological Protection. Oxford: Pergamon Press; ICRP Publication 103. Annuals of the ICRP 37 (2–4).

Kaufman, S.R. 1997. The World War II plutonium experiments: Contested stories and their lessons for medical research and informed consent. Culture, Medicine and Psychiatry 21:161–97.

Keens, T.G., Oki, G.S.F., Gilsanz, V., et al. 1994. Does radiation research in healthy children pose greater than minimal risk? IRB: a review of human subjects research. 16: 5–10.

Larson, D.B., Rader, S.B., Forman, H.P., et al. 2007. Informing parents about CT radiation exposure in children: it's OK to tell them. AJR. 189:271–275.

Lee, C.I., Haims, A.H., Monico, E.P., et al. 2004. Diagnostic CT scans: assessment of patient, physician, and radiologist awareness of radiation dose and possible risks. Radiology 231:393–398.

Lee, C.I., Flaster, H.V., Haims, A.H., et al. 2006. Diagnostic CT scans: institutional informed consent guidelines and practices at academic medical centers. AJR. 187:282–287.

Lock, S. 1995. Research ethics - a brief historical review to 1965. J Intern Med 238:513–20.

Martin, C.J. 2007. Effective dose: how should it be applied to medical exposures? Br J Radiol. 80(956):639–647.

Malone, J., Guleria, R., Craven, C., et al. 2012. Justification of diagnostic medical exposures: some practical issues. Report of an International Atomic Energy Agency consultation. Br J Radiol. 85:523–38.

McCollough, C.H., Christner, J.A., and Kofler, J.M. 2010. How effective is effective dose as a predictor of radiation risk? AJR. 194:890–896.

Merck, L.H., Hauck, M.G., Houry, D.E., et al. 2011. Informed consent for computed tomography. Letters to the Editor. Am J Emerg Med. 29(2):230–232. McColNational Commission for the Protection of Human Subjects of Biomedical and Behavioral Research. 1975. Reports and Recommendations: Research on the Fetus. DHEW Publication No. (OS) 76–127 https://repository.library.georgetown.edu/bitstream/handle/10822/559372/research_fetus.pdf?sequence=4 (Accessed October 27, 2014).

National Commission for the Protection of Human Subjects of Biomedical and Behavioral Research. 1977. Reports and Recommendations: Research Involving Children. DHEW Publication No. (OS) 77–0004. https://repository.library.georgetown.edu/bitstream/handle/10822/559373/Research_involving_children.pdf?sequence=1 (Accessed October 27, 2014).

National Commission for the Protection of Human Subjects of Biomedical and Behavioral Research. 1978. The Belmont Report: ethical principles and guidelines for the protection of human subjects in research. DHEW Publication No. (OS) 78–0014. http://videocast.nih.gov/pdf/ohrp_appendix_belmont_report_vol_2.pdf (Accessed October 27, 2014).

NCRP. 2009. National Council on Radiation Protection and Measurements. Ionizing radiation exposure of the population of the United States: NCRP Report No. 160. Bethesda, MD: National Council on Radiation Protection and Measurements.

Picano, E. 2004a. Sustainability of medical imaging. BJM. 328:578–580.

Picano, E. 2004b. Informed consent and communication of risk from radiological and nuclear medicine examinations: how to escape from a communication inferno. BMJ 329:849–851.

Semelka, R.C., Armao, D.M., Elias, Jr., J., et al. 2007. Imaging strategies to reduce the risk of radiation in CT studies, including selective substitution with MRI. J. Magn. Reson. Imaging 25:900–909.

Semelka, R.C., Armao, D.M., Elias Jr., J, et al. 2012. The information imperative: is it time for an informed consent process explaining the risks of medical radiation? Radiology 262(1):15–18.

Shapiro, H.T. and Meslin, E.M. 2001. Ethical issues in the design and conduct of clinical trials in developing - countries. NEJM 345(2):139–42.

Shiralkar, S., Rennie, A., Snow, M., et al. 2003. Doctors' knowledge of radiation exposure: questionnaire study. BMJ 327:371–372.

Sparks. J. 2014. Timeline of laws related to the protection of human subjects. http://history.nih.gov/about/timelines_laws_human.html (Accessed October 30, 2014).

Stecker, M.S. 2010. Patient radiation management and preprocedure planning and consent. Tech Vasc Interventional Rad 13:176–182.

Sugarman, J., Mastroianni, A.C., and Kahn, J.P. 1998. Ethics of research with human subjects: selected policies and resources. Frederick, Maryland: University Publishing Group, Inc.

Terry, P.B., 2007. Informed consent in clinical medicine. Chest 131(2):563–568.

Thomas, K.E., Parnell-Parmley, J.E., Haidar, S., et al. 2006. Assessment of radiation dose awareness among pediatricians. Pediatr Radiol 36(8):823–832.

Tubiana, M., Feinendegen, L.E., Yang, C., et al. 2009. The linear no-threshold relationship is inconsistent with radiation biologic and experimental data. Radiology 251(1):13–22. UKDOH, 2009 United Kingdom Department of Health 2009. Reference guide to consent for examination or treatment, 2nd ed.

United States Department of Health and Human Services. 1979. The Belmont Report. http://www.hhs.gov/ohrp/policy/belmont.html (Accessed December 23, 2014).

United States Department of Health and Human Services. 2009. Code of Federal Regulations. Washington, DC: U.S. Government Printing Office; 45 CFR Part 46.

United States Holocaust Memorial Museum. 2002. The Nuremberg Code.

http://www.nus.edu.sg/irb/Articles/Nuremberg.pdf (accessed October 27, 2014).

Varmus. H. and Satcher. D. 1997. Ethical complexities of conducting research in developing countries (letter to editor). NEJM 337:1003–6.

Walters, E.A., Weinstein, N.D., Colditz, G.A., et al. 2006. Formats for improving risk communication in medical tradeoff decisions. Journal of Health Communication: International Perspectives, 11(2):167–182.

Weindling, P. 1996. Human guinea pigs and the ethics of experimentation: The BMJ's correspondent at the Nuremberg medical trial. BMJ 313:1467–70.

WHO. 2014. World Health Organization. 2014. Patients' rights. http://www.who.int/genomics/public/patientrights/en/ (Accessed October 23, 1014).

WMA. 2009. World Medical Association. 2009. Medical Ethics Manual - Chapter 2: Physicians and Patients. http://www.wma.net/en/ 30publications/30ethicsmanual/ pdf/chap_2_en.pdf (Accessed July 11, 2014).

WMA. 2013. World Medical Association. 2013. World Medical Association Declaration of Helsinki: Ethical principles for medical research involving human subjects. JAMA 310:2191–4.

20.5 APPENDIX A: EUROPEAN UNION LEARNING OUTCOMES IN RADIATION PROTECTION FOR DIAGNOSTIC RADIOLOGISTS

Taken from European Commission, Radiation Protection No. 175, Guidelines on Radiation Protection Education and Training of Medical Professionals in the European Union, Luxembourg: Publications Office of the European Union, 2014

Table 20.6: European Union learning outcomes in radiation protection for diagnostic radiologists

	Knowledge (facts, principles, theories, practices)	Skills (cognitive and practical)	Competence (responsibility and autonomy)
Radiation Physics	K1. List sources and properties of ionising radiation K2. List and explain mechanisms of interaction between ionising radiation and matter/tissues K3. List and explain mechanisms of radioactive decay K4. Explain the phenomena of X-ray interaction with matter and the consequences for image generation, image quality and radiation exposure K5. List and explain definitions, quantities and units of kerma, absorbed energy dose (Gy), organ and effective doses (Sv), as well as exposure rate and dose rate	S1. Apply radiation physics to optimally select the best imaging modality S2. Apply radiation physics to optimise the protocols, using minimal exposure to reach the image quality level needed for the task S3. Use the laws of physics to minimise scatter and optimise contrast S4. Use the correct terms to characterise exposure in daily radiograph fluoroscope and CT examinations and define organ risk, and estimate the genetic and cancer risk	

Continued on next page

Table 20.6 — *Continued from previous page*

	Knowledge (facts, principles, theories, practices)	Skills (cognitive and practical)	Competence (responsibility and autonomy)
Equipment	K6. Explain the mechanism of X-ray production K7. List the components of an X-ray unit and explain the process of X-ray generation K8. Explain the function of filters and diaphragms K9. List the common analogue and digital detectors, explain their function and their relative pros and cons K10. Explain the role of screens (in analogue radiography) and grids and their effect on image quality and exposure	S5. Continuously check image quality to recognise and correct technical defects S6. Demand the best in image quality, technical innovation and exposure reduction for the lowest cost S7. Coordinate the commissioning of new equipment with the other members of the core team (radiographer, medical physicist) S8. Use the technical features of the specific equipment and take advantage of all quality-improving and dose-reducing capabilities while recognising the limits of the machine	C1. Choose the best equipment for your patient spectrum based on the resources available

Continued on next page

Table 20.6 — *Continued from previous page*

	Knowledge (facts, principles, theories, practices)	Skills (cognitive and practical)	Competence (responsibility and autonomy)
Radiobiology	K11. Describe radiation effects on cells and DNA K12. Describe cellular mechanisms of radiation response, repair, and cell survival K13. Describe radiation effects on tissues and organs K14. Explain differences in radiation response between healthy tissue and tumours as basis for radiation treatment K15. Define and explain stochastic and teratogenic radiation effects and tissue reactions K16. Describe types and magnitudes of radiation risk from radiation exposure in medicine	S9. Inform patients of their health problems and the planned procedure S10. Communicate the radiation risk to the patient at an understandable level, whenever there is a significant deterministic or stochastic risk, or when the patient has a question	

Continued on next page

Table 20.6 — *Continued from previous page*

Knowledge (facts, principles, theories, practices)	Skills (cognitive and practical)	Competence (responsibility and autonomy)
General radiation protection	K17. Describe the basic principles of radiation protection, as outlined by the ICRP K18. Specify types and magnitudes of radiation exposure from natural and artificial sources K19. Describe concepts of dose determination and dose measurement for patients, occupationally exposed personnel and the public K20. Explain the nature of radiation exposure and the relevant dose limits for the worker, including organ doses and dose limits for pregnant workers, comforters, careers, and the general public	S11. Communicate with referrer regarding justification; if necessary, suggest a different test S12. Apply the three levels of justification in daily practice, with respect to existing guidelines, but also to individual cases (e.g., polymorbidity)	C2. Take responsibility for choosing the best imaging modalities for the individual patient (radiography, CT, alternatives such as ultrasound or MRI) by taking into consideration the risk of the disease, patient, age and size, the dose level of the procedure, and exposure of different critical organs C3. Consult both the patient and staff on pregnancy related concerns in radiation protection C4. Take responsibility for patient dose management in different imaging modalities being aware of specific patient dose levels

Continued on next page

Table 20.6 — *Continued from previous page*

	Knowledge (facts, principles, theories, practices)	Skills (cognitive and practical)	Competence (responsibility and autonomy)
Radiation protection in radiology (X-rays)	K21. Define ALARA and its applicability to diagnostic radiology settings K22. Explain the concepts and tools for dose management in diagnostic radiology with regard to adult and paediatric patients K23. Explain the factors influencing image quality and dose in diagnostic radiology K24. Describe the methods and tools for dose management in diagnostic radiology: radiography, fluoroscopy, CT, mammography, and those for paediatric patients K25. Explain the basic concepts of patient dose measurement and calculation for the different modalities in diagnostic radiology	S13. Optimise imaging protocols by using standard operating procedures (SOPs) and by adapting these to the specific patient's size S14. Use specific paediatric protocols, by taking into consideration the physics of small size, but also the elevated risk, vulnerability and specific pathology of each age group S15. Choose the best compromise between risk-benefit ratios, image quality and radiation exposure on a case-by-case basis S16. Supervise the use of personal protective equipment. Support monitoring of the workplace and individuals. Support exposure assessment, investigation and follow up, health surveillance, and records	C5. Advise patients on the radiation-related risks and benefits of a planned procedure C6. Take responsibility for justification of radiation exposure for every individual patient, with special consideration for pregnant patients C7. Take responsibility for choosing and performing the diagnostic procedure with the lowest dose for a given referrer's request C8. Take responsibility for optimising the radiographic technique/protocol used for a given diagnostic procedure based on patient-specific information

Continued on next page

Table 20.6 — *Continued from previous page*

Knowledge (facts, principles, theories, practices)	Skills (cognitive and practical)	Competence (responsibility and autonomy)
K26. Describe the key considerations relevant to radiation protection when designing a diagnostic radiology department	S17. Apply radiation protection measures in diagnostic radiology (radiography, fluoroscopy-intervention, CT, mammography and paediatric patients) and advise on their use	C9. Take responsibility for applying the optimal size-adapted and problem-adapted individual protocol for high-dose procedures (CT, fluoroscopy-intervention)
K27. List diagnostic procedures performed outside radiology department with relevant radiation protection considerations	S18. Stay within guidance/reference levels in daily practice	C10. Implement the concepts and tools for radiation protection optimisation
K28. List expected doses (reference person) for frequent diagnostic radiology procedures	S19. Set up size-specific protocols for high-dose procedures	
K29. Explain quantitative risk and dose assessment for workers and the general public in diagnostic radiology	S20. Estimate organ doses and effective doses for diagnostic radiology examinations, based on measurable exposure parameters (KAP, DLP)	
K30. Explain the concepts and tools for radiation protection optimization		

Continued on next page

Table 20.6 — *Continued from previous page*

	Knowledge (facts, principles, theories, practices)	Skills (cognitive and practical)	Competence (responsibility and autonomy)
Quality	K31. Define QA in radiology, QA management and responsibilities, outline a QA and radiation protection programme for diagnostic radiology K32. List the key components of image quality and their relation to patient exposure K33. Explain the principle of diagnostic reference levels (DRLs)	S21. Apply standards of acceptable image quality. Perform retake analyses. Understand the effects of poor-quality images S22. Avoid unnecessary radiation exposure during pregnancy by screening the patient before examination (warning signs, questionnaire, pregnancy test, etc.) S23. Double check the appropriate protection measures when exposing a pregnant woman (size and positioning of the X-ray field, gonad shielding, tube-to-skin distance, correct beam filtration, minimising and recording the fluoroscopy time, excluding non-essential projections, avoiding repeat radiographs) S24. Develop an organisational policy to keep doses to the personnel ALARA	C11. Supervise QC procedures on all equipment related to patient exposure C12. Take responsibility for the establishment of formal systems of work (SOPs) C13. Take responsibility for organisational issues and implementation of responsibilities and local rules C14. Take responsibility for organising the radiological workflow in order to avoid accidental/unintended exposures and for adequate handling of such an event

Continued on next page

Table 20.6 — *Continued from previous page*

	Knowledge (facts, principles, theories, practices)	Skills (cognitive and practical)	Competence (responsibility and autonomy)
Laws and regulations	K34. List national and international bodies involved in radiation protection regulatory processes K35. Specify the relevant regulatory framework (ordinances, directives, regulations, etc.) governing the medical use of ionising radiation in your country and the EU K36. Specify the relevant regulatory framework governing the practice of diagnostic radiology in your country	S25. Find and apply the relevant regulations and guidance for any clinical situation in radiology	C15. Take responsibility for compliance with regulatory requirements concerning occupational and public radiation exposures C16. Take responsibility for compliance with ALARA principles concerning occupational and public radiation exposure C17. Take responsibility for conforming with patient protection regulations (DRLs, where applicable)

20.6   APPENDIX B: THE NUREMBURG CODE

1. Consent must be voluntary.

2. The study must be sound, justified, and scientifically based; results should benefit society.

3. The expected results of the study should justify performance of the study and be based on prior animal study results along with a knowledge of disease progression.

4. The study should be performed in a manner that eliminates or minimizes physical and mental harm.

5. A study should not be conducted if it is known in advance that death or disabling injury will occur.

6. The level of risk associated with the study shall be below the level of significance given to the societal problem the study is looking to solve.

7. Subjects shall be protected from injury, disability, or death by assuring the study is properly executed by preplanning and having suitable facilities available.

8. The study must be conducted by persons qualified to do so.

9. The study participant shall be able to discontinue the study if they deem it necessary.

10. The study investigator must cease the study if it is determined that continuing could result in injury, disability or death of a subject.

*Adapted from "Trials of War Criminals before the Nuremberg Military Tribunals" under Control Council Law No. 10. Nuremberg, October 1946–April 1949. Washington, D.C.: U.S. G.P.O, 1949–1953.

20.7 APPENDIX C: BELMONT REPORT ETHICAL PRINCIPLES AND CIOMS GENERAL ETHICAL PRINCIPLES

Respect for persons*

- Dignity and freedom of every person
- Informed consent from all potential research subjects or their legally authorized representatives

Beneficence

- Researchers maximize benefits and minimize harm
- Research-related risks must be reasonable in light of expected benefits

Justice

- Equitable selection, recruitment, and fair treatment of research subjects

Autonomy

- Patient's right to information
- Patient's right to reject or accept treatment

*CIOMS incorporates "Autonomy" with "Respect for Persons".

20.8   APPENDIX D: CIOMS ELEMENTS OF CONSENT (ADAPTED FROM CIOMS 2002)

A. Participation is voluntary;

B. Participation can be stopped at any time without penalty;

C. Reason for the study; what is different from current practice;

D. An explanation of how the study will be conducted;

E. Length of participation and number of visits to research institution;

F. Level of reimbursement;

G. Participants will be informed of the results and how that might relate to their situation;

H. Study data is available to participants if requested;

I. Risks;

J. Benefits for the participant;

K. Benefits for society;

L. Whether the drug/device/intervention being studied will be available if found to be safe and effective;

M. Alternative treatments or therapies;

N. How confidentiality of data will be ensured;

O. Consequences if confidentiality is violated;

P. How results of genetic testing will be used for current research and future research;

Q. Who is funding the research and who is conducting the research;

R. Other research uses of the subjects' medical data;

S. Biological specimen destruction or storage;

T. Whether a participant will be compensated if biological specimens will be used in the development of a commercial product;

U. Whether the investigator is only overseeing the research study or is actively participating as a physician for the participant;

V. The degree to which the investigator will provide medical services;

W. If and how participants will be compensated if there are complications or injury;

X. If and how the participant's family will be compensated if the participant dies due to participation;

Y. Whether or not compensation is guaranteed;

Z. The study has been reviewed and approved by appropriate committees (ethical, scientific, IRB, etc.)

Glossary

This glossary contains technical terms used in this book. Most definitions were obtained from the e-Encyclopaedia for Lifelong Learning (EMITEL), an e-encyclopedia of medical physics and multilingual dictionary of terms (http://www.emitel2.eu). For a more complete definition, consult EMITEL or another medical encyclopedia or dictionary.

AAPM

American Association of Physicists in Medicine

Absorbed Dose

The mean energy absorbed within a mass of absorber (tissue).

AFOMP

Asian-Oceania Federation of Organizations for Medical Physics

ALARA

A basic principle of radiation protection set down by the ICRP in their 1977 Recommendations (Publication 26). ALARA is an acronym for "as low as (is) reasonably achievable," which means making every reasonable effort to maintain exposures to ionizing radiation as far below the dose limits as practical.

ALFIM

Latin American Medical Physics Association

Basic Safety Standards

A primary legal tool for radiation protection purposes, published by the IAEA.

Brachytherapy

Radiation therapy where a sealed radioactive source is placed close to or inside the target tissue.

Controlled Area

Any area in which specific protection and safety measures are or may be required to control exposures.

Deterministic Effects

The detrimental biological effects of exposure to ionizing radiation seen at higher doses/dose rates. These effects occur above a threshold in all persons exposed and the severity increases with the dose received.

Diagnostic Reference Level

Defined by the ICRP in Publication 73 as a form of investigation level, applied to an easily measured quantity, usually the absorbed dose in air, or tissue-equivalent material at the surface of a simple phantom or a representative patient.

Effective Dose

Defined by ICRP (Publication 103) as the tissue-weighted sum of the equivalent doses in all specified tissues and organs of the body, and representing the stochastic risk.

EFOMP

European Federation of Organizations for Medical Physics

Equivalent Dose

The product of the absorbed dose D to an organ or tissue and a factor, WR, called the radiation weighting factor, a dimensionless quantity that characterizes that damage associated with the relative biological effectiveness of different types of radiation.

FAMPO

Federation of African Medical Physics Organizations

Gray (Gy)
This is the SI unit for absorbed dose and kerma of ionizing radiation, and is abbreviated to Gy. 1 Gray = 1 Joule/Kilogram.

ICRP
The International Commission on Radiological Protection, the principal body to provide an appropriate international standard of protection for man without unduly limiting the benefit of the practices using ionizing radiation.

IFMBE
The International Federation for Medical and Biological Engineering (IFMBE) is primarily a federation of national and transnational societies. These professional organizations represent interests in medical and biological engineering.

Image Guided Radiotherapy, IGRT
The process of frequent two- and three-dimensional imaging used during a course of radiation treatment to direct radiation therapy using the imaging coordinates of the actual radiation treatment plan.

Intensity Modulated Radiation Therapy, IMRT
A type of three-dimensional radiation therapy in which narrow beams of radiation of different intensities are aimed at the tumor from many angles to maximize dose to the tumor and minimize the damage to healthy tissue near the tumor.

IOMP
International Organization for Medical Physics. IOMP is charged with a mission to advance medical physics practice worldwide by disseminating scientific and technical information, fostering the educational and professional development of medical physics, and promoting the highest quality medical services for patients.

IUPESM
International Union for Physics and Engineering Sciences in Medicine. The IUPESM represents the combined efforts of more than 40,000 medical physicists and biomedical engineers working on the physical and engineering science of medicine.

Justification
The first principle of protection against ionizing radiation for workers, patients, and members of the public, specified by the ICRP. Justification is a Cost/Benefit Analysis shows that the net benefit outweighs the risks associated with the potential adverse radiation effects of the exposure.

Lifetime Attributable Risk
The estimated lifetime rate of cancer that could, in theory, be prevented if all exposures to a particular causative agent (such as medical radiation) were eliminated.

Medical Exposure
Exposure incurred by patients as part of their own medical or dental diagnosis or treatment; by persons, other than those occupationally exposed, knowingly while voluntarily helping in the support and comfort of patients; and by volunteers in a program of biomedical research involving their exposure.

Medical Physicists
Medical Physicists apply knowledge and methodology of science of physics to all aspects of medicine, to conduct research, develop or improve theories, and address problems related to diagnosis, treatment, and rehabilitation of human disease. They are directly involved with patients and people with disabilities.

MEFOMP
Middle East Federation of Organizations for Medical Physics

NMO
National Member Organization

Optimization
Defined by the ICRP in Publication 103 as the process to keep the likelihood of incurring exposures, the number of people exposed, and the magnitude of individual doses ALARA, taking economic and societal factors into account.

Patient Safety
The prevention of avoidable risks and adverse effects to patients, associated with healthcare in terms of radiation protection.

Personal Protective Equipment
An element of the BSS protection and safety program. Includes, but not limited to, lead aprons, thyroid shields, leaded glasses, and lead gloves.

PET Radionuclides
Radionuclides used in Positron Emission Tomography, an imaging technology that detects photons generated by positrons emitted by specific radionuclides used to image tumors and other abnormalities.

Radiation Therapy, External Beam
In external beam radiotherapy the radiation source is at a certain distance from the patient and the target within the patient is irradiated with an external radiation beam. Most external beam radiotherapy is carried out with photon beams, some with electron beams, and a very small fraction with more exotic particles such as protons, heavy ions, or neutrons.

Radiation Therapy, Treatment Planning
Using computer workstations networked with imaging equipment, e.g., CT machines, to generate a plan to deliver a specific absorbed dose to a target, e.g., a tumor, while minimizing doses to surrounding healthy tissue.

Radionuclide Therapy
Radionuclide therapy uses radionuclides, administered either orally or intravenously, to deliver highly targeted therapy for cancer or noncancerous diseases, enabling the delivery of a high dose to the target while minimizing normal-tissue toxicity.

Safety Culture
The ways in which safety is managed in the workplace, often reflecting the attitudes, beliefs, perceptions, and values that employees share in relation to safety.

SEAFOMP
Southeast Asian Federation for Medical Physics

Sievert (Sv)

The unit given to the radiation protection quantities Equivalent Dose, and Effective Dose, as defined by the ICRP. One Sv is equivalent to 1 Joule of energy from incident ionizing radiation absorbed in each kilogram of human tissue.

Simulation

Simulating a radiation therapy treatment through use of treatment planning.

Stochastic Effects

Effects that are statistically detectable only in populations because of their random nature. As opposed to deterministic effects, the probability rather than the severity of the effect is a function of radiation dose.

Supervised Area

Areas in which occupational exposures may reach the levels that would require the area to be controlled.

Treatment Vault

A room constructed with thick walls, usually concrete, in which a radiation therapy machine, such as a linear accelerator, is housed and patient treatments take place.

Weighting Factor, Radiation, w_R

The effectiveness of a type of radiation to cause damage is used to convert Absorbed Dose to a tissue or organ to the Equivalent Dose to that tissue.

Weighting Factor, Tissue, w_T

Values chosen to represent the contributions of individual organs and tissues to overall radiation detriment from stochastic effects.

Index